COMPSTAT

Proceedings in
Computational Statistics

8th Symposium held
in Copenhagen 1988

Edited by
D. Edwards and N. E. Raun

With 65 Figures

Physica-Verlag Heidelberg

David Edwards, NOVO, Clinical Research Department,
Novo Allé 9Q, DK-2880 Bagsværd, Denmark

Niels E. Raun, UNI-C, Danish Computing Centre for Research and
Education, Vermundsgade 5, DK-2100 Copenhagen Ø, Denmark

ISBN-13: 978-3-7908-0411-9 e-ISBN-13: 978-3-642-46900-8
DOI: 10.1007/978-3-642-46900-8

Cover: Erich Kirchner, Heidelberg
Offsetprinting: Kiliandruck Grünstadt
Bookbinding: T. Gansert GmbH, Weinheim-Sulzbach
7120/7130-543210

Preface

The papers assembled in this volume were presented at COMPSTAT 1988, the 8th biannual Symposium in Computational Statistics held under the auspices of the International Association for Statistical Computing.

The current impact of computers on the theory and practice of statistics can be traced at many levels: on one level, the ubiquitous personal computer has made methods for explorative data analysis and display, rarely even described in conventional statistics textbooks, widely available. At another level, advances in computing power permit the development and application of statistical methods in ways that previously have been infeasible. Some of these methods, for example Bayesian methods, are deeply rooted in the philosophical basis of statistics, while others, for example dynamic graphics, present the classical statistical framework with quite novel perspectives.

The contents of this volume provide a cross-section of current concerns and interests in computational statistics. A dominating topic is the application of artificial intelligence to statistics (and vice versa), where systems deserving the label "expert systems" are just beginning to emerge from the haze of good intentions with which they hitherto have been clouded. Other topics that are well represented include: nonparametric estimation, graphical techniques, algorithmic developments in all areas, projection pursuit and other computationally intensive methods.

COMPSTAT symposia have been held biannually since 1974. This tradition has made COMPSTAT a major forum for advances in computational statistics with contributions from many countries in the world. Two new features have been introduced at COMPSTAT '88. Firstly, the category of keynote papers has been introduced to highlight contributions judged to be of particular importance. Secondly, tutorial sessions in dynamic graphics (R. Becker), artificial intelligence in statistics (W. Gale) and graphical modelling (N. Wermuth) have been arranged, to satisfy the widespread interest in these new topics.

The programme committee, which consisted of E. B. Andersen, H. Caussinus, D. Edwards (chairman), D. Hand, T. Havránek, N. Lauro, F. van Nes and B. Streitberg, had the painful task of choosing 60 papers for publication in these proceedings, out of several hundred received. The criteria used were originality, accuracy and that the topics should have bearing on both statistics and computation.

The scientific programme consisted of the contributed, invited and keynote papers (collected in this volume) and short communications, posters and tutorials (collected elsewhere). Moreover, presentations and demonstrations of non-commercial software, an exhibition of commercial software, a book exhibition, and not least important an exhilarating social programme were arranged by the organizing committee, which consisted of P. Allerup, I. A. Larsen, A. Milhøj and N. E. Raun (chairman). The assistance of Lone Cramer is also gratefully acknowledged. The meeting was arranged by UNI•C, Danish Computing Centre for Research and Education and was sponsored by the Danish Research Council.

David Edwards • Niels E. Raun

Copenhagen

July 1988

Contents

Keynote Papers

Non-Parametric Estimation

Projection Pursuit

Graphical Techniques

Expert Systems

Statistical Methods

Time Series

Contributors

Addresses of authors will be found at the end of this volume

Keynote Papers

Parallel Linear Algebra in Statistical Computations

G. W. Stewart*, Maryland

1. Introduction

The main problem in parallel computation is to get a number of computers to cooperate in solving a single problem. The word "single" is necessary here to exclude the case of processors in a system working on unrelated problems. Ideally we should like to take a problem that requires time T to solve on a single processor and solve it in time T/p on a system consisting of p processors. We say that a system is *efficient* in proportion as it achieves this goal.

In some statistical applications, like bootstrapping or simulations, this goal is easy to achieve. The reason is that the problems divide into independent subtasks, which can be run separately with the results being collected at the end. Although this should be gratifying to statisticians, such problems are not very interesting to people doing research in parallel computing.

Fortunately, there are large problems in regression analysis, signal processing, geodetics, etc. that could potentially benefit from efficient parallelization. For many of these, the heart of the computations is the numerical linear algebra. Consequently this paper is devoted to some of the issues in implementing parallel matrix algorithms.

Just as there is no single general architecture for parallel computers, there is no general theory of parallel matrix algorithms. The same sequential algorithm will be programmed one way on one system and in a completely different way on another. Since the number of potential architectures is very large [1,16], I have chosen to restrict this paper to three, for which commercial systems are available. They are SIMD systems, shared-memory systems, and message-passing systems. We shall treat each of these in the next three sections.

Given this paper's title, its focus on computer architectures requires explanation. When I sat down to write, I intended to stress statistics and parallel matrix computations. But as I proceeded, it became clear that the key to current research and practice lay in the machines themselves. Hence the change in emphasis.

2. SIMD Systems and Systolic Arrays

A *single-instruction, multiple-data* (SIMD [8]) system is a group of (usually simple) processors that execute the same sequence of instructions in lockstep under a global control. This results in a nontrivial computation because the instructions are executed with different data, which can pass from processor to processor to be combined with other data.

*This work was supported in part by the Air Force Office of Sponsored Research under grant AFOSR-82-0078.

Compstat 1988

Originally a *systolic array* meant an array of special processors, acting in lockstep, through which data was pumped like blood through the heart [20]. Since the processors are not conceived to be programmable but do have the ability to perform different functions, systolic arrays are not, strictly speaking, SIMD systems. But neither system exists in a pure form, and the distinction has become blurred. Thus the term "systolic algorithm" is now used for algorithms that can be implemented on either system.

As an example, let us consider a systolic array to accumulate the cross-product matrix $A = X^{\mathrm{T}}X$ of an $n \times k$ regression matrix X. If we write A in the form

$$a_{ij} = x_{1i}x_{1j} + x_{2i}x_{2j} + \cdots + x_{ni}x_{nj} \tag{1}$$

we see that the problem is to extract the ith and jth elements from each row of X and add their products to a_{ij}. If we assign a processor to each element of A, then the problem becomes one of making sure that x_{ki} and x_{kj} arrive at the processor responsible for a_{ij} at the same time.

A systolic array for accumulating the upper half of A might be organized as follows.

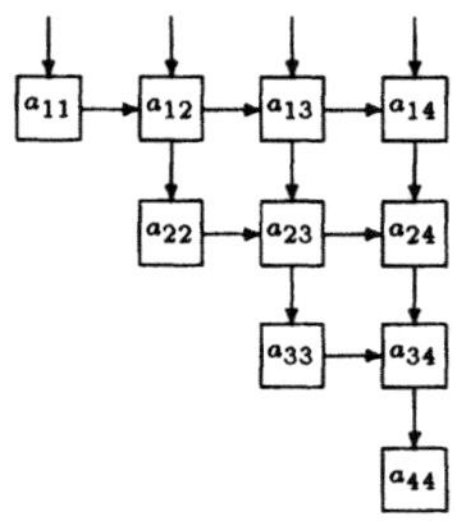

Here the boxes stand for processors and the arrows indicate how data flows through the array. Each processor is associated with an element of the upper half of the cross-product matrix as shown in the figure. The rows of x are streamed through this processor array as follows. Each element x_{ij} enters the jth column of the array at the top. At each step it moves down one processor until it gets to the diagonal, at which point it begins moving across the jth row and eventually out of the computer.

Figure 1 shows the flow of data in greater detail. The numbers associated with the arrows are the subscripts of the elements of X that are about to enter the processor. When they enter, the processor multiplies them and adds them to the element of A for which it is responsible. Note that the flow of data is such that the appropriate elements of X end up at a processor at the right time.

It is instructive to look at this system from the point of view of an individual processor, say processor `(I,J)`. If we refer to the values in the communication links to the north, south, east and west by **northx**, **southx**, **eastx**, and **westx**, and **a** denotes the current value of a_{IJ}, then a program for the `(I,J)`-processor might read as follows.

```
if (I == J)
   westx = northx;
a = a + northx*westx;
eastx = westx;
southx = northx;
```

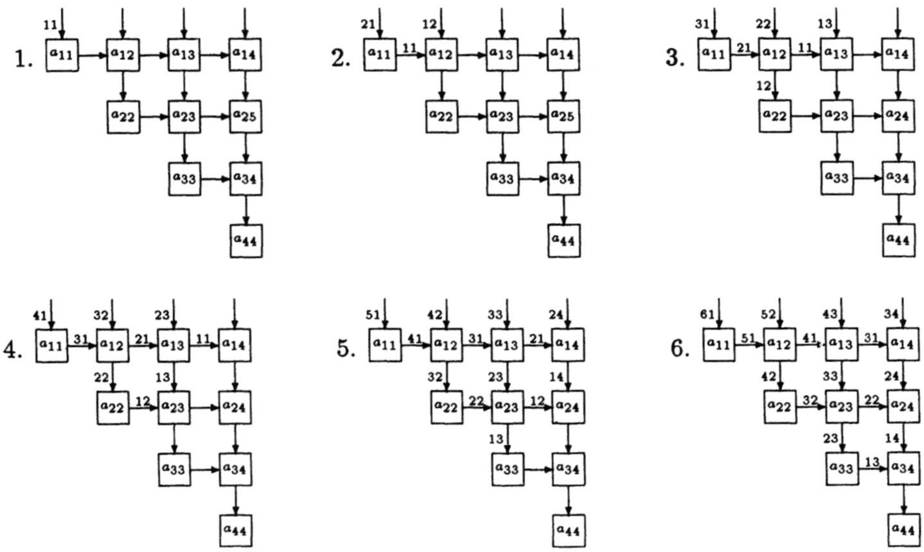

Figure 1: Flow of Data for $A = X^{\mathrm{T}}X$

Here it is understood that one iteration of the code is performed each time the controller signals the system to advance a step.

There are three things to note about this code. First, it is not strictly SIMD, since the processors on the diagonal behave differently from the others. However, a single program suffices for all processors, a situation sometimes tagged SPMD (single program, multiple data). Second, the code is local. Each processor knows only about its own variables and its input and output. Third, communication is explicit; the code specifies where the data comes from and where it goes to. These characteristics, which programs for systolic systems share with message-passing systems, make for code that is not obviously linked to its task. Certainly it is not easy to recognize equation (1) in the above program. Nonetheless, there is a certain satisfaction in designing systolic algorithms for matrix computations to judge from the number that have been published (e.g., see [3,5,21,29,30]).

When systolic arrays were first proposed, it was hoped that they would provided inexpensive, special-purpose processing for a variety of applications. Things have not worked out this way. The array above is a toy that solves a 4×4 problem. To accumulate a 100×100 matrix one would require 5,050 processors—nontrivial processors that can perform floating-point arithmetic. Moreover, one cannot afford to build such a big system for a single application; the processors must also be programmable, which increases their complexity. The end result of these considerations is the WARP computer, a linear systolic array of high-performance processors [2]. By all accounts it is effective, but it is neither simple nor cheap.

On the other hand, general purpose SIMD machines have been built and run on a variety of problems. Their main advantage is that they can bring large numbers of simple processors to bear on single problems. They are very effective with simple algorithms that proceed in short, repetitive bursts of computations. Their main disadvantage is their inflexibility. They are tedious to code, even for highly structured problems like computing

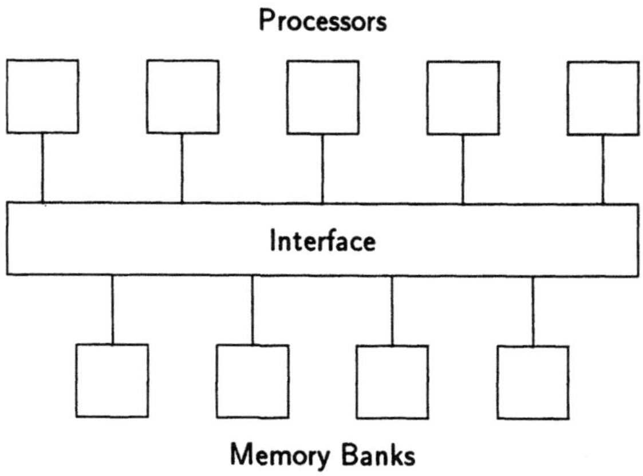

Figure 2: A Shared-Memory System

the FFT [6], and complicated, irregular computations are beyond them.

3. Shared-Memory Systems

We now turn to systems of general purpose processors that run asynchronously, the MIMD (multiple processor, multiple data) systems. These systems are commonly divided into two categories: shared-memory systems and message-passing systems.

A *shared-memory system* is one in which each processor can access a common, global memory. Figure 2 illustrates the essential features of such a system. The processors at the top are general purpose processors. In current systems they range from fairly powerful micro-computers to full-fledged super-computers. At the bottom are banks of memory each of which can be accessed in parallel. In between is an interface which routes memory requests to banks. In principal, every processor has equal access to all the data of a given problem.

Whenever processors run asynchronously, their actions must be coordinated. For example, if one processor is responsible for computing a number `x`, another processor must not attempt to use that value before it has been computed. Shared-memory systems can synchronize their actions through shared variables. For example, two processors, one of which generates x and the other which consumes it, might synchronize their actions by means of a shared variable `xready` (initially `FALSE`) as follows:

Processor A

```
while (TRUE){
   /* generate x */
   xready = TRUE;
   while (xready){}
}
```

Processor B

```
while (TRUE){
   while (!xready){}
   /* use x */
   xready = FALSE;
}
```

Processor A generates `x` and sets `xready` to tell Processor B that it may use `x`. In the meantime Processor B spins, testing the `xready` until it turns `TRUE`, after which it uses `x`.

It then resets **readyx** to tell Processor A, which itself has been spinning on **xready**, that it may generate another value of **x**.

This, of course, is an artificial example. In practice the processors would be doing other useful work, only spinning on **xready** when there is nothing else to do. Moreover, we have ignored the tricky question of how to initialize **xready**. However, people who write operating systems have long ago developed techniques for doing this [18, Ch. 8].

Actually, the applications programmer may never see this kind of synchronization. Shared-memory systems lend themselves to elegant extensions of existing programming languages that make the implementation of parallel matrix algorithms easy [13]. For example our problem of forming the normal equations might be coded in an extension of the language C as follows.[1]

```
for (i=1; i<=n; i++){
   /* Get the i-th row of X. */
   parfor (j=1; j<=k; 1)
      for (l=j; l<=k; l++)
         a[j][l] = a[j][l] + x[j]*x[l];
}
```

The effect of this code is the following. The `parfor` statement informs the operating system that there are independent tasks to be done, one for each value of a variable `j` ranging from one to `k` by increments of one. The system then starts distributing these jobs to the processors, which can begin executing right away, since all the data they require is in the shared memory.

The ability to implement parallel constructions like `parfor` make shared memory systems very attractive. In particular, it offers hope that existing sequential code can be effectively parallelized by altering and inserting commands at suitable places. Given the large investment in sequential code (the dusty deck problem), this is a strong argument for developing shared-memory systems.

However, shared memory has its difficulties. In the first place, at most one processor at a time can access a memory bank. This means that when two processors must access the same location, one must wait. The formation of the cross-product matrix is an example, since the processes spanned by the `parfor` are all contending for the same row of X. It is very difficult to model contention in shared-memory systems, since the processors run asynchronously, but it is a serious problem [22,31].

Moreover, as the number of processors grows, it becomes impractical to connect every processor to every bank; the interface in the diagram above becomes a switching network. This has three consequences. First, the complexity, and hence the cost of the network increases with the number of processors. Second, the time to access memory also increases with the number of processors. Finally, programs must contend not just for memory access but for a path through the network.

For these reasons, the most successful shared memory systems, for example the CRAY XM-P, the ALLIANT FX/8, the SEQUENT, all have relatively few processors. Moreover, many shared-memory systems are being run, not as true parallel machines, but as multitasking

[1] For those not familiar with C, the statement `for (i=j; i<=k; i++)` is equivalent to the FORTRAN `do <xx> i=j,k,1`.

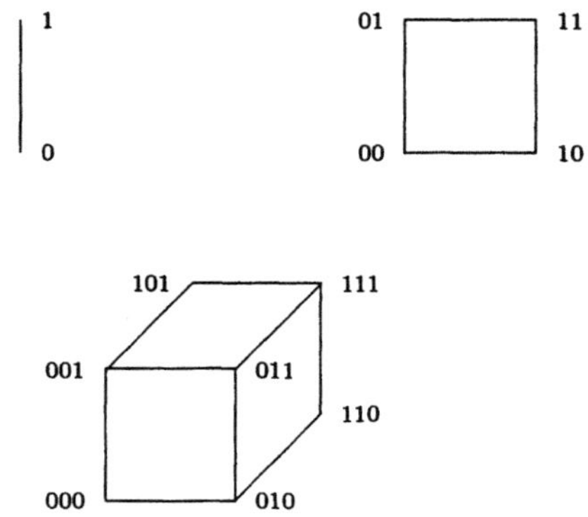

Figure 3: One, Two, and Three Cubes

systems with more than one processor. Such systems can process a large number of jobs, but the individual user sees no speedup.

A recent development in matrix computations on shared-memory systems is worth noting. To avoid contention, some systems equip their processors with a fast memory local to the processor. This *cache memory* is loaded with blocks of contiguous data in the hope that once a block is loaded its data will be used repeatedly before it is necessary to access the shared memory again. A current research problem is to develop matrix algorithms that take advantage of a cache. The idea is to work in blocks and take advantage of the fact that although an $k \times k$ block requires k^2 memory accesses to load it into the cache, we can often perform $O(k^3)$ operations on it once it is there. For more, see [7,4].

4. Message-Passing Systems

A *message-passing* system is one in which processors communicate by passing packets of data between themselves. The medium may be something as simple as an ethernet, or a slotted ring [27], in which case any processor can talk to any other processor. For economy's sake, however, processors are usually connected to one another in a fixed pattern or geometry. In such a system a processor can talk only to its neighbors.

In a system with a fixed geometry, the time required to broadcast data throughout the system will be proportional to the number of processors lying between the most widely separated processors—the diameter of the system. Thus systems with small diameter are at a premium. Currently the most popular geometry is the hypercube [28], which is commercially available in the I-PSC and the NCUBE. Geometrically, an n-dimensional hypercube is a system of 2^n processors connected in an n-dimensional cube. Figure 3 shows a 1-cube, a 2-cube, and a 3-cube. From the numbering, it is evident that an $(n+1)$-cube can be formed from two n-cubes by connecting pairwise nodes having the same number. It is also evident that the diameter of an n-cube is about $\log_2 n$.

The communication advantages of the n-cube are somewhat offset by the fact that each processor must communicate directly with $\log_2 n$ neighbors. This increasing number of connections per processor must ultimately limit the size of hypercubes that can actually be constructed. Consequently people have become interested in geometries that have only a constant number of connections per processor, for example a rectangular grid or a torus of processors.

We have mentioned that algorithms for message-passing systems have much in common with systolic algorithms. The chief differences are granularity and synchronization. Message-passing algorithms work with chunks of data larger than a single item, and it is the flow of the data that synchronizes the processors.

To see this let us return to our example of forming the cross-product matrix. We will consider a $\mathtt{P} \times \mathtt{P}$ triangular array of processors connected as the systolic array in Section 2. The algorithm is based on a block variant of (1). Namely, let X be partitioned in the form

$$X = \begin{pmatrix} x_{11}^{\mathrm{T}} & x_{12}^{\mathrm{T}} & \cdots & x_{1\mathrm{P}}^{\mathrm{T}} \\ x_{21}^{\mathrm{T}} & x_{22}^{\mathrm{T}} & \cdots & x_{2\mathrm{P}}^{\mathrm{T}} \\ \vdots & \vdots & & \vdots \\ x_{n1}^{\mathrm{T}} & x_{n2}^{\mathrm{T}} & \cdots & x_{n\mathrm{P}}^{\mathrm{T}} \end{pmatrix},$$

and let A be partitioned conformally in the form

$$A = \begin{pmatrix} A_{11} & A_{12} & \cdots & A_{1\mathrm{P}} \\ A_{21} & A_{22} & \cdots & A_{2\mathrm{P}} \\ \vdots & \vdots & & \vdots \\ A_{\mathrm{P}1} & A_{\mathrm{P}2} & \cdots & A_{\mathrm{PP}} \end{pmatrix}.$$

Then

$$A_{ij} = x_{1i}x_{1j}^{\mathrm{T}} + x_{2i}x_{2j}^{\mathrm{T}} + \cdots + x_{ni}x_{nj}^{\mathrm{T}}.$$

Figure 4 contains a message-passing implementation of this algorithm as seen by processor `(I,J)`. The dimensions of $x_{k\mathtt{i}}$ and $x_{k\mathtt{j}}$ are `ni` and `nj`, and they will be contained in the arrays `xw` and `xn`. The variables `north`, `south`, `east`, and `west` refer to the processors to the north, south, east, and west.

The *send-request* protocol by which data is passed is typical of message passing systems (e.g., see [10,25]). There is no direct link between the program that sends the message and the program that receives it. Instead the sending program calls a routine to send the message, the receiving program calls a routine to request the message, and the operating system does the rest. Data arriving early is queued in anticipation of a request, and a program requesting data is blocked from further execution until the data it has requested arrives. For uniformity in the above program, input to the first row of processors is achieved by requesting data from an imaginary processor to the north.

The program itself is a block analogy of the systolic algorithm. It divides into three parts: an input section (which is implicit in the systolic algorithm), a computation section, and an output section. The pieces of the rows x_{ij}^{T} flow through the processor array in the same way as the elements x_{ij} flowed in the systolic algorithm (cp. Figure 1).

Although it is troublesome to specify communication explicitly, the reward is free synchronization. The send-request protocol described above is sufficient to keep each processor performing the right operations in the right order, no matter how fast or slowly the individual processors run [19,24].

```
for (row=1; row<=n; row++){

   /* Get the pieces of the current
      row. */

   request(north, xn);
   if (I != J)
      request (west, xw);
   else
      for (i=1; i<=ni; i++)
        xw[i] = xn[i];

   /* Update A. */

   for (i=1; i<=ni; i++)
      for (j=1; j<=nj; j++)
         a[i,j] = a[i,j] + xw[i]*xn[j];

   /* Pass on the pieces of the row. */

   if (I != J)
      send(south, xn);
   if (J != P)
      send(east, xw);
}
```

Figure 4: Message-Passing Algorithm for $A = X^{\mathrm{T}}X$

At the present time, message-passing systems offer the greatest hope for connecting very large numbers of general purpose processors in an MIMD system. Unfortunately, communication delays can easily cancel out arithmetic parallelism. To see how this comes about, let us analyze the communication costs for the above algorithm. The basic assumption underlying the analysis is that it takes time

$$\sigma + m\tau$$

to transmit a packet of m data items between processors. The constant σ can be regarded as a startup time, while τ^{-1} is a transmission rate. This model has been widely used to analyze communication in message passing systems.

For our algorithm, let us assume that X has k columns, so that the packets being transmitted are of length $k/\mathtt{P}$. Consulting Figure 1, we see that it will take roughly time $2\mathtt{P}(\sigma + \frac{k}{\mathtt{P}}\tau)$ for the first packet to reach the $(\mathtt{P},\mathtt{P})$-processor. The job will be completed after all n packets arrive, each requiring time $\sigma + \frac{k}{\mathtt{P}}\tau$. Since we necessarily have $\mathtt{P} \leq k$, for $n \gg k$, we have a communication time of

$$T_{\mathrm{comm}} \cong n(\sigma + \frac{k}{\mathtt{P}}\tau).$$

Unfortunately, most message-passing systems are unbalanced in that the startup time is large compared to the inverse transmission rate (e.g., see [9]). Typically $\sigma = 10^{-3}$ while τ can be as small as 10^{-6}. In this case if $k = 100$ (say), the transmission time is insignificant compared to the communication time and

$$T_{\text{comm}} \cong n\sigma,$$

independent of the number of processors.

It is instructive to compare this with the time to do the arithmetic operations. This can be estimated by estimating how much work the (P, P)-processor must do. Specifically, it must perform $(k/\text{P})^2$ floating-point operations for each row of X. If floating-point operations can be performed at a rate of α^{-1} per second, then

$$T_{\text{arith}} \cong n \left(\frac{k}{\text{P}}\right)^2 \alpha.$$

Thus if $\text{P} > k\sqrt{\alpha/\sigma}$, the communication time will dominate, and we can expect little improvement from adding processors.

If we assume that $\sigma = 10^{-3}$ as above and $\alpha = 10^{-5}$, then for $k = 100$ the break-even point is $\text{P} = 10$. In other words, with these values of the parameters, we can effectively use no more than 55–60 processors!

The phenomena described here is real and has been observed in practice. In some applications there is even a point where adding processors *increases* the running time. The difficulty here is the large startup time σ. There are three cures.

The first is to take great care in the implementation of parallel algorithms. Seemingly simple changes in an algorithm can have an inordinate effect on its efficiency. For example, what efficiency there is in the algorithm for forming the cross product matrix depends on waves of computation and communication, one for each row, passing behind each other in the array. In more complicated algorithms, say Gaussian elimination with partial pivoting, it is possible to inadvertently arrange the computations so that waves do not form. (Unfortunately, eigenvalue algorithms seem to resist the formation of waves.) For examples of well considered implementation see [12,11,14]

The second is to run large jobs. In many applications, the ratio of arithmetic to communication grows with the size of the problem, so that ultimately communication costs are insignificant. For a highly publicized example of this phenomenon see [15].

The second is to reduce σ. Unfortunately, this is not easily done, since the send-receive protocol requires costly calls to the operating system. At the University of Maryland, we are experimenting with a hybrid architecture in which neighboring processors can access one-another's memory [23]. The idea is to give local communication the efficiency of shared-variable systems while preserving the limited number of connections of message-passing systems. But this work is in an early phase.

5. Summary

We have just completed a survey of three important classes of architectures. I chose them because they are the most extensively developed and have been tested in the field. But they are by no means the only possible architectures; nor have I described all possible variants. For more see [1,17,18,26].

The statistician contemplating using parallel computers to solve a problem should be prepared to do research in parallel computations. The area is in a state of flux. Machines are proposed and become obsolete even before they are built. Models for predicting performance are crude, and every new application has its surprises. The literature is scattered in technical reports and conference proceedings.

Moreover, most parallel computers are new—with all that implies. The hardware is often flakey; the software invariably so. Documentation is confused and ,incomplete on important points. Support from the developers often leaves much to be desired.

Nonetheless, I have seen enough to convince me that these problems and many of the ones discussed in the body of the paper are transient. There will probably never be a general purpose parallel computer; but I expect machines to emerge in the next decade that can utilize many processors to solve broad classes of problems, problems that could be solved in no other way. When that begins to happen, the statistical community should be prepared to take advantage of them.

References

[1] G. S. Almas. Overview of parallel computing. *Parallel Compting*, 2:191–203, 1985.

[2] M. Annoratone, E. Arnould, T. Gross, H. T. Kung, O. Menzicioglu, and J. A. Webb. The WARP computer: architecture, implementation, and performance. *IEEE Transaction on Computers*, C-36:1523–1538, 1987.

[3] J. L. Barlow and I. C. F. Ipsen. Scaled Givens rotations for the solution of linear least squares problems on systolic arrays. *SIAM Journal on Scientific and Statistical Computing*, 8:716–734, 1987.

[4] C. Bischof and C. Van Loan. The *WY* representation for products of Householder transformations. *SIAM Journal on Scientific and Statistical Computing*, 8:s2–s13, 1987.

[5] A. Bojanczyk and R. P. Brent. Parallel solution of certain toeplitz least-squares problems. *Linear Algebra and Its Applications*, 77:43–60, 1986.

[6] A. Brass and G. S. Pawley. Two and three dimensional FFTs on highly parallel compters. *Parallel Computing*, 3:167–184, 1986.

[7] Jack J. Dongarra, Jeremy Du Croz, , Iain Duff, and Sven Hammarling. A proposal for a set of level 3 basic linear algebra subprograms. *SIGNUM Newsletter*, 22:2–14, 1987.

[8] M. Flynn. Very high speed computing. *Proceedings of the IEEE*, 1901–1909, 1966.

[9] G. C. Fox, S. W. Otto, and A. J. G. Hey. Matrix algorithms on a hypercube I: matrix multiplication. *Parallel Computing*, 4:17–31, 1987.

[10] W. M. Gentleman. *Using the Harmony Operating System.* Technical Report ERB-966, Natonal Research Council of Canada, 1985.

[11] A. George, M. T. Heath, J. Liu, and E. NG. Sparse cholesky factorization on a local-memory multiprocessor. *SIAM Journal on Scientific and Statiscal Computing*, 9:327–340, 1988.

[12] Alan George and Eleanor Chu. *Gaussian Elimination with Partial Pivoting and Load Balancing on a Multiprocessor.* Technical Report ORNL/TM-10323, Mathematical Sciences Section, Oak Ridge National Laboratory, 1987.

[13] Carlo Ghezzi. Concurrency in programming languages: a survey. *Parallel Computing*, 2:224–241, 1985.

[14] L. Guangye and T. F. Coleman. A parallel triangular solver for a distributed memory multiprocessor. *SIAM Journal on Scientific and Statistical Computing*, 9:485–502, 1988.

[15] J. L. Gustafson, G. R. Montry, and R. E. Benner. Development of parallel methods for a 1024-processor hypercube. *SIAM Journal on Scientific and Statistical Computing*, 9:, 1988. To appear in July.

[16] R. W. Hockney. MIMD computing in the USA—1984. *Parallel Computing*, 3:119–136, 1985.

[17] R. W. Hockney and C. R. Jesshope. *Parallel Computers.* Adam Hilger Ltd., Bristol, 1981.

[18] K. Hwang and F. A. Briggs. *Computer Architecture and Parallel Processing.* McGraw-Hill, New York, 1984.

[19] R. Karp and R. Miller. Properties of a model for parallel computation: determinacy, termination, queuing. *SIAM Journal on Applied Mathematics*, 14:1390–1411, 1966.

[20] H. T. Kung. Why systolic architectures? *IEEE Computer*, 15:37–46, 1982.

[21] F. T. Luk. A triangular processor array for computing singular values. *Linear Algebra and Its Applications*, 77:259–274, 1986.

[22] W. Oed and O. Lange. Modelling, measurement, and simulation of memory interference in the CRAY XM-P. *Parallel Computing*, 3:343–358, 1986.

[23] D. P. O'Leary, Roger Pierson, G. W. Stewart, and Mark Wieser. *The Maryland* CRAB*: A Module for Building Parallel Computers.* Technical Report TR-1660, Department of Computer Science, University of Maryland, 1986.

[24] D. P. O'Leary and G. W. Stewart. From determinacy to systaltic arrays. *IEEE Transactions on Computers*, C-36:1355–1359, 1987.

[25] D. P. O'Leary, G. W. Stewart, and Robert van de Geijn. DOMINO*: A Message Passing Environment for Parallel Computation.* TR-1648, University of Maryland, Computer Science, 1986.

[26] J. M. Ortega. *Introduction to Parallel and Vector Computing.* Plenum Press, New York, 1988.

[27] C. Rieger. Zmob: hardware from a user's viewpoint. In *Proceedings of the IEEE Computer Society Conference on Pettern Recognition and Image Processing*, pages 399–408, 1981.

[28] C. Seitz. The cosmic cube. *Communications of the Association for Computing Machinery*, 28:22–33, 1985.

[29] G. W. Stewart. A Jacobi-like algorithm for computing the Schur decomposition of a non-Hermitian matrix. *SIAM Journal on Scientific and Statistical Computing*, 28:853–864, 1985.

[30] Robert A. van de Geijn. *Implementing the QR-Algorithm on an Array of Processors.* Technical Report 1897, Department of Computer Science, University of Maryland, 1987.

[31] Pen-Chung Yew, Niau-Feng Tzeng, and D. H. Lawrie. Distributing hot-spot addressing in large scale multiprocessors. *IEEE Transactions on Computers*, C-36:338–395, 1987.

Non-Parametric Estimation

Efficient Nonparametric Smoothing in High Dimensions Using Interactive Graphical Techniques

W. Härdle[1], Bonn

Abstract

Smoothing techniques are used to reduce the variability of point clouds. There is great interest not only among applied statisticians but also among applied workers in biostatistics, economics and engineering to model the data in a nonparametric fashion. The benefits of this more flexible modeling come at the cost of greater computation, especially in high dimensions. In this paper several possibilities of smoothing in high dimensions are described using additive models. The algorithms for solving the nonparametric smoothing problems are based on WARPing, i.e. Weighted Averaging using Rounded Points. Interactive graphical techniques are a *conditio sine qua non* for tuning and checking the structure of lower dimensional projections of the data and of smooths produced by the algorithms. Applications of the WARPing technique to a side impact study are shown by smoothing in Projection-Pursuit-type models using Average Derivative Estimation.

[1] This research was supported by the Deutsche Forschungsgemeinschaft, Sonderforschungsbereich 303.

Smoothing means parameter free re-expression of data points in a form that is easier to understand than the raw point cloud itself. In a ***regression smoothing*** problem a $(d+1)$-dimensional point cloud is observed consisting of response variables $\{Y_i\}_{i=1}^n \in \mathbb{R}$ at predictor variables $\{X_i\}_{i=1}^n \in \mathbb{R}^d$. One is then interested in smoothing the data in order to recover the mean function $m(x) : \mathbb{R}^d \to \mathbb{R}$. In a ***density smoothing*** problem a $d-$dimensional point cloud is observed and data smoothing yields an estimate of the density of the joint density $f(x)$.

There exists a wide variety of smoothing methods, see e.g. Silverman (1986), Härdle (1988b). Correctly compared and tuned these different methods have very similar asymptotic properties, but may exhibit quite different small sample behavior, see Müller (1987), Silverman (1984) and Härdle, Hall and Marron (1988). This is a perfect context in which the interactive graphics tools of an intelligent computing environment are very helpful. From a mathematical viewpoint smoothing involves function and functional estimation and therefore the various results can only be studied in the form of comparing graphs or tables of values. Another scenario where the power of an interactive computing environment is useful is the analysis of residuals : Several algorithms for finding *additive models* are based on iterative schemes based on "intermediate residuals". The ***back-fitting algorithm*** for Generalized Additive Models described in Hastie and Tibshirani (1986) uses the residuals iteratively.

Smoothing in dimensions bigger than 4 creates problems on the computational and the statistical side. Since parameter free smoothing techniques are based on the idea of ***local data pooling*** one is usually faced with the problem of sparseness of point clouds. There is a nice illustrating example in the introduction of Friedman and Stuetzle (1981). The sparseness of observations in higher dimensions is sometimes called *curse of dimensionality*, see Huber (1985). A consequence of this curse of dimensionality is the lower rate of convergence of nonparametric smoothers, see Ibragimov and Khas'minskii (1981) and Stone (1982). On the presentation side the problem is that one cannot see the full fit for dimensions greater than three. In four dimensions one can introduce an artificial time dimension (Scott, 1986) but for higher dimensions one has to rely on studying several interesting projections. A complimentary approach is to perform a Principal Components Analysis (PCA) for dimensionality reduction, see Caussinus(1986). ***Additive models*** reduce the dimensionality problem on the statistical side by imposing more structure on the functions to be estimated. Despite this the computational burden is still present if not increased since several iterative steps may have to be computed. The necessity for interactive control of smoothing algorithms becomes evident if one summarizes the typical operations. These operations might be performed several times in some sort of "smoothing analysis cycle" until one or several satisfactory results are obtained, see McDonald and Pederson (1986), Härdle (1988a).

SCATTERPLOT SMOOTHING. An elementary building block for higher dimensional model fitting is an efficent scatterplot smoother. Breiman and Friedman (1985) in their ACE-algorithm use a symmetrized k-nearest neighbor algorithm to iteratively compute the "transformations" for Alternating Conditional Expectations.

OPTIMIZATION. Often a measure of "interestingness" is computed and optimized by choice of a tuning parameter. Jones and Sibson (1986) study several indices of "interestingness" for projection pursuit and state that it may be sometimes necessary to run a preliminary PCA in order to reduce computational efforts, see also Caussinus (1987). In the projection pursuit regression setting the optimization is performed over linear combinations of the predictor variables together with one dimensional scatterplot smoothers.

ITERATION. The process of extracting features from residuals might involve changing the "interestingness" functional or a transformation in order to find good approximations to additive models. The CART algorithm for instance iterates recursively to find a Classification And/or Regression Tree of a prescribed complexity, see Breiman, Friedman, Olshen and Stone (1984).

CALIBRATION. A desirable operation in density or regression smoothing is to calibrate the smoothing parameter or to construct confidence bands. Cross validation and related methods (Scott and Terrell, 1987, Härdle, Hall and Marron, 1988) have been used to find good smoothing parameters. Confidence bands have been constructed using the bootstrap in order to calibrate the variability bands, see Härdle and Bowman (1988).

The above elementary operations involve computations that are typically linear in the squared sample size if one uses straightforward implementations.

Improved computation may be obtained by either reducing the number of arithmetic calculations or using techniques that eliminate one or more of the above four steps. In this paper it is demonstrated how WARPing achieves this goal of improved computational efficiency. Furthermore some requirements of a computing environment for performing efficient smoothing in high dimensions are given. It is seen that *XploRe - a computing environment for eXploratory Regression and density smoothing* - meets some of these requirements even on a "small scale" machine like an IBM AT.

2. Additive Models and WARPing

Additive models have been studied in order to reduce the dimensionality problem. It is probably easiest to explain the development of this model class starting from binary regression, i.e. the Y-observations are in $\{0,1\}$. A quite common approach to model the conditional probability $m(x) = P(Y = 1|X = x)$ is to assume that there is a known *link function* $G : \mathbb{R} \rightarrow \mathbb{R}$ such that

$$m(x) = G(x^T \beta) \tag{2.1}$$

for some parameter vector β. A well known example is the logistic regression model where G is the distribution function of the logistic disctribution. This model is additive but may be unsatisfactory in several ways. First it is rather narrow in the sense that the link function is assumed to be known. Secondly one might argue that it is too restrictive to assume that the regression function depends on a one dimensional link function of certain projections of the predictor variables. The first criticism can be overcome by so called projection pursuit models which have the form

$$m(x) = \sum_{l=1}^{D} g_l(x^T \beta_l), \tag{2.2}$$

where $g_l : \mathbb{R} \rightarrow \mathbb{R}$ are nonparametric functions operating on projections $x^T \beta_l$. An algorithm to fit such models is described in Friedman and Stuetzle (1981). The second criticism can be overcome by leaving the link function in a suitable known form but extend the linear projection $x^T\beta_l$. This leads to Generalized Additive Models

$$m(x) = G(\sum_{l=1}^{d} g_l(x_l)), \tag{2.3}$$

where as above the $g_l : \mathbb{R} \to \mathbb{R}$ are nonparametric functions but now operating on the l-th coordinate $x_l = (0, .., 1, .., 0)^T x$ of x. Hastie and Tibshirani (1987) discuss this model and show how the functions g_l can be estimated by the *local scoring* algorithm. The ACE-model by Breiman and Friedman (1985) is yet another extension where also the function G is nonparametrically estimated.

The ACE-algorithm attempts to find nonparametric functions $G^*, g_l^*, l = 1, .., d$ such that

$$e^2(G, g_1, ..., g_d) = \frac{E\{(G(Y) - \sum_{j=1}^{d} g_j(X_j))^2\}}{EG^2(Y)} \tag{2.4}$$

is minimized. The ACE-algorithm is highly iterative and makes extensive use of an elementary scatterplot smoother. The computational burden is even bigger if a "supersmoother" - a symmetrized k−nearest neighbor smoother with locally optimized bandwidth choice - is implemented, see Friedman (1986). Similar computational costs occur for the other additive models. In order to make additive model fits accessible for the average user one should start optimizing the scatterplot smoothing routines. The WARPing approach is a promising technique to achieve computational efficieny.

The WARPing technique is based on *Weighted Averaging using Rounded Points.* It is easiest to demonstrate this approach in the univariate density smoothing context. It is well known that the histogram may show quite different shapes if the origin of the histogram defining classes is changed, see Scott (1986). A natural way of eliminating this effect is to construct an average of histograms over a collection of origin choices. In particular, if there are q origins $\{x_{0,k} = kh/q, k = 0, ..., m-1\}$ for a histogram $HG_h(x, x_{0,k})$ with bin width h, then the WARPed histogram is simply

$$\hat{f}_{h,q} = q^{-1} \sum_{k=0}^{q-1} HG_h(x, x_{0,k}). \tag{2.5}$$

It is straightforward to see by partial summation that this density estimate is simply a weighted average of points aggregated into smaller bins of width $\delta = h/q$. If the parameter q tends to infinity the density estimate (2.5) approaches the kernel estimate with triangular kernel. This motivates a generalization of this technique to other weighting schemes. More specifically one constructs a weighting sequence $w_q(k)$ that sum to one and estimates the density by

$$(WARP) \qquad \hat{f}(x) = q^{-1} \sum_{k=1-q}^{q-1} RP_{i(x)+k},$$

where $i(x)$ is the bin in which x falls and RP_i is the frequency of *Rounded Points* in the i-th bin. This technique is also applicable to scatterplot smoothing and is described in detail in Härdle and Scott (1988).

Consider the ACE algorithm again. In this context the WARPed kernel smoother has to perform $2\delta^{-1}q + n$ operations for each of the elementary scatterplot smoothing fits. The number n comes from discretizing the data into rounded points. The effective cost on the rounded points is equal to $2\delta^{-1}q$. Depending on δ and q it can be made highly efficient compared to ordinary kernel smoothing or symmetrized k−nearest neighbor smoothing. The latter will take $nlog(n) + 2n$ operations, the $nlog(n)$ cost coming from sorting the one dimensional X variables for recursive updating of local linear fits, see Friedman and Stuetzle (1982). Also the Projection Pursuit Regression (PPR) algorithm requires an efficient scatterplot smoother as a basic tool. The PPR algorithm searches first for the best pair (β, g) such that with β suitably normalized the residual sum of squares $\sum_{i=1}^{n}(Y_i - g(X_i^T\beta))^2$ is minimized. Again this task is done iteratively. In a further step residuals are fitted and the same procedure is applied to the set of estimated residuals. Note that if β is standardized such that $E_X(dg/d(x^T\beta) = 1$ then the terms in the projection pursuit model (2.2) can be estimated by the average derivate

$$(2.6) \qquad \delta = E_X(m'(X)),$$

where m' denotes the vector of partial derivatives. *Average Derivative Estimation (ADE)* is a technique for estimating δ. An estimator for δ is given by

$$(2.7) \qquad \hat{\delta} = n^{-1} \sum_{i=1}^{n} Y_i(-\hat{f}_h'(X_i)/\hat{f}_h(X_i)).$$

Again the density estimates in this formula can be estimated using WARPing. The estimator $\hat{\delta}$ is then used to construct a two dimensional scatterplot of $\{(\hat{\delta}^T X_i, Y_i)\}$. A smoother applied to this scatterplot yields an estimate of the nonparametric function g in (2.2). This technique also applies to partial linear models, see Spiegelman (1976), Rice (1986), Heckman (1986). Theoretical properties of the ADE technique are given in Härdle and Stoker (1988).

3. The Desirable Computing Environment

The computing environment necessary to perform smoothing techniques falls into three layers (Chambers, 1986) : The individual computer, the operating system and the special logical structures designed for effective smoothing. It is clear that all three parts interact with each other and that the degree of interaction varies from machine to machine. There is no useful purpose here to discuss the interaction between hardware and operating system here since sophisticated machines become affordable nowadays even on the desktop level. However, the minimal requirement on the hardware side for good performance of smoothing techniques seems to be a Personal Computer with color display and at least 200*640 pixels. Most important is the user interface which should allow various graphical controls desirably mouse oriented. The logical structures of such an intelligent computing environment should meet the following basic requirements in order to allow a high degree of interactiveness with the data or intermediate stages of an analysis.

(3.1) The interactive system must enable fast and informative comparison of different fits, preferably in a graphical way (colors, lineystyle,...).

(3.2) Since a data set might show an interesting structure only from a certain angle, certain viewpoints or snapshots of the data and its current fits should be recordable.

(3.3) Intermediate stages of the analysis should be storable and evocable in a further analysis step.

(3.4) The system should be open to new programs and macros. The macro language or the new programs should have well defined system interfaces.

Oldford and Peters (1986) propose the object oriented approach in order to meet such requirements. In the object oriented approach data is not only seen as a data matrix but as a more complex object that not only carries the pure numbers but also display information, linestyle, masks and other data analytic features. The system S by Becker and Chambers (1984) is a good example for an open analysis system. Also in XploRe (1987) the object oriented approach is taken. Moreover the *inheritance principle* is employed in order to facilitate graphical interpretation of the results. This principle allows the user of the system to inherit certain attributes from existing objects (data) to offspring objects (data).

Suppose for instance one has decided to mask part of a certain data set with a marker "1". (This can be done interactively through mouse control defining a rectangle.) This mask information can be inherited to any smoother applied to that data. The smooth fit as an offspring object can thus show the marker "1" at the same data points where it was in the raw data. This technique makes it simple to trace influences of points or to understand certain local structures of a nonparametric fit.

In XploRe there exist four different object types: vector-, workunit-, text- and picture-objects. Workunits are collections of vector objects and represent point clouds. Workunits are established by pointers to existing vector objects and include various data analytic attributes, for details see Härdle (1987). Text objects are containing ASCII information and have been included for formatted input/output and for display of intermediate numerical results and summary statistics. Picture objects have been defined to allow handling of different projections or aspects of data in different viewports or sections of the screen. Figure 1 shows two different picture objects (viewports). Three workunit objects are shown in two picture objects. The big plot (picture object *picture 1*) shows two vector objects of a ten dimensional workunit as dots. The smooth curve is another workunit created by a k−nearest neighbor. The second picture object in the right upper corner shows a workunit object containing the marginal density of the X variable of first workunit.

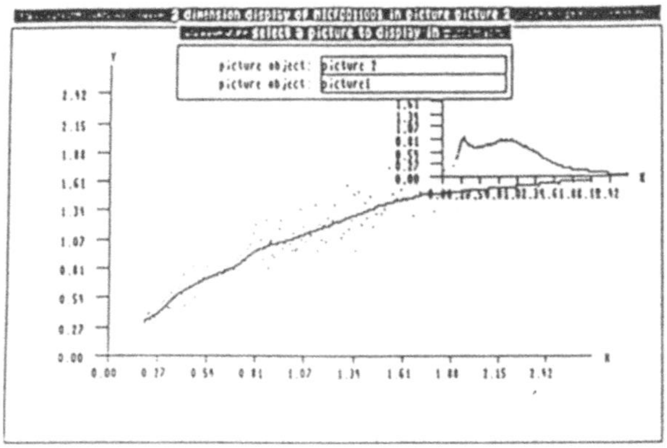

Figure 1. Three workunit objects in two picture objects.

The smoothing parameters for these plots have been selected by cross-validation.

4. An Application to Side Impact Data

The above described techniques have been applied to some side impact data. The object of the study was to compare the behavior of dummies and Post-Mortem-Human-Subjects (PMHS), see Kallieris, Mattern and Härdle (1986). The WARP technique was used to compute a three dimensional density estimate of three biomechanical variables. The object was to see the distribution of these variables for dummies and for PMHS. There were 31 dummie variables and 58 PMHS variables. Figure 2 shows the three dimensional density of these variables for dummies with a contour at level 5 percent of the modal level.

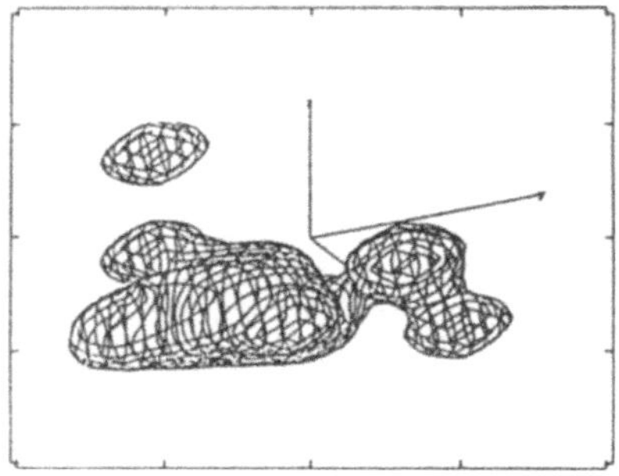

Figure 2. WARPed density smoother for dummies

Figure 3 shows the density of the same variables for PMHS at the same contour level.

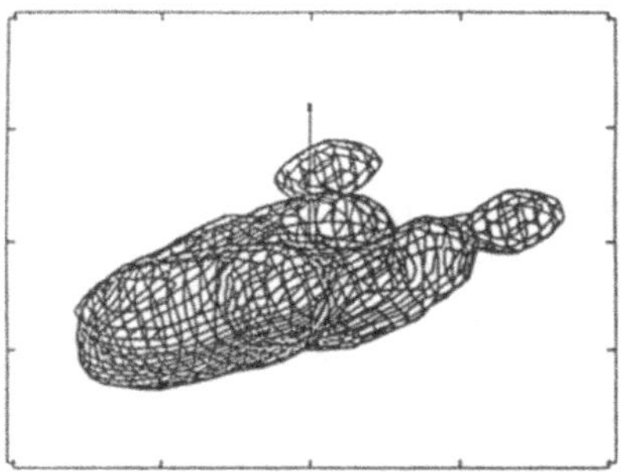

Figure 3. WARPed density smoother for PMHS

Interestingly there are two clusters at the "drumsticks" of the "frozen duck" shape in Figure 3. Both density contours are plotted on the same scale and show a very different shape due to the fact that the frozen duck has higher z–values and lower x–values than the "flying duck" (Figure 2).

The average derivative technique has been applied to this data set as well. Let Y be the random variable with $Y = 1$ indicating fatal injury and $Y = 0$ indicating survival. Figure 4 shows the density contours for the variables $(AGE, VEL, T12RM)$ for the two groups with $Y = 0$ or 1 respectively. The contour in the foreground corresponds to $Y = 1$.

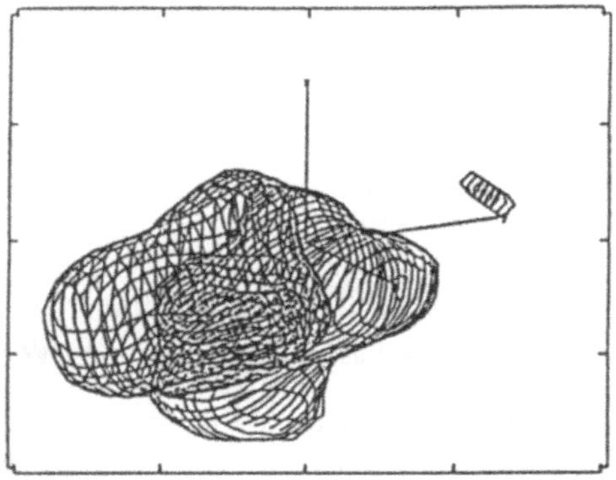

Figure 4. WARP density countours for PMHS

Setting up the above PPR model with unknown link function g we have applied the ADE technique to find projections $\hat{\delta}$. Figure 5 displays the projected predictor variables $\hat{\delta}^T X$ versus the reponse Y. The solid line indicates the WARP kernel smoother with biweight kernel.

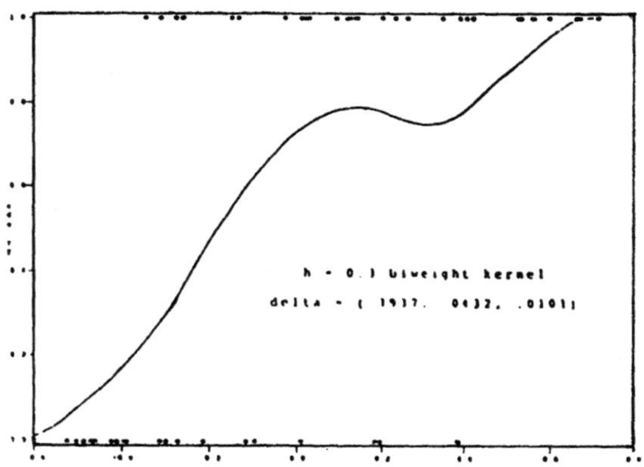

Figure 5. WARPed ADE technique and WARPed kernel smoother.

The group of outlying observations could be identified as observations with low AGE variable.

References

Becker, R.A. and Chambers, J.M. (1984). S: An Interactive Environment for Data Analysis and Graphics. *Wadsworth, Belmont.*

Breiman, L. ; Friedman, J. ; Olshen, R. and Stone, C.J. (1984). Classification and Regression Trees. *Wadsworth, Belmont.*

Breiman, L. and Friedman, J. (1985). Estimating Optimal Transformations for Multiple Regression and Correlation (with discussion). *J.Amer.Stat.Assoc., 80, 580-619.*

Caussinus, H. (1986). Models and Uses of Principal Component Analysis. *Proc. Multidimensional Data Analysis, Cambridge, DSWO Press.*

Caussinus, H. (1987). Discussion of "What is Projection Pursuit ?" by Jones, M.C. and Sibson, R. *J.Royal Statist. Soc.(A), 150, 26.*

Chambers, J.M. (1986). Computing Environments for Quantitative Applications. *ATT Bell Labs Stat. Research Reports 17.*

Friedman, J. and Stuetzle, W. (1981). Projection Pursuit Regression. *J.Amer.Stat.Assoc., 76, 817-823.*

Friedman, J. and Stuetzle, W. (1982). Smoothing of Scatterplots. *Tech. Report Orion 3. Dept. Statistics, Stanford University.*

Härdle, W. (1988a). Interactive Smoothing Techniques. *in: Proceedings of the Interface Conference, 1988, Reston Virginia, E. Wegman, Ed.*

Härdle, W. (1988b). Applied Nonparametric Regression. *Cambridge University Press, to appear.*

Härdle, W. and Bowman A. (1988). Bootstrapping in Nonparametric Regression : Local Adaptive Smoothing and Confidence Bands. *J. Amer. Statist. Assoc., 83, 102-110.*

Härdle, W. ; Hall, P. and Marron, J.S. (1988). How Far are Automatically Chosen Regression Smoothing Parameters from Their Optimum ? (with discussion). *J. Amer. Statist. Assoc., 83, 86-101.*

Härdle, W. and Scott, D.W. (1988). Smoothing in Low and High Dimensions by Weighted Averaging using Rounded Points. *Statistical Science, submitted.*

Härdle, W. and Stoker, T. (1988). Investigating Smooth Multiple Regression by the Method of Average Derivatives. *J. Amer. Statist. Assoc., submitted.*

Hastie, T. and Tibshirani, R. (1986). Generalized Additive Models. *Statistical*

Science, 1, 297-318.

Heckman, N. (1986). Spline Smoothing in a Partly Linear Model. *J.Royal Statist. Soc.(B), 48, 244-248.*

Huber, P.J. (1985). Projection Pursuit (with discussion). *Ann. Statist., 13, 435-375.*

Ibragimov, I.A. and Khasm'insksii, R.Z. (1981). Asymptotic Quality Boundaries of Regression Estimation in L_p. *Zap. Nauch. Sem. Lomi., 97, 88-101.*

Jones, M.C. and Sibson, R. (1987). What is Projection Pursuit ? (with discussion). *J.Royal Statist. Soc.(A), 150, 1-39.*

Kallieris, D. ; Mattern, R. and Härdle, W. (1986). Belastbarkeitsgrenye und Verletzungsmechanik des angegurteten PKW-Insassen beim Seitaufprall. Phase II: Ansätze zur Verletzungsprädiktion. *FAT Schriftenreihe 60, Forschungsvereinigung Automobiltechnik e.V. (FAT).*

McDonald, J. and Pederson, J. (1986). Computing Environments for Data Analysis, Part 3: Programming Environments. *Laboratory for Computational Statistics, Stanford Technical Report 24.*

Müller, H.G. (1987). Weighted Local Regression and Kernel Methods for Nonparametric Curve Fitting. *J. Amer. Statist. Assoc., 82, 231-238.*

Oldford, R.W. and Peters, S.C. (1985). DINDE : Towards more Statistically Sophisticated Software. *MIT, Technical Report 55.*

Rice, J.A. (1986). Convergence Rates for Partially Splined Models . *Statistics and Probability Letters, 4, 203-208.*

Scott, D.W. (1986). Data Analysis in 3 and 4 Dimensions with Nonparametric Density Estimation. *in: Statistical Image Processing, E. Wegman and D. DePriest, eds., Marcel Dekker, New York, 291-305.*

Scott, D.W. and Terrell, G.R. (1987). Biased and Unbiased Crossvalidation in Density Estimation. *J. Amer. Statist. Assoc., 82, 1131-1146.*

Silverman, B.W. (1984). Spline Smoothing: The equivalent Variable Kernel Method. *Ann. Statist., 12, 898-916.*

Silverman, B.W. (1986). Density Estimation for Statistics and Data Analysis. *Chapman and Hall, London.*

Stone, C.J. (1982). Optimal Global Rates of Convergence for Nonparametric Regression. *Ann.Statist., 10, 1040-1053.*

XploRe (1987). XploRe - a computing environment for eXploratory Regression and density smoothing. *Wirtschaftstheorie II, Universität D-5300 Bonn.*

A Boundary Modification of Kernel Function Smoothing, with Application to Insulin Absorption Kinetics

P. Hougaard, Bagsværd

Summary

Kernel function smoothing of observations lying around a regression function is a procedure with marked boundary effects and therefore specific consideration of boundary problems is important. A traditional solution is the construction of kernels designed specifically for use at the boundary. These are derived assuming that the design is continuous. The problem is solved for minimum variance kernels, but not for optimal kernels. In practice, the design is discrete and the boundary kernels may give large variance and biased results even for linear functions. A new procedure, the so-called time-adjustment is proposed. This procedure can be applied to estimates both with and without boundary modifications and is an improvement in both cases.

Introduction

A smooth curve and its derivatives can be estimated non-parametrically from observations at some design points (for example different times) by means of kernel function smoothing. With this technique the function estimate at a time point is a weighted average of observations in a neighborhood, given by the window b.

Suppose we have observations $Y_1,\ldots,Y_n$ at time points $t_1,\ldots,t_n$ placed in an interval, say $t_1=0$, $t_n=\tau$. The model which is assumed is that $Y_j=\eta(t_j)+\varepsilon_j$, where $\varepsilon_1,\ldots,\varepsilon_n$ are independent and identically distributed with zero mean and variance σ^2. The weights are determined by a kernel function, say $w(\cdot)$, which for convenience has support on $[-1;1]$. Following Gasser et al (1984) each observation is assigned the weight of an interval given by the interpolating sequence s_j, $j=0,\ldots,n$, with

$$s_0=-\infty,\ t_j\le s_j\le t_{j+1},\ j=1,\ldots,n-1,\ s_n=+\infty.$$

It is natural to choose $s_j=0.5(x_j+x_{j+1})$, $j=1,\ldots,n-1$. That gives the following expression for the estimate at time t

$$\hat{\eta}(t) = b^{-1}\sum_{j=1}^{n} Y_j \int_{s_{j-1}}^{s_j} w\{(t-u)/b\}\,du.$$

The kernels w are chosen such that the procedure is exact (unbiased) for certain polynomials, when the design is continuous, i.e. sampling infinitely often, $b>0$ and $\sigma^2=0$. A kernel is of order (ν,k) if it estimates the νth derivative and is exact for polynomials of order $0,1,..,k-1$. It is natural to study kernels with some kind of optimality properties, see Gasser, Müller and Mammitzsch (1985). At the boundary (i.e. within the intervals $[0,b]$ and $[\tau-b,\tau]$) special consideration is needed, because with the above procedure a severe bias is obtained. An approach to avoid this is to use specific boundary kernels, see Gasser, Müller and Mammitzsch (1985). Theorem 3 of that paper describes how to find the minimum variance kernels. There is one kernel function for each value of t/b respectively $(t-\tau)/b$. It is the polynomial of smallest order among those satisfying the side conditions concerning exactness for certain polynomials, in the case of continuous design. For optimal kernels this problem is not completely solved.

The time-adjustment procedure

In practice the boundary kernels procedure is biased even for linear functions, because the design is discrete. This paper suggests to adjust the time-scale to remove the bias for linear functions. To show how this works, consider an example. Let the first observation times be 0, 1, 2 and 3. Using the simple weights obtained by not accounting for boundary effects, the weights of the first three observations for estimating the function value at 0 are approximately 0.68, 0.27 and 0.05 using a window of 2 and Epanechnikovs kernel $0.75-0.75x^2$. This is biased and cannot represent the function value at time 0. Rather it could represent the function at time 0.37, the corresponding weighted average of time values. The suggestion is to adjust the time scale, such that the estimate at time 0 is moved to 0.37. For the minimum variance boundary kernel of order (0,2) the weights are 0.8125, 0.5 and -0.3125. In this case the corresponding average of time values is -0.125, better than the simple estimate, but not perfect. The variance of this estimate is approximately 1.008 times the variance on a single observation, implying that it is worse than using the observation at 0 as the estimate. To calculate the time-adjustment in practice we determine $\hat{i}$, the kernel estimate of the discrete identity, that is the identity restricted to the design points. This is done by inserting $t_1,...,t_n$ both as observations and design points. The time-adjusted kernel estimate is then the function $\hat{\eta}\circ \hat{i}^{-1}$, where $\hat{\eta}$ is the ordinary kernel estimate. Values outside the observed time range are neglected. Of course the same kernel function and window size is used for $\hat{\eta}$ and $\hat{i}$. With continuous sampling this gives no difference with the usually applied kernels, because these are exact for linear functions. This implies that the order of the kernel is not influenced by the adjustment. Asymptotically the adjustment disappears.

As described the time-adjustment procedure can be applied both to the simple estimate and to the boundary kernels estimate. In the case of the simple estimate it is easy to differentiate the estimate, provided the kernel is differentiable. We find

$$\frac{d}{dx}\hat{\eta} \circ \hat{i}^{-1} = \frac{d\hat{\eta}}{dt}\left(\frac{d\hat{i}}{dt}\right)^{-1},$$

letting x denote the adjusted time scale. For kernels to calculate derivatives we have first to find the corresponding kernel for the function value in order to calculate the time-adjustment, then it can be differentiated. For boundary kernels the relation between $\hat{\eta}$ and $d\hat{\eta}/dt$ is less clear because the kernel functions change with t. However, a simple approach which seems to work well is to estimate $d\hat{\eta}/dt$ and $d\hat{i}/dt$ by a kernel for estimation of derivatives and to estimate the time-adjustment for each t by the kernel which is the integral of the kernel used at that time point.

Comparison

For any design the actual variance is easily calculated as $c \cdot \sigma^2$ and can be integrated over the observation period. For any specific function the actual bias can also be calculated and the integrated mean square bias is simply found. We examine as an example a specific design with observations times i/n, $i=0,1,\dots,n$ for $n=10$ and using a window of 0.4, such that boundary effects are present over practically the whole interval. We find the root integrated variance (IV) described as the factor to σ and the root integrated mean square bias for the polynomials t (IMSB(1)) and t^2 (IMSB(2)).

Method	IV	IMSB(1)	IMSB(2)
Simple	0.431	0.0398	0.0588
Adjusted	0.496	0.0000	0.0413
Boundary	0.483	0.0019	0.0363
Adjusted boundary	0.458	0.0000	0.0360

This picture is repeated for most of a number of designs and kernels examined. The time-adjusted estimate has higher variance but smaller bias than the simple estimate. The boundary kernel procedure is better than the time-adjusted unless the window is fairly small. The best procedure is the time-adjusted boundary kernels estimate.

Application to insulin absorption

The absorption of insulin after subcutaneous injection of radioactively labelled insulin (Actrapid, dose 0.1 U/kg) was studied in seven pigs. The insulin depot was measured by external γ-counting at 0, 0.25, 0.5, 0.75, 1, 1.5, 2, 2.5, 3 and 4 hours after injection. The measurements are expressed in percentage of the first measurement. The aim was to estimate the rate of absorption, i.e. the transfer from the depot to the blood, which is minus the derivative of the residual amount of insulin in the depot. The figure shows the kernel estimates of the absorption rate, all based on the minimum variance kernel of order (1,3) and a window of 1.25 h. The simple estimate is too low near the boundaries. The boundary estimate shows strange fluctuations, in particular at the right boundary, due to the sparse design at that time. The window was chosen to get a good fit at the left boundary, which is the most important from a clinical point of view. The adjusted curves show that it is possible to remove most of the design-created fluctuations and that it is possible to estimate the whole curve non-parametrically after only 7 measurement points. Thus the proposed time-adjustment can in practical applications be expected to reduce bias and stabilize the boundary estimates.

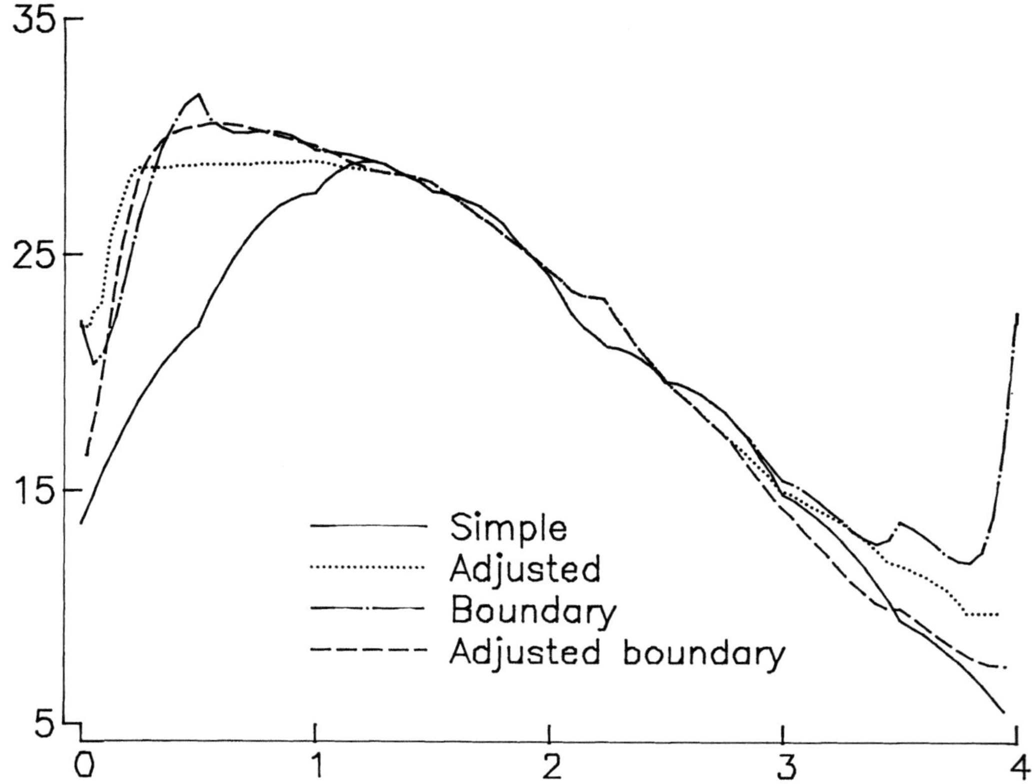

Estimates of insulin absorption (% of dose/h) as function of time since injection (hours).

References

Gasser, T., Müller, H.-G., Köhler, W., Molinari, L., Prader, A. (1984). Nonparametric regression analysis of growth curves. Annals of Statistics 12, 210-229.

Gasser. T., Müller, H.-G., Mammitzsch, V. (1985). Kernels for nonparametric curve estimation. Journal of the Royal Statistical Society B47, 238-252.

A Roughness Penalty Regression Approach for Statistical Graphics

M. G. Schimek, Graz and Bath

ABSTRACT: A discrete version of the spline smoothing technique is developed to deal with curve estimation problems when the errors have stationary, but not necessarily independent stochastic structure. Both autoregressive and moving average error structures are considered. The use of band matrix manipulations makes it possible to construct linear time algorithms in both cases.
KEYWORDS: AR errors, Cholesky decomposition, MA errors, nonparametric regression, roughness penalty, spline smoothing, statistical graphics

INTRODUCTION

In practice statisticians are often confronted with data sets relating a design variable to a response variable. Their main interest is the functional relationship between the design points and the observations. Usually only noise measurements are available. Assuming this, a parametric relationship needs the support of background information or past empirical experience. If prior knowledge is not obtainable or the pattern of the data is too complex, a nonparametric roughness penalty regression approach is a good choice. Most of these approaches make use of smoothing splines (WAHBA 1975, 1978, SILVERMAN 1984,1985). As long as the true function is known to be smooth and the assumption of independent identically distributed errors holds, nonparametric spline regression is a technique of high value for the data analyst (e.g. SCHIMEK and KATSCHNIG 1987). If the errors depart from the usual independence assumption there is little known about the consequences. There are two spline smoothing approaches for autocorrelated error sequences (WECKER and ANSLEY 1983, DIGGLE and HUTCHINSON 1987). Both are of limited practical value To overcome most of their drawbacks we propose an alternative roughness penalty approach, which does not make use of the spline concept, although being functionally related to it. It allows for both autoregressive (AR) and moving average (MA) error sequences. Our aim is a good graphical representation of potential functional relationships and a better understanding of the effects of independence assumption violations when smoothing with splines.

THE REGRESSION MODEL AND ITS LEAST SQUARES ESTIMATOR

Let us have an univariate regression problem

$$y_i = g(t_i) + e_i \; , \qquad i = 1,2,\ldots,n, \tag{1}$$

with observations y_i, equally spaced design (time) points t_i and errors e_i determined by an AR or MA process of any order. $g(t_i)$ is assumed to be some smooth regression function which we want to know at the discrete design points t_i. Our intention is to identify an underlying trend model while having measurements which are corrupted by dependent error sequences or white noise as a special case (i.e. independent errors).

We propose a nonparametric regression model with a discrete roughness penalty based on a second order difference quotient

$$g''(t_i) = \frac{g(t_i+h)-2g(t_i)+g(t_i-h)}{h^2} = \frac{\Delta_h}{h^2} = g(t_i)$$

with Δ_h a linear operator and h a scaling integer. The roughness penalty takes the form

$$P(g) = \alpha h^{-3} \sum_{i=h+1}^{n-h} (\Delta_h g(t_i))^2 \tag{2}$$

with a smoothing parameter $\alpha>0$. Without loss of generality we assume h=1 and (2) defines a (n-2)xn linear operator matrix $\underline{\Delta}$ for $i= 1,2,\ldots,n$.

Under the assumptions made, our regression model is specified by a discrete sum of squares

$$S(\underline{g}) = (\underline{y}-\underline{g})^T \underline{D}^{-1} (\underline{y}-\underline{g}) + \alpha \underline{g}^T \underline{Q}\ \underline{g} \; . \tag{3}$$

$\underline{g}$ is a n-dimensional column vector of curve estimates $g(t_i)$ and $\underline{y}$ a n-dimensional column vector of observations y_i. $\underline{D}^{-1}$ denotes the inverted nxn dispersion matrix $\underline{D}$ as usually defined and $\underline{Q}$ is a symmetric and banded nxn matrix with $\underline{Q} = \underline{\Delta}^T\underline{\Delta}$. The roughness penalty of our approach is as simple as a quadratic form $\underline{Q}\,(\underline{g})$

multiplied by the scalar α, representing the smoothing parameter. For a given error process the vector $\underline{g}$ of the curve estimates does not only depend on the vector $\underline{y}$ of the observations but also on the choice of the parameter α which compromises between the smoothness and the goodness of fit of the regression function. This is what makes roughness penalty approaches so flexible in terms of modelling trend corrupted by noise.

For the derivation of a least squares estimator we do some algebra on our sum of squares equation (3) and end up with

$$S(\underline{g}) = \underline{g}^T (\underline{D}^{-1} + \alpha\underline{Q})\underline{g} - 2\underline{g}^T\underline{D}^{-1}\underline{y} + \underline{y}^T\underline{D}^{-1}\underline{y} .$$

As can be seen, this equation has a known minimal solution

$$\hat{\underline{g}} = (\underline{D}^{-1} + \alpha\underline{Q})^{-1}\underline{D}^{-1}\underline{y} . \tag{4}$$

We can write this least squares result as

$$\hat{\underline{g}} = \underline{A}\ \underline{y}$$

where

$$\underline{A} = (\underline{D}^{-1} + \alpha\underline{Q})^{-1}\underline{D}^{-1} .$$

Whether we can solve the equation above dependens on the covariance structure of the error process involved.

A LEAST SQUARES SOLUTION FOR AR ERRORS

Let us define an autoregressive error sequence of order p (AR(p)) according to BOX and JENKINS (1976, p.51)

$$e_t = \emptyset_1 e_{t-1} + \emptyset_2 e_{t-2} + \dots + \emptyset_p e_{t-p} + \varepsilon_t$$

with autoregressive parameters $\emptyset_i$ and a $N(0, \sigma_\varepsilon^2)$ distributed random variable ε_t. Due to WISE (1955) the inverse of the semi-infinite covariance matrix of an AR(p) process is symmetric and banded with bandwidth 2p+1. Applying some results from GALBRAITH and GALBRAITH (1974) we are in the position to calculate directly those elements of $\underline{D}^{-1}$ we need for our purpose. Under the assumption of weak stationarity and positive definiteness Cholesky decomposition can be applied to the inverse $\underline{D}^{-1}$ of the dispersion matrix $\underline{D}$ (HARVEY 1981, p.206). After some matrix manipulations equation (4) takes the form

$$\underline{B}\,\hat{\underline{g}} = \underline{D}^{-1}\underline{y} \qquad \text{with} \qquad \underline{B} = (\underline{D}^{-1} + \alpha\underline{Q}),$$

where $\underline{B}$ is a symmetric positive definite band matrix, which can be decomposed by Cholesky in $\underline{B} = \underline{L}_B\underline{L}_B^T$. Decomposing $\underline{D}^{-1} = \underline{L}_D{-1}\ \underline{L}_D^T{-1}$ yie

$$\underline{B}\,\hat{\underline{g}} = \underline{L}_D{-1}\ \underline{L}_D^T{-1}\ \underline{y}.$$

The substitutions $\underline{u} = \underline{L}_D^T{-1}\ \underline{y}$ and $\underline{v} = \underline{L}_D{-1}\ \underline{u}$ result in $\underline{B}\,\hat{\underline{g}} = \underline{v}$ which can be solved since

$$\underline{L}_B\ L_B^T\ \hat{\underline{g}} = \underline{v}.$$

Another substitution $\underline{L}_B\ \underline{x} = \underline{v}$ allows us to solve for $\underline{x}$ and finally for the curve estimator $\hat{\underline{g}}$ due to the triangular structure of $\underline{L}_B^T$ in

$$\underline{L}_B^T\ \hat{\underline{g}} = \underline{x}.$$

Thus we have obtained an exact least squares solution for known AR error structures.

A LEAST SQUARES SOLUTION FOR MA ERRORS

Let us define a moving average error sequence of order q (MA(q)) consistent with BOX and JENKINS (1976, p.52)

$$e_t = \varepsilon_t - \theta_1\ \varepsilon_{t-1} - \theta_2\ \varepsilon_{t-2} - \dots - \theta_q\ \varepsilon_{t-q}$$

with moving average parameters θ_i and idenpendent identically $N(0,\ \sigma_\varepsilon^2)$ distributed random variables ε_t. Under the zero mean assumption and $k = \dots, -2,-1,0,1,2,\dots$ the autocovariance function of a MA(q) process is given by

$$k = \begin{cases} 0 & \text{for} \quad k > q \\ \sigma_\varepsilon^2 \sum_{i=0}^{q-k} \theta_i\ \theta_{i+k} & \text{for} \quad k = 0,1,\dots,q \\ \gamma_{-k} & \text{for} \quad k < 0 \end{cases}$$

Due to the symmetry of the function γ_k and the fact that the number of different non-zero values is a function of the order q, the dispersion matrix $\underline{D}$ has the Laurent property, is positive definite and banded with bandwidth $2q + 1$. The Cholesky decomposition can be applied and we have $\underline{D} = \underline{L}_D\ \underline{L}_D^T$. Under the symmetry of $\underline{D}$ follows $\underline{D}^{-1} = \underline{L}_D^{-T}\ \underline{L}_D^{-1}$. We transform equation (4) to

$$(\underline{D}^{-1} + \alpha\underline{Q})\ \hat{\underline{g}} = \underline{D}^{-1}\ \underline{y}$$

and the decomposition of $\underline{D}^{-1}$ derived above can be applied resulting in

$$(\underline{L}_D^{-T}\ \underline{L}_D^{-1} + \alpha\underline{Q})\ \underline{\hat{g}} = (\underline{L}_D^{-T}\ \underline{L}_D^{-1})\ \underline{y}$$

and after some algebra

$$\underline{B}\ \underline{L}_D^{-1}\ \underline{\hat{g}} = \underline{L}_D^{-1}\ \underline{y} \quad \text{with} \quad \underline{B} = (\underline{I} + \alpha\underline{L}_D^{T}\ \underline{Q}\ \underline{L}_D). \tag{5}$$

We substitute $\underline{u} = \underline{L}_D^{-1}\ \underline{y}$ and $\underline{v} = \underline{L}_D^{-1}\ \underline{\hat{g}}$.
Then we solve $\underline{L}_D\ \underline{u} = \underline{y}$ for $\underline{u}$ and substitute back into equation (5), obtaining

$$\underline{B}\ \underline{v} = \underline{u} \tag{6}$$

with $\underline{v}$ unknown. As $\underline{B}$ is symmetric, positive definite and banded, we decompose in equation (6) to

$$\underline{L}_B\ \underline{L}_B^{T}\ \underline{v} = \underline{u}\ .$$

Finally we have

$$\underline{L}_B^{T}\ \underline{v} = \underline{L}_B^{-1}\ \underline{u},$$

substitute $\underline{w} = \underline{L}_B^{-1}$ and multiply with $\underline{L}_B$ from left to avoid the calculation of $\underline{L}_B^{-1}$. We successively solve $\underline{L}_B\ \underline{w} = \underline{u}$, $\underline{L}_B^{T}\ \underline{v} = \underline{w}$ and $\underline{L}_B\ \underline{v} = \underline{\hat{g}}$ instead of $\underline{v} = \underline{L}_B^{-1}\ \underline{\hat{g}}$. Due to the triangular structure of $\underline{L}_B$ the last equation can be solved for $\underline{\hat{g}}$ and yields an exact least squares solution for known MA error structures.

IMPLEMENTATIONAL AND NUMERICAL FEATURES

The roughness penalty regression approach has been implemented in PASCAL as a structured program. The basic concept is a tree structure with an AR and a MA path. There is an interactive user surface and a graphics interface (FORTRAN formated ASCII files) for output. Throughout the program we made use of the specific features of the pattern matrices involved to keep the number of operations low. We do not apply numerical inversions of matrices to save costs and to increase numerical stability. From the algorithmic point of view the most important property of our approach is that $\underline{D}^{-1}$ and $\underline{B} = (\underline{D}^{-1} + \alpha\underline{Q})$ for AR errors and $\underline{D}$ and $\underline{B} = (\underline{I} + \alpha\underline{L}_D^{T}\ \underline{Q}\ \underline{L}_D)$ for MA errors are all positive definite, symmetric and bandlimited. As the bandwidth is a function of the order of the linear stochastic error process, thus small in practice, we can apply a special case of Cholesky decomposition for band matrices (MARTIN and WILKINSON 1965). Other algorithms we use are Gaussian elimination for triangular equation systems (back and forward substitution) and non-standard multiplication of triangular matrices by vectors.

The numerical evaluation of the algorithms for the calculation of $\underline{D}^{-1}$ (AR) and $\underline{D}$ (MA) and for the above computations in terms of operations (multiplications, divisions and square roots) yields the following approximate results, given n observations and error process order m:

algorithm	operations
$\underline{D}^{-1}$ (AR)	$(n-m)(m+1)(m+2)$
$\underline{D}$ (MA)	$(n-m)(m+1)+(m+1)^2$
band decomposition	$1/2n(m+1)(m+2)$
triangular equation	$1/2n(n-1)$
triangular multiplication	$1/2n(n+1)$

Since all other computations are less complex, for $n \gg m$ the overall calculation costs of our procedure are of $O(n)$. Thus the roughness penalty regression approach is fast and efficient, even for large n, in spite of its potential to model a wide variety of error structures

STATISTICAL GRAPHICS APPLICATION

For our regression problem (1) we simulated n=50 observations y_i based on a trend model $g(t_i)=2.5\sin(5\pi t_i)$ with $t_i=(i-0.5)/n$ and additive errors e_i determined by an AR(1) process with $\emptyset = 0.5$ and ε_t distributed N(0,1). We applied our procedure for a set of values α with and without prior knowledge of the error process. Graphical comparisons of the estimated regression function $\hat{g}(t_i)$ and the trend model $g(t_i)$ indicate that the latter can be identified properl even with little prior information on the error structure. For a given α the correct error specification reduces the error sum of squares of the concerned model. Moreover, it makes the functional relationship more reliable in some data ranges. For the purpose of illustration see the following figure:

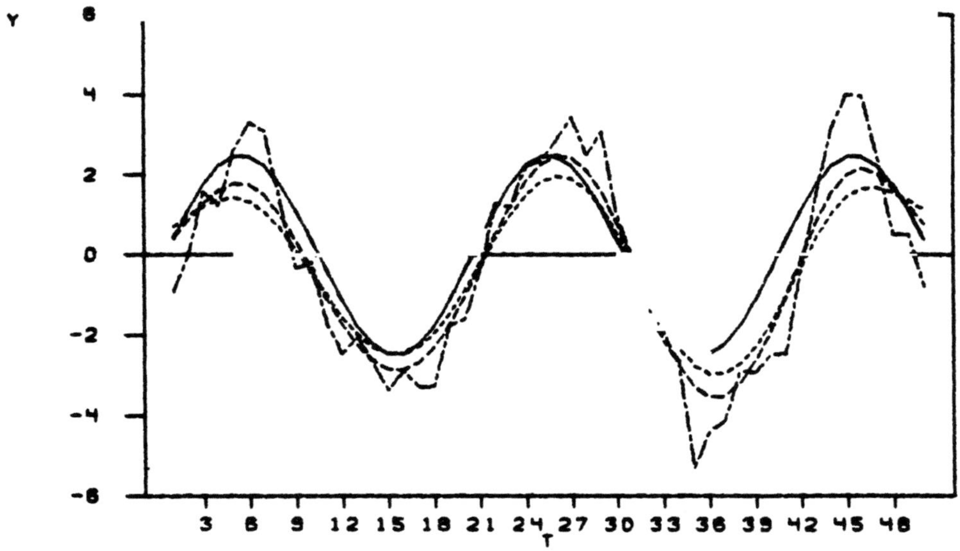

Sine function with positive autocorrelated errors

Observations y_i .... -·-
True trend $g(t_i)$....................................... ——
Estimated trend $\hat{g}_1(t_i)$ under AR(0) error assumption ... ---
Estimated trend $\hat{g}_2(t_i)$ under AR(1) error assumption ... — —

BOX,G.E.P. and JENKINS,G.M. (1976), Time series analysis. Forecasting and control. 2nd edition, Holden-Day, San Francisco.

DIGGLE,P.J. and HUTCHINSON,M.F. (1987), On spline smoothing with autocorrelated errors. CSIRO Technical Report.

GALBRAITH,R.F. and GALBRAITH,J.I. (1974), On the inverses of some patterned matrices arising in the theory of stationary time series. J Appl Prob 11, 63-71.

HARVEY,A.C. (1981), The econometric analysis of time series. Allan, Oxford.

MARTIN,R.S. and WILKINSON,J.H. (1965), Symmetric decomposition of positive definite band matrices. Numer Math 7, 355-361.

SCHIMEK,M.G. and KATSCHNIG,H. (1988), Die Entgiftung des Wiener Stadtgases - eine quasiexperimentelle Evaluation einer suizidpräventiven Maßnahme mittels nichtparametrischer Splineregression. ZSI 17, 293-320.

SILVERMAN,B.W. (1984), Spline smoothing: The equivalent variable kernel method. Annals Stat 12, 898-916.

SILVERMAN,B.W. (1985), Some aspects of the spline smoothing approach to non-parametric regression curve fitting. JRSS B 47, 1-52 (with discussion).

WECKER,W.E. and ANSLEY,C.F. (1983), The signal extraction approach to non-linear regression and spline smoothing. JASA 78, 81-89.

WHABA,G. (1975), Smoothing noisy data with spline functions. Numer Math 24, 383-393.

WHABA,G. (1978), Improper priors, spline smoothing and the problem of guarding against model errors in regression. JRSS B 40, 364-372.

WISE,J. (1955), The autocorrelation function and the spectral density function. Biometrika 42, 151-159.

Projection Pursuit

Detecting Structures by Means of Projection Pursuit

I. S. Yenyukov, Moscow

Summary

In this paper, the consideration of the projection pursuit for testing the presence of clusters is based on the model of the ellipsoidally symmetric unimodal densities mixture. It is shown that under this model the use of projections indices based on Renyi entropy or on third or fourth moments results in obtaining an estimate of the discriminant subspace. For estimating the Renyi indices values some forms of the order statistics are used. For detecting outliers the ratio of the standard variance estimate to a robust one is proposed as projection index. Indeces for discriminant analysis problem are introduced.

Keywords: projection pursuit, discriminant subspace estimating, cluster analysis, outliers detecting, discriminant analysis.

1. Introduction

The projection pursuit (PP) technique (Friedman, Tukey (1974), Buchshtaber, Maslov (1975, 1977), Huber (1985), Jones, Sibson (1987), Friedman (1987)) can be used to reveal any singularities of multivariate data and to find out a geometric structure of it, for instance the presence of clusters, outliers, concentration of points in the vicinity of nonlinear manifolds. The PP approach offers the visual analysis of "interesting" linear projections of the initial data for this goal.

More formally, let us have a sample $\mathbf{X}^{(n)}=(X_1,\ldots,X_n)$ of p-variate observations and let $\mathbf{U}$ be a linear orthogonal projection operator from R^p into R^q ($q \ll p$) and $Q(\mathbf{U},\mathbf{X}^{(n)})$ be the sampling value of some projection index (PI). The PI is disigned in such a way that its value is the larger the more expressed the explored structure in R^q is. We select the q-variate projection $Z=\mathbf{U}_1 X$ with the help of the solution of the problem

$$\mathbf{U}_1 = \underset{\mathfrak{P}(\mathbf{U})}{\operatorname{argmax}}\, Q(\mathbf{U},\mathbf{X}^{(n)}) \qquad (1)$$

where $\mathfrak{P}(\mathbf{U})$ is the set of orthogonal linear operators from R^p

into R^q .
Then we can do the visual analysis of q-variate points $Z_t=U_1X_t$.

The problem (1) is usually rather difficult to solve if q>1.That is why a step-wise approach is often used for obtaining an approximation of solution of (1). In this case we determine the sequentional of the projections vectors $U_1,\ldots,U_q$. At first the vector U_1 is determined by the solution of the problem

$$U_1 = \underset{U}{\operatorname{argmax}}\, Q(U,\mathbf{X}^{(n)}) \qquad (1')$$

Each following vector U_j is defined from (1') but on additional condition that the influence of all the before obtained vectors is excluded in a suitable way. So the q-dimensional projection $Z = \mathbf{U}_2'X$ can be defined, where the columns of matrix $\mathbf{U}_2$ are just these vectors. Certainly,generally speaking,the projectors $\mathbf{U}_2$and $\mathbf{U}_1$ result in the different projections. But the problem (1') is more simple and it may be shown that under some assumptions the use of the step-wise approach does not lead to loss of the information about the explored structure. Such a probability model for cluster structure we discuss in Sections 2. In Section 3 we consider the PI family based on Rényi entropy (Rényi (1970)). The main result obtained in Section 3 is that the subspace spanned on the projection vectors selected by the step-wise maximization of such PI is the discriminant subspace introduced by Rao (1965). In Section 4 we consider the problem of estimating the PI values.Forms based on the order statistics are proposed. Sections 5, 6 contain some results connected with the use of the PI based on high degree moments, including asymptotical results when $p \to \infty, n \to \infty$, $p/n=\eta>0$.In Section 7 some consequences of violation of the probability model assumptions are discussed.It is shown by example that PP procedure can select the projection with visual pattern like "dense kern with outliers" even if there are clearly discernable clusters and there are no actual outliers. Detecting the outliers is discussed in Section 8. In Section 9 we consider PP for discriminant analysis.

2. Probability model for describing the cluster structure.

We consider that the underlying distribution of the sample $\mathbf{X}^{(n)}$ is a k-component mixture with the density

$$p(X)=c(d,p,\mathbf{W})\sum_{i=1}^{k} a_i d((X-M_i)'\mathbf{W}^{-1}(X-M_i)), \qquad (2)$$

where

$d(y)$ is a postitive monotonously decreasing function of y $(y>0)$, that has finite moment of $(p/2+1)$- degree;

$c(d,p,\mathbf{W})$ is the normalizing constante;

$a_i > 0$ is the weight of the i-th component;

M_i is the means vector of the i-th component;

$\mathbf{W}$ is the within-component scatter (covariance) matrix.

So the density of i-th component is the unimodal ellipsoidally symmetric function. For instance, if $d(y)=\exp(-y/2)$ it is the normal distribution density. The mixture (2) may be considered as some model of the cluster structure. The density (2) has k modes (if the mixture components are enough scattered). And the points surrounding some mode may be regarded as the members of the same cluster.

Now let z be an one-dimensional projection $z=U'X$. The density of the z is the k-components unimodal symmetric densities mixture

$$f(z)=\sum_{i=1}^{k} a_i e_i(z), \quad e\ (z)=e((z-m_i)/w)/w,$$

where

$$e(z)=c(d)\int \rho^{p-2}d(z^2+\ \rho^2\ )d\rho; \quad m_i=U'M_i; \quad w^2=U'\mathbf{W}U.$$

The variance of the z is $s^2=U'\mathbf{S}U$.

Now define the discriminant subspace (Rao (1965)). It is good known that the covariance matrix $\mathbf{S}$ of the X may be presented as the sum the whithin-components scatter matrix $\mathbf{W}$ and between-components scatter matrix $\mathbf{B}$, i.e. $\mathbf{S}=\mathbf{B}+\mathbf{W}$, where $\mathbf{B}=\sum a_i(M_i-M)(M_i-M)'$, $M=\sum a_i M_i$ is means vector of the X. In according to it the variance $s^2(U)$ of the one-dimensional projection is $s^2(U)=b^2(U)+w^2(U)$, where $b^2(U)=U'\mathbf{B}U$. Regarde ratio

$$t^2(U)=b^2\ (U)/w^2\ (U) \qquad (3)$$

The value $t^2(U)$ is a measure of dissimilarity of the mixture components for the projection with vector U. Maximizing the $t^2(U)$ leads to the solution of the generalized eigenvectors problem

$$(\mathbf{B}-t\mathbf{W})V=0.$$

There are q' eigenvectors $V_1, \ldots, V_{q'}$ with corresponding positive eigenvalues $t_1, \ldots, t_{q'}$ $(q'< \min(p,k-1),\ t_i=t^2(V_i))$.

The discriminant subspace is defined as subspace

$\Omega = \text{span}(V_1, \dots, V_{q'})$.
The another definition of the Ω is $\Omega = \text{span}(\mathbf{W}^{-1}M_1, \dots, \mathbf{W}^{-1}M_k)$.
The q' depends on the configuration of the vectors $M_1, \dots, M_k$.

The Ω contains the complete information about the diffrences among the mixture (2) components. The PP gives the way to get the Ω without any information about the matrix $\mathbf{W}$ unlike discriminant analysis problem where we can estimate $\mathbf{W}$ and $\mathbf{B}$. Further without loss generality we will consider the case when M =0 (centered data).

3. PI suitable for testing the presence of clusters.

We consider the one-parametric family of the PI (Yenyukov (1986b), Jones, Sibson (1987)) based on Rényi entropy

$$Q_\beta(U,X)=s^\beta E_f f^\beta(z) = s^\beta \int f^{1+\beta}(z)dz \quad , \quad (\beta>0) . \tag{4}$$

Here the "theoretical" quantity of PI is designated by $Q_\beta(U,X)$ in contrast to the sampling quantity $Q_\beta(U,\mathbf{X}^{(n)})$.

We give without the proof the inequalities connecting the $Q_\beta(U,X)$ with the ratio $t^2(U)$ (3)

$$g(e,\beta)(\sum a_i^{1+\beta})(1+t^2(U))^{\beta/2}<Q_\beta(U,X)<g(e,\beta)(1+t^2(U))^{\beta/2} \tag{5}$$

where $g(e,\beta)=E_e e^\beta(z)$ does not depend on the U. When $t^2(U)=0$ then $Q_\beta(U,X)=g(e,\beta)$, it is the minimal value of this PI. If the Mahalanobis distances between all the mixture component pairs increase then the low bound is reached asymptotically. And so one may expect that if there are projections with the large values of the t^2 then the maximazing procedure of the Q_β selects these The last reasoning is heuristic rather than exact. However, it is possible to get exact results. We formulate it as Lemma 1.

Lemma 1. Let the model (2) be true. Assume that the vectors $U_1, \dots, U_{q'}$ are found by the step-wise (sequentional) maximitation of the $Q_\beta(U,X)$, and so that the vector U_i is S-orthogonal to the subspace $\text{span}(U_1, \dots, U_{i-1})$. Then every vector U_i belongs to the discriminant subspace Ω and what is more $\Omega = \text{span}(U_1, \dots, U_{q'})$.

Proof. The PI $Q_\beta(U,X)$ is some function G of the $s, w, m_1, \dots, m_k$, i.e. $Q_\beta(U,X)=G(s,w,m_1,\dots,m_k)$. Differentiating the G with respect to U and equating the result to zero gives

$$(G_s'/s)\mathbf{S}U + (G_w'/w)\mathbf{W}U + \sum G_{m_i}' M_i = 0$$

Multiplplying this equality by the $\mathbf{W}$ (from the left) gives after little algebra the equation for the vector U_1

$$(h(U_1)I_p + \mathbf{W}^{-1}\mathbf{B})U_1 - V = 0 \ , \qquad (6)$$

where $V = \sum G'_{m_i}\mathbf{W}^{-1}M_i$; $h(U) = G'_W + G'_S$.

The vector V is a linear combination of the vectors $\mathbf{W}^{-1}M_i$ from the Ω and therefore, itself belongs to the Ω . Now show that the maximazing vector U_1 belongs to the Ω .

Assume that it is not true, i.e. $U_1 = c_1U^+ + c_2U^-$, where $U^+ \in \Omega$, $U^- \in A$ (the subspace A is S-orthogonal to the subspace Ω) and $c_2 \neq 0$. Then the substitution of the U_1 into the (6) yields $c_2h(U_1)U^- = V - c_1(h(U_1)I_p + \mathbf{W}^{-1}\mathbf{B})U^+$.
The right side of this equality is a vector from Ω .Further,the value $h(U)=0$ only if $t^2(U)=0$, i.e. if $U \in A$(it can be checked by direct calculating the derivatives G'_S and G'_W).

Therefore, the last equality can be true either if $U_1 \in A$(and then $V=0, c_1=0$, and U_1 is not maximazing vector) or if $U^-=0$.Thus, the maximazing vector U_1 belongs to Ω . By analogy it is proved that the vectors $U_2, \dots, U_{q'}$ belong to the Ω (under the condition S-orthogonality). Since rank$(\Omega)=q'$ and these vectors are S- orthogonal,the $\Omega = \text{span}(U_1, \dots, U_{q'})$ and they are the basis of Ω .■
But, generally saying, this basis is other than canonical basis.

The following example shows that the PI (4) are not, generaly saying, monotonic functions of the varianaces ratio t^2 and it explains some behaviour details of the step-wise optimisation of PI (4).

Example 1. Take two one-dimensional mixtures with the densities

$$f_1(z) = a(2n_1(0,1;z) + n_1(\delta_1,1;z));$$

$$f_2(z) = a(n_1(-\delta_2,1;z) + n_1(0,1;z) + n_1(\delta_2,1;z)),$$

where $n_1(m,\sigma^2;z)$ is normal density with parametres m and σ^2, $a=\frac{1}{3}$.

We use PI (4) with $\beta=1$, i.e. $Q_1 = s\int f^2(z)dz$. The values of the t_i^2 , s_i^2 and $E_i = \int f_i^2(x)dx$ $(i=1,2)$ are given below

i	E_i	t_i^2	s_i^2
1	$g_0(5+4\exp(-\delta_1^2/4))/9$	$2\delta_1^2/9$	$1+2\delta_1^2/9$
2	$g_0(3+4\exp(-\delta_2^2/4)+2\exp(-\delta_2^2))/9$	$2\delta_2^2/3$	$1+2\delta_2^2/3$

When the values δ_1^2 and δ_2^2 are large enough the values $q_1 = s_1 E_1 \simeq (5\backslash 9) g_0 (1+2\delta_1^2/9)^{1/2}$, $q_2 = s_2 E_2 \simeq (3/9) g_0 (1+2\delta_2^2/3)^{1/2}$. Here $g_0 = \int n_1^2(0,1;x)dx$.

It is possible to find after some algebra that the inequality $(\sqrt{27/5})\delta_2 < \delta_1 < \sqrt{3}\,\delta_2$ entails inequalitis: $t_1^2 < t_2^2$ but $q_1 > q_2$.

One of the consecquences of the example is as follows. If some densities mixture (2) admits projections of both types - $f_1()$ and $f_2(\ )$, then a procedure of selecting the projections from maximazing PI (4) will at first select the bimodal projections . The effect has been noted as experimental fact Sibson, Jones (1987) (see too Friedman, Tukey(1974)).

4. Estimating of the PI values.

In practice it is necessary to be able to estimate the PI values for any onedimensional projection. Here we consider the estimate based on the high-order gaps.This approach have been used for estimating usual entropy functional in Vasicek (1976),Cressie (1978). We use it to estimating the (4).
Let $z_1,\ldots,z_n$ be the projections of the initial observations by a vector U, i.e. $z_i = U'X_i$ and let $z_{(1)},\ldots,z_{(n)}$ be the corresponding order statistics. Then we can design the following estimate of $E_f f^\beta(z)$ (we denote it by $E(\beta,U)$)

$$nE(\beta,U)=(2r/n)^\beta \sum_{i=1}^{n} (z_{(i+r)_-} - z_{(i-r)_+})^{-\beta}$$

where the r is integer value ($r<n/2$,$r=O(n^d)$,$d>0$);

$$(i+r)_- = \min(n,i+r)\ ,\ (i-r)_+ = \max(n,i-r).$$

It is possible to prove that this estimate is consistent and asymptotically normal. Finally, the estimate of the PI (4) is

$$Q_\beta(U,X^{(n)})=\hat{s}^\beta E(\beta,U)\ .$$

The estimate $E(\beta,U)$ depends on the value of r. But we have found experimentally that the influence of r is not so great. In practice we use the values $r=cn^{1/2}$ ($0.5< c <1.5$).

As the estimates of $\hat{s}^2$ we use the routine estimate and the weighted robust estimate with weights v_i ($i=1,\ldots,n$) defined in Section 9.

5. Discriminant subspace estimate based on high degree moments.

The third and fourth moments were proposed as PI by Huber (1985), Jones,Sibson (1987).We also used them (Yenyukov,Neuschtadt (1978)). Here we discuss the simple way to obtain with the help of these moments some families of vectors from Ω . We consider here only the case of third moment. But this results can be applied to the moments of order higher than the third one, and to the characteristic function(about the fourth moment see Yenyukov,Neuschtadt (1978)).

At first we note that if a vector $U \in \mathbb{M}=\mathrm{span}(M_1, \dots M_k)$ then the vectors $\mathbf{W}^{-1}U$ and $\mathbf{S}^{-1}U$ belong to the Ω . So, it is enough to obtain the estimate of vector from $\mathbb{M}$ and then to multiplay it by an estimate of the matrix $\mathbf{S}$.

The further results are based on the following expressions fo the third central moment of the projection of the mixture (2) by some vector V

$$\mu_3(V) = \sum_{i=1}^{k} a_i (M_i'V)^3 \tag{7}$$

and its estimate

$$\hat{\mu}_3(V) =(1/n)\sum_{j=1}^{n} (V'X_j)^3 \tag{7'}$$

Evidently

$$E\, \hat{\mu}_3(V)=c\mu_3(V) \tag{8}$$

where constant c=c(n) and Ez is the average of random value z.

Lemma 2. Let V be a p-dimensional vector,**H** is any positively defined matrix with sizes p*p. Consider the vectors

$$\hat{U}(V) =(1/n)\sum_{j=1}^{n} (V'X_j)^2 X_j \tag{9}$$

$$\hat{U}(\mathbf{H}) =(1/n)\sum_{j=1}^{n} (X_j'\mathbf{H}X\ )X_j \tag{10}$$

Then

$$E\hat{U}(V) =c\sum_{=1}^{k} a_i (M_i'V)^2 M_i \tag{11}$$

$$E\hat{U}(\mathbf{H}) =c\sum_{=1}^{k} a_i (M_i'\mathbf{H}M_i)M_i=cU(\mathbf{H}) \tag{12}$$

Thus the means of the vectors of (9) and (10) belong to the $\mathbb{M}$.

Proof. Let us take the derivatives from both parts of expression (7') with the respect to V and then use averaging E for them.Since differentiation is a linear operation we can change the order of operations of averaging and differentiation for the left part of expression (7') that entails equality (11). Equality (12) may be proved if we additionally consider that the vector $V \sim N_p(0,\mathbf{H})$ and use proper average. ■

6. The case when the means of components are placed on the same direct line (q'=1).

This case is interesting in several applications. On the other hand, we can obtain some results about the proposed estimates when the dimension p is great. In this case we can write $\mathbf{M}_t=b_t D$. ($\sum a_t b_t=0$) and the vectors U(**H**) and D are collinear. Suppose that the mixture (2) components densities are normal and matrix $\mathbf{W}=\mathbf{I}_p$. Let the ϕ be the angle between $\hat{U}(\mathbf{H})$ and D and $\gamma=\cos(\phi)$.

Lemma 3. Let $n \to \infty$, $p \to \infty$, $\eta=p/n>0$. Then

$$E\gamma=\mu_3(L)/[\mu_3^2(L)+ (p/n)(15+\mu_4(L)-(L'SL)^2 +2(L'SL))]^{1/2}+O(1/r),$$

$$E(\gamma-E\gamma)^2 =O(1/r),$$

where $L=D/\|D\|$;

r is n or p ; $\mu_4(L)$ is the fourth moment of the z=L'X.

Remark: The case when $\mathbf{W}=\mathbf{I}_p$ may be reduced by the linear transformation of the variables to the one considered above. The scalar product of D and U becomes $D'\mathbf{W}^{-1}U$.

If k=2 the main term of $E\gamma$ transforms in

$$E\gamma= A/(A +(p/n)B))^{1/2}$$

where $A=a(1-a)(1-2a)\tau^3$, $B=19+6a(1-a)\tau^2+a(1-a)(1-2a)^2\tau^4$

τ is Machalanobis' distance between the mixture components. Values γ for several sets (a, τ) given in the table below (the ratio η=0.1)

a	τ	$E\gamma$
0.2	2.5	0.67
0.2	2	0.45
0.05	4	0.83

7. Some consequences of violating the probabilty model assumptions.

Let us discuss some consequences of violating the condition of equality of the whithin-covariances matrices **W** of mixture (2) components. Consider the simple example of the density of two-variates vector X

$$p(X)=a\, n_2(M_1,W_1;X) + (1-a)n_2(M_2,W_2;X),\ (0<a<1)$$

where $n_q(M,W;X)$ is the normal density of q-variate vector with means vector M and covariances matrix W;

$$M_1'=(0\ ,-m)\quad ,\qquad M_2'=(0,\ m);$$

$$W_1=\begin{bmatrix}1/a-(1-a)/c & 0\\ 0 & 1/æ\end{bmatrix}\quad ,\ W_2=\begin{bmatrix}a/c & 0\\ 0 & 1/æ\end{bmatrix};$$

$$m^2=(1-1/æ)/2a(1-a),\ (æ>1).$$

It is easy to check that the X has the means vector M'=(0,0) and covariances matrix $S=I_2$. For the first coordinate axis the induced density is

$$f_1(z)=a\, n_2(0,1-a(1-a)/c;z) + (1-a)n_2(0,a(1-a)/c;z);$$

and for the second

$$f_2(z)=a\, n_2(-m,1/æ;z) + (1-a)n_2(m,1/æ;z).$$

Let us calculate the values of PI (4) under β =1. Note that $f_1(z)$ does not depend on æ and $f_2(z)$ does not depend on the value of c. Then

- for any value of c there exists such $æ_0=æ_0(c)$ that if $æ > æ_0$ the projection on the second coordinate axis gives maximum to $Q_1(U_2,X)$, i.e. optimal projection vector $U_2' = (0,1)$
- for any value of æ there exists such $c_0=c_0(æ)$ that if $c > c_0$ the projection on the first coordinate axis gives maximum to $Q_1(U_1,X)$, i.e. optimal projection vector $U_1' = (1,0)$

Therefore, the value of PI (4) will increase also due to the increase (or decrease) of the variances ratio $w_1^2(U)/w_2^2(U)$.

The projection with the vector U_2 gives visual pattern of two clearly discernable clusters if the m is large enough. At the same time if $c > c_0$ (m) the projection U_1 will be selected by condition of the PI maximum . But this projection gives visual pattern rather like a dense kern with outliers.

8. Choosing the projection for detecting outliers.

As the PI suitable for choosing the projections vectors for detecting outliers we propose the ratio

$$Q(U,X^{(n)})=s^2(U)\ /s^2(U) \qquad (13)$$

where $s^2(U)$ is the usual estimate of the variance of the projection with the vector U, and $s_r^2(U)$ is a robust estimate of it.

It is known that the usual estimate of s^2 is rather sensitive to outliers and if they are present its value increases as a rule. So, the projection where the value of PI (13) is maximal may be reasonably considered as the one where the influence of outliers is the most expressed. An approximate maximization of the PI (13) can be achieved by the solution of the generalized eigenvectors problem

$$(\mathbf{S} - h\, \mathbf{S}_r)U=0 , \qquad (14)$$

where $\mathbf{S}$ is the routine estimate of covariances matrix and $\mathbf{S}_r$ is some robust estimate of it.
For visual analysis it is necessary to use the projection with maximal eigenvalues.

The index (14) may be improved if we take into account the difference between the location parameter estimates (usual and robust). For example, we can use index

$$Q(U,\mathbf{X}^{(n)}) = (s^2(U) + \| \mathbf{M} - \mathbf{M}_r \|^2)/s_r^2(U)$$

where $\mathbf{M}$ is the usual estimate of the means vector and $\mathbf{M}_r$ is some robust estimate of it.

Approximative solution may be again defined as solution of the generalized eigenvalues (vectors) problem

$$(\mathbf{S} + (\mathbf{M} - \mathbf{M}_r)(\mathbf{M} - \mathbf{M}_r)' - h\mathbf{S}_r)U=0$$

The estimates $\mathbf{S}_r$, $\mathbf{M}_r$ that we use have been proposed by Meshalkin (1970) and they are solution of the following set of equations system

$$\mathbf{M}_r = \sum_{i=1}^{n} w(t_i)X_i \Big/ \sum_{i=1}^{n} w(t_i)$$

$$\mathbf{S}_r = (1 + \lambda) \sum_{i=1}^{n} w(t_i)(X_i - \mathbf{M}_r)(X_i - \mathbf{M}_r)' \Big/ \sum_{i=1}^{n} w(t_i) , \qquad (15)$$

where $t_i^2 = (X_i - \mathbf{M}_r)' S_r^{-1} (X_i - \mathbf{M}_r)$,

$$w(t_i) = \exp(- \lambda\, t_i^2 / 2), \quad (\lambda > 0).$$

It is possible to use an iterative procedure for seeking the solution of equations (15).

The factor ($1+ \lambda$) affords the unbiasedness of the estimate S_r when the underlying distribution of the X is in fact normal (Meshalkin (1970)). Therefore, in such case the eigenvalues are close to 1. The values λ that we use are $c/(p+1)$, $1<c<4$.

This approach gives the weights $v_i = w_i/\sum w_i$ for every observation X_i. These weights can be used for several purposes: automatical removal of the observations with the low weights (suspected as outliers), marking such ones on visual display , obtaining robust scale estimates for calculation of PI values (Section 4)

10. Projection pursuit for discriminant analysis (DA).

Assume that the p-variate sample X is divided into the k subsets $\mathbf{X}_1,\ldots,\mathbf{X}_k$ with the sizes $n_i (i=1,\ldots,k)$, $\sum n_i = n$.

The classical DA is based on the assumption that the mixture model (2) is true and the vectors from the group $\mathbf{X}_i$ are vectors corresponding to the i-th component of the mixture.

On these assumptions and if $k=2$, the PI for selecting the most interesting one-dimensional projection is Fisher ratio

$$Q(U,\mathbf{X}^{(n)})=(U'X_1 - U'X_2)^2/ w^2(U) \quad , \qquad (16)$$

where X_i is means vector of the $\mathbf{X}_i$

Robust variants of the PI may be obtained if in (16) we replace the values X_i, $w^2(U)$ by some robust analogs (Huber (1985)).

If $k>2$ we can use the canonical vectors V_i (see Section 2) as the vectors of interesting projections.

When the assumptions of classical model of DA are violated even in robust variants, it is necessary to use PI that is based on more complete information about distributions of vectors in $\mathbf{X}_i$.

If $k=2$ such a PI is for example

$$Q(U,\mathbf{X}^{(n)})= \sqrt{(w_1 w_2)} \int \hat{f}_1(z)\hat{f}_2(z)dz$$

where w_i^2- is variances estimate of one-dimensional projection of i-th component and $\hat{f}_i(z)$ is induced density estimates.

Now, the most interesting projection will be obtained as the solution of minimization problem

$$U=\underset{Y\in R^p}{\operatorname{argmin}}\, Q(Y,\mathbf{X}^{(n)})$$

References.

Buchschtaber V.M., Maslov V.K.(1975) Factor analysis on manifolds and problem of selecting the variates in pattern recognition. Proceedings of Ac.Sc.of the USSR,ser."Engineering Cibernetika", N 6, p. 194-200 (in russian).
Buchschtaber V.M., Maslov V.K. (1977) Factor analysis and extremal problems on Grassman manifolds. In "Mathemathical methods of solving the economical problems". Moscow:Science,N 7, p. 85-102 (in russian).
Cressie N.(1978) The power results for tests based on high-order gaps. Biometrika, 65, p. 214-218.
Friedman J.H., Tukey J.W.(1974) A projection pursuit algorithm for exploratory data analysis. IEEE Trans. Comput., C-23,p.881-889.
Friedman J.H.(1987) Exploratory projection pursuit. J. Amer. Stat. Ass.,82, p.249-266.
Huber P.J.(1985) Projection pursuit. Ann. Statist., 12, p.433-474.
Jones M.C., Sibson R.(1987) What is Projection Pursuit. J.Royal Stat. Soc.,150,ser.A,p.1-38
Meshalkin L.D. (1970) Approximation of multidimensional densities by normal distributions.-7th International Biometric Conference,Hannover.
Maronna R.A.(1976) Robust M-estimators of multivariate location and scatter. Ann. Statist. 4, p.51-56.
Rao C.R.(1965) Linear statistical inferences and its applications. New-York: Wiley.
Renyi A (1970) Probability Theory . Amsterdam :North-Holland
Vacicek O.(1976) A test of normality based on Sample Entropy. J. Royal Stat. Soc.,ser.B, 38, p.54-59.
Yenyukov I.S.,Neuschtadt(1978) Estimating the discriminant subspace when the sample is unclassificated. In "Applied Multivariate Statistical Analysis",Moscow: Science,p.326-333.
Yenyukov I.S.(1986a) Methods, algorithms and programmes of the multivariate statistical analysis (in russian). Moscow:Financy & Statistica.
Yenyukov I.S.(1986b) Detecting outliers and clusters in multivariate data based on projection pursuit.In "Proceedings of the World Congress of Bernoulli Society,Tashkent,USSR,Sept. 1987. Utrecht:VHU Science Press, 2, p.137-141.

Confidence Regions for Projection Pursuit Density Estimates

E. Elguero and S. Holmes-Junca, Montpellier

Keywords: Projection Pursuit, Multivariate Density Estimation, Bootstrap, Pivot, Confidence Regions, Micro-computers.

Abstract: Multivariate Projection Pursuit Density Estimation (PPDE) does not suffer from the "curse of dimensionnality" as the more classical kernel density estimation does, however a means of evaluating its stability and precision is needed, and this paper shows how the bootstrap can provide certain useful confidence intervals, the method used for constructing them starts by a pre-pivoting process.

1 Projection Pursuit Density Estimation

Multivariate density estimation has been done using the "Projection Pursuit" approach as introduced by FRIEDMAN AND AL. [4], however certain modifications have been necessary especially as the PPDE programme was originally written for a micro-computer using Pascal(cf. ELGUERO [3]). As in FRIEDMAN [4] a sequence of nested models is defined by

$$g^{(k)}(x) = g_o \prod_{i=1}^{k} g_i(a_i'x) \quad \text{where}$$

- g_o is a density belonging to a parametric family
- a_i is a unit vector
- g_i is a smooth function called an *augmenting function.*

So that for each $k = 1..K$, :

- $g^{(k)}$ is a density, i.e. $g^{(k)} \geq 0$ and $\int g^{(k)} = 1$
- the Kullback-Leibler information $I(f, g^{(k)})$ is minimum.

Compstat 1988

This procedure comes down to going through the following steps K times:

- Look for a direction a_i that minimizes the Kullback-Liebler distance between the marginals of f and $g^{(i-1)}$ upon this direction.
- Construct $g^{(i)}$ as the ratio of these projections who are the marginal densities f_{a_i} and $g_{a_i}^{(i-1)}$.

The ratio $f_{a_i}/g_{a_i}^{(i-1)}$ is easy to estimate because :

1. as $g^{(i-1)}$ is known, monte-carlo simulations are sufficient for defining its marginals,
2. a sample drawn from f_{a_i} is available by projecting the data upon a_i and then simply applying a kernel.

This method breaks down the problem to a sequence of one dimensional estimations, thus sparing the worry about "the curse of dimensionality".

The complexity of the methods used to estimate the Projection Pursuit Density Approximation (PPDA) $g^{(K)}(x)$ makes it difficult to attain sufficient knowledge about the PPDE $\widehat{g}^{(K)}(x)$'s distribution.

2 Using the Bootstrap

The Bootstrap (EFRON [2]) provides a way of constructing confidence intervals, even in this non-parametric framework. The estimate whose variation we are interested in is $\widehat{\theta} = \widehat{g^{(K)}}(x_o)$, an exact pivot is not of course available, but it seems plausible to use a pre-pivoting method as suggested by BERAN [1].

The idea behind BERAN's method is the following: through consideration of a potential pivot called the root $R_n(\theta, x_n)$ whose distribution can be estimated consistently from the sample, one can suppose that the confidence set

$$C_n = \{T \in \mathcal{T} : R_n(T, x_n) \leq \widehat{c_n(\alpha)}\}$$

is such that it's level should converge to α as n increases.

If $H_n(., F)$ denotes $R_n(\theta, x_n)$'s c.d.f., and its limit is denoted by $H = H(., F)$ then writing

$$C_n = \{T \in \mathcal{T} : R_n(T, x_n) \leq \widehat{H}^{-1}(1 - \alpha)\}$$

indicates that using the usual bootstrap procedure one will be able to estimate C_n through estimation of $\widehat{H}$ by $\widehat{H}_n$. But BERAN pushes his idea further, and what is even more interesting is the fact that $R_{n,1}(\theta, x_n) = \widehat{H}_n(R_n(\theta, x_n))$ has a distribution that is less dependent on the real c.d.f. F than is the original root. So one could re-iterate the bootstrap procedure on this new root $R_{n,1}(\theta, x_n)$ whose c.d.f. noted $H_{n,1}$ will thus be estimated by $\widehat{H}_{n,1} = H_{n,1}(., \widehat{F}_n)$.

So the simulations run as follows:

- **STEP 1 : Resampling in x_n generates** B new samples $\{x^*_{n,b}, b = 1..B\}$ **that provide the empirical c.d.f. of $R_n(\hat{\theta}, x^*_n)$ allowing approximation of $\hat{H}_n$.**

- **STEP 2 : Resampling P times in each of the B samples $\{x^*_{n,b}, b = 1..B\}$ provides an approximation of $\hat{H}_{n,1}$ by the empirical c.d.f. of**

$$Z_j = \frac{\#\{j, j = 1..P | R_n(x^{**}_{n,b,j}, \hat{\theta}^*_j) \le R_n(x^*_{n,b}, \hat{\theta})\}}{P}.$$

HALL [5] shows how the number of bootstrap replications does not have any notable effect on the precision with which the confidence intervals are estimated as long as $1/(B+1) \le \alpha$.

In our case we have taken $R_n(\theta, x_n) = |\widehat{g^{(K)}}(x_0) - g^{(K)}(x_0)|$.

3 Testing the Procedure through Simulation

The first stage of our test was done by considering a sample of n=300 observations drawn from a $d = 2$ dimensional mixture of three normal distibutions centred on $(0,5)$, $(3,0)$ and $(-3,0)$ respectively. In this case $K = 5$ terms proved to be largely sufficient for a good quality estimation of f, as shows fig.1.

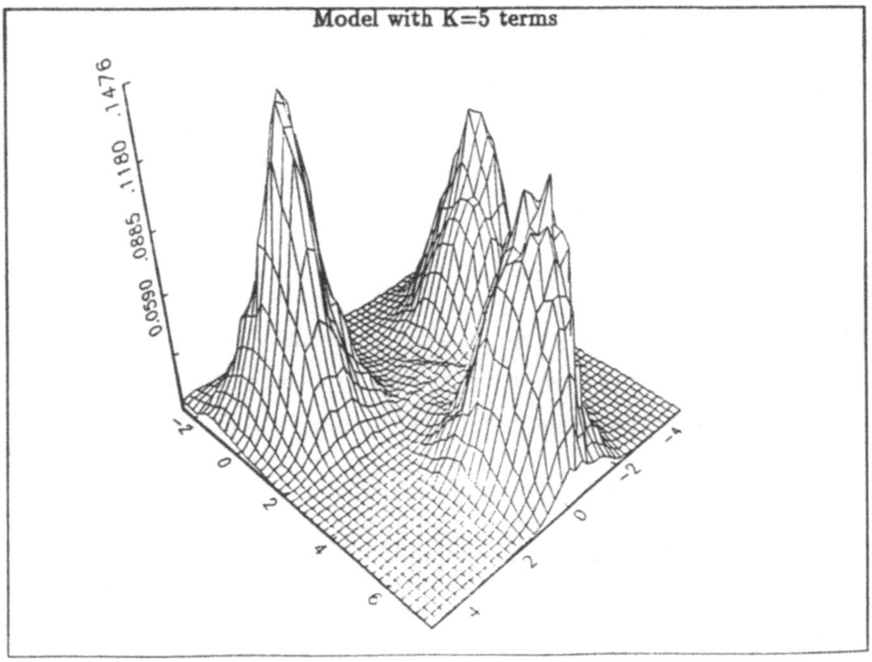

fig.1

The pre-pivoting procedure thus provided the following confidence intervals for B=50 and P=50:

x_0	$\widehat{g^{(K)}}(x_0)$	*left bound*	*right bound*	*width*
$(0,5)$	0.0304	0.0193	0.0415	0.0222
$(-3,0)$	0.0109	0.00731	0.01449	0.00718
$(3,0)$	0.0092	0.00500	0.01341	0.0084
$(-1,-1)$	$1.5\ 10^{-4}$	0	$3\ 10^{-4}$	$3\ 10^{-4}$
$(-1,1)$	$3.2\ 10^{-4}$	0	$11.2\ 10^{-4}$	$8\ 10^{-4}$

4 Computational considerations

We have chosen NELDER & MEAD 's *Simplex* method for maximizing $I(f_{a_i}, g_{a_i}^{(i-1)})$ because it doesn't include computing of derivatives.

The *Acceptance-Rejection* method was used for generating observations according to $g^{(i-1)}$ becaus it enables generation from any density p, even if p is unknown as long as generation from another density q is possible, q can be unknown as well, as long as the ration p/q is known and bounded.

As the above procedure needed the use of more than $n \times d \times K \times B \times P$ random numbers, we had to be careful about the generators used, and we have implemented a Tausworthe generator as exposed in KENNEDY & GENTLE [7].

In the future, it seems preferable to use larger data sets thus increasing needs for memory space as well as for computational speed.

One of the feasible answers to these growing requirements seems to be use of a vectorial or parallel computer architecture. Bootstrapping only calls for repetition, a large number of times of the same instructions thus making a simple SIMD architecture very useful, research in this direction is actually being carried out at our Laboratory.

5 Theoretical Considerations

Apart from PPDE's consistency , no asymptotic theory is available for this particular density estimate, although we know that for K large enough this estimate does better than the kernel one. A good starting root seems to be $R_n = n^{\frac{2}{5}}|\widehat{g^{(K)}}(x_0) - g^{(K)}(x_0)|/\sigma$, but this is yet to be proved.

6 An application to a pollution problem

A network of automatic stations measures SO_2 concentrations every quarter of an hour, as well as some other meteorological variables. What is needed is a way for predicting pollutant overflow.

The method proposed herewith supplies a way of predicting sudden pollution-growth by taking into account the meteorological variables. First we have classified the pollutant's increase (large, null and negative), then we have applied a decisionnal discriminant analysis to the data. Conditional densities that are needed to compute the minimal risk have been estimated by the non-parametric PPDE method.

References

[1] Beran R. , 1987, *"Pre-pivoting to reduce level error of confidence sets"*, Biometrika, vol.74 ,pp. 457-468.

[2] Efron B. , 1987, *"Better Bootstrap Confidence Intervals"*, J.A.S.A., vol. 82, pp. 171-185.

[3] Elguero E. ,(to appear in 1988), *"Contribution à l'Elaboration d'une Méthodologie pour l'Analyse Statistique d'un Réseau de Télémesure"* , Thèse à l'U.S.T.L. , Montpellier.

[4] Friedman J.H., Stuetzle W., and Schroeder A., 1984, *"Projection Pursuit Density Estimation"*, J.A.S.A., vol. 79.

[5] Hall P. ,1986, *"On the number of Bootstrap Simulations Required to Construct a Confidence Interval"*, Ann. Stat., vol. 14, pp.1453-1462.

[6] Huber P.J. , 1985, *"Projection Pursuit"*, Ann. Stat., vol. 13, pp.435-475.

[7] Kennedy W.J. and Gentle J.E., 1985, STATISTICAL COMPUTING,Ed. M. Dekker, New York.

A Robustness Property of the Projection Pursuit Methods in Sampling from Separably Dependent Random Vectors

B. Baldessari and F. Gallo, Roma

1. INTRODUCTION

A purpose of this paper is to point out that the dependence of the observations may improve the capability of a P.P.M. (Projection Pursuit Method) in detecting clusters in the projected observation values. To this purpose, the mean vector and the covariance matrix of the r.v. (random variables) representing the projected observation values are calculated under the assumptions of the Intrinsic Inference Model for Finite Populations of Separably Dependent Random Vectors. Comparison with the mean vector and the covariance matrix under the assumption of a population of independent r.vt. (random vectors) shows the effect of the dependence. It is proved, at least for the case of normal r.v., that the probability that the projected observation values are separeted into two destinct clusters is greater in the case of dependent r.vt. than in the case of independent r.vt..

We present also a necessary and sufficient condition under which the covariance matrices of the projected observation r.v. in the cases of dependent and independent r.vt. are equal. This condition is very simple as it ammounts to requiring that the sums of correlations among the r.vt. are null.

Our results are true for whatever P.P.M. but, for simplicity, we consider the case of a unidimensional projections only. Extensions to multidimensional projections are relatively simple.

In section 2, the inference models and the P.P.M. are presented; in section 3, the results of the covariance matrices are proved; and, in section 4, the probabilities of distinct clusters of the projected observation values are calculated.

2. MODELS

The Intrinsic Inference Model for Finite Populations of Separably Dependent Random Vectors is defined as follows.

(1) Suppose that there exists a population of N random p vectors, i.e., that there exists a random Np vector $X = (X(1)',\dots,X(N)')'$; whose j-th component $X(j) = (X_1(j),\dots,X_p(j)')$, $j = 1,\dots,N$, represents p variables studied on the j-th unit of a statistical population. The X(j) are denominated population r.vt..

(2) Let $\mu(j)$, $j = 1,\dots,N$, be the random p vector of the

expected value of $X(j)$, $j = 1,\dots,N$, let Σ be a covariance $p \times p$ matrix and let R be a correlation $N \times N$ matrix such that the covariance $Np \times Np$ matrix $\Sigma(x)$ of x is given by

$$\Sigma(X) = R \otimes \Sigma, \qquad (2.1)$$

where $\otimes$ denotes the Kronecker's product. The matrix R may be interpreted as an aspect of the dependence among the population r.vt. $X(j)$. The matrix Σ may be interpreted as an aspect of the dependence of the population r.v. $X_h(j)$, $h = 1,\dots,p$, within each population r.vt. $X(j)$. With this interpretation, assumption (2) (introduced by Aragon and Caussinus, 1980) corresponds to the intuitive idea that the dependence among the population r.v. may be separated into a dependence among the population r.vt. $X(j)$ and a dependence of the population r.v. $X_h(j)$ within the population r.vt. $X(j)$.

(3) A sample $Y(1),\dots,Y(n)$, i.e., a set of $n(< N)$ random p vectors, is selected without replacement from the population r.vt. $X(1),\dots,X(N)$, with selection probabilities $p(Y)$ where $Y = (Y(1)',\dots,Y(n)')'$ and

$$\Sigma p(Y) = 1, \qquad (2.2)$$

in which Σ is extended over the $N!\, n![(N-n)!]$ possible selections of $Y(1),\dots,Y(N)$. The r.vt. $Y(1),\dots,Y(n)$ are denominated observation r.vt..

(4) The observation r.vt. $Y(1),\dots,Y(n)$ are realized; they take on the vt. (vectors) $y(1),\dots,y(n)$ and the realization process of $y(1),\dots,y(n)$ is independent of the selection process of $Y(1),\dots,Y(n)$. The vt. $y(1),\dots,y(n)$ are denominated observation vt.. Assumptions (1), (3) and (4) have been introduced by Baldessari (1987) as the Intrinsic Inference Model for Finite Populations of Dependent Random Vectors.

To connect the above inferential model with the P.P.M., we assume that the observation $n \times p$ matrix $y = (y(1),\dots,y(n))'$, over which the P.P.M. has to be applied, is composed by the observation vt. $y(1),\dots,y(n)$. A P.P.M. for y (see: Huber, 1985 and Jones and Sibson, 1987) may be sintecally defined as follows.

(I) The observation vt. $y(1),\dots,y(n)$ are preliminarily centered, scaled and sphered by the transformation

$$u = n^{-1} s^{-\frac{1}{2}} y(I_n - n^{-1} 1_n 1_n')s^{-\frac{1}{2}}, \qquad (2.3)$$

where I_n is the identity $n \times n$ matrix, 1_n is the n vt. of ones,

$$s = n^{-1} y(I_n - n^{-1} 1_n 1_n')y', \qquad (2.4)$$

is the covariance matrix of the observation vt., and s is defined in the usual way.

(II) For any $q \times p$ matrix A of rank $q(\leq p)$ and with orthonormal row vt., the observation vt. $y(1),...,y(n)$ are projected and the projected observation $q \times n$ matrix is given by

$$z_A = A\,u\,. \tag{2.5}$$

For any kernel function $\phi(.)$, for any $0 < w$, and for any $z \in R^q$ the kernel density estimate, i.e., the function

$$f_A(z) = (n\,w)^{-1} \sum_{i=1}^{n} \phi[(z - z_A(i))/w]\,, \tag{2.6}$$

is calculated. Here: $z_A(i)$ (denominated projected observation vt.) is the i-th colum of z_A; $z \in R^q$; and, $\phi(.)$ is any non negative function whose argument is a q' vt. The function $\phi(.)$ is even with respect to each component of the argument, decreases monotonically to zero as any component of the argument tends to $\pm\infty$ and has unit integral over the R^q space.

(iv) For any A, $\phi(.)$ and w, a projection index, i.e., an index designed for detecting clusters in the projected observation vt., is defined as follows

$$d(A) = g\,[f_A(z)], \tag{2.7}$$

where $g(.)$ is any convenient function such that $d(A)$ does not depend on z. For example, when $A = a'$ and a is an n vector of lenght one, Friedman and Tukey (1974) suggest the projection index

$$d(a') = \int_{-\infty}^{+\infty} f_{a'}(z)\; d\,F_{a'}(z), \tag{2.8}$$

where $F_{a'}(z)$ is the empirical distribution function of $z_{a'}(1),...,z_{a'}(n)$.

Jones and Sibson (1987) propose

$$d(a') = \int_{-\infty}^{+\infty} f_{a'}(z)\;\ln f_{a'}(z)\; d\,z, \tag{2.9}$$

and, an index based on the third and forth moments of the projected observation values.

(v) By numerical evaluation, the absolute and the local maxima of $d(A)$ are determined; the corresponding scatters of the projected observation vt. are examined for identification of clusters and, finally, clusters of the unprojected observation vt. are surmised.

Assumption (i) is discussed by Gover (1987) which points out that scaling and centering may alter the structure of the unknown clusters. He states that centered or uncentered observation vt. may be scaled by other ways, for example by variance-stabilizing logaritmic transformations.

We assume that a convenient preliminarily transformation has been applied; thus, we consider the observation vt. $y(1),...,y(n)$ only.

Our argument is based on the r.vt. $Y(1),\dots,Y(N)$ and does not involve neither the kernel density estimate $f_A(z)$ nor the projection index $d(A)$. . Therefore, only the assumptions (1)-(4) and (II) are used in this paper.

Let us end this section by remarking that our argument may be extended to the centered observation vt

$$v(i) = y(i) - n^{-1} \sum_{i=1}^{n} y(i), \quad i = 1,\dots,n, \tag{2.10}$$

but that this aspect is not discussed, for brevity.

3. RESULTS

We consider the observation r.vt. $Y(1),\dots,Y(n)$ in their lexicographical order, i.e., we consider the random np vector $Y = (Y(1)',\dots,Y(n)')'$. Also, with $A = a'$, we consider the random n vector $Z(a') = (Z_{a'}(1),\dots,Z_{a'}(n))$, where $Z_{a'}(i) = a'Y(i)$, $i = 1,\dots,n$, is the projected observation r.vt..

Theorem. Under the assumptions (1)-(4) and (II) of section 2, the mean n vector $\mu(a')$ and the covariance $n \times n$ matrix $\Sigma(a')$ of $Z(a')$ are given by

$$\begin{aligned} \mu(a') &= a' \nu 1_n , \\ \Sigma(a) &= a' \Sigma a \,[(1-\rho)I_n + \rho 1_n 1_n'] + (N-1)^{-1} a'\Delta a(N I_n - 1_n 1_n'), \end{aligned} \tag{3.1}$$

where: $\nu = (\nu_1,\dots,\nu_n)'$, with

$$\nu_h = N^{-1} \sum_{j=1}^{N} E X_h(j), \ h = 1,\dots,p; \tag{3.2}$$

Δ is the covariance $p \times p$ matrix of the random p vector $\eta = (\eta_1 + \dots + \eta_p)'$ that takes on the p vt. $(E X_1(j),\dots,E X_p(j))'$, $j=1,\dots,N$, with equal probabilities $1/N$; and,

$$\rho = [N(N-1)]^{-1} [1_N' R 1_N - N]. \tag{3.3}$$

Proof. The proof follows from an application of some results of Baldessari (1987) under the assumptions (1), (3) and (4) of section 2. For reasons of brevity, the proof is omitted.

The two cases of separably dependent and independent r.vt. are compared in the following corollary.

Corollary. Under the assumptions (1), (3), (4) and (II) of section 2, the projected observation r.v. $Z_{a'}(i)$ in the two cases of dependent and independent r.vt. have:

- the same expected values:

$$E Z_{a'}(i) = a' \nu, \qquad i = 1,\dots,n ; \tag{3.4}$$

- and the same variances:

$$\sigma^2[Z_{a'}(i)] = a'\,\Sigma\, a + a'\,\Delta\, a\ , \quad i = 1,\ldots,n. \tag{3.5}$$

Moreover, the covariances

$$\mathrm{Cov}\ [Z_{a'}(i),\ Z_{a'}(s)] = \rho\, a'\Sigma\, a - (N-1)^{-1}\, a'\Delta\, a,\ i \neq s = 1,\ldots,n, \tag{3.6}$$

of the case of dependent r.vt. is smaller, equal or greater than the covariances

$$\mathrm{Cov}\ [z_{a'}(i),\ z_{a'}(s)] = -\ (N-1)^{-1}\ a'\ \Delta\ a, \quad i \neq s = 1,\ldots,n, \tag{3.7}$$

of the case of independent r.vt. iff ρ is smaller, equal or greater then zero.

Proof. If the population r.vt. are independent, $R = I_n$ Therefore,

$$\mu(a') = a'\ \nu\ I_n\ ,$$
$$\Sigma(a') = a'\ \Sigma\ a\ I_n + (N-1)^{-1}\ a'\ \Delta\ a\ (N\ I_n - 1_n\ 1_n'). \tag{3.8}$$

Comparison of the equations (3.1) and (3.8) proves the corollary.

4. DEPENDENCE OF THE OBSERVATIONS AND CAPABILITY OF A P.P.M. TO DETECT CLUSTERS

The results of section 3 support the statement that, at least for some population r.vt. $X(1),\ldots,X(N)$, when $\rho < 0$, the probability that a P.P.M. detects separate clusters of the projected observation values is greater in the case of dependent than in the case of independent population r.vt.

The general, but obviously not rigorous, argument runs as follows:

(a) If $\rho < 0$, the negative covariances of the projected observation r.v. of the dependency case are smaller than the negative covariances of the projected observation r.v. of the independency case;

(b) since the expected values and the variances are equal in both the dependency and the independency cases, statement (a) suggests that large projected observation values are more likely to be followed by small projected observation values in the dependency than in the independency case;

(c) therefore, at least for some population r.vt., the capability of a P.P.M. to detect clusters in the projected observation values is larger in the dependency than in the independency case.

The above statement can be proved true, at least when the $Z_{a'}(i)$ r.v. are normally distributed.

Lemma. If $U \sim N[0, I_n + \theta(1_n 1_n' - I_n)]$, i.e., if the random n vector U is normally distributed with null mean vector and correlation matrix $I_n + \rho(1_n 1_n' - I_n)$, then

$$I(\rho) = p(U_1 < 0,\ 0 < U_2) = \int_{-\infty}^{+\infty} [2\pi(1-\theta^2)]^{-1} \exp\{-2^{-1} t^2\} [1 - \Phi(t\sqrt{2(1+\theta)})]\, dt\ , \qquad (4.1)$$

where $\Phi(.)$ denotes the cumulative distribution function of the standard normal r.v. .

Proof. The proof follows from the double normal integral extended over the upper-right quarter of R^2 by linear substitutions. For brevity, the proof is omitted.

Since $Z_{a'}(i) \sim N(a'\nu,\ a'\Sigma a + a'\Delta a),\ i = 1,\ldots,n$,

$$P[Z_{a'}(1) < a'\nu,\quad a'\nu < Z_{a'}(2)] = P(U_1 < 0,\ 0 < U_2) = I\left(\frac{\rho a'\Sigma a - (N-1)^{-1} a'\Delta a}{a'\Sigma a + a'\Delta a}\right). \qquad (4.2)$$

Now, equation (4.1) shows that, for $\theta < 0$, $I(\theta)$ is a decreasing function of θ , i.e., that the probability that the projected observation values $Z_{a'}(1)$ and $Z_{a'}(2)$ are separeted by their expected value $a'\nu$ is greater for the dependency than for the independency case.

REFERENCES

ARAGON, Y. and CAUSSINUS, H. (1980). Une analyse en composantes principales pour les unités statistiques correlées. Data Analysis and Informatics, E. Diday et al., eds., North Holland, 121-131.

BALDESSARI, B. (1987). Intrinsic Inference Model for Finite Populations of Dependent Random Vectors. Metron, XLV, 169-186.

HUBER, P.J. (1985). Projection Pursuit. The Ann. of Statist., 13, 2, 435-475.

JONES, M.C. and SIBSON, R. (1987). What is Projection Pursuit?, J.R. Statist. Soc., 150, 1, 1-36.

Graphical Techniques

Graphical Modelling with Large Numbers of Variables: An Application of Principal Components

J. Whittaker, Lancaster, A. Iliakopoulos, Fontenay-aux-Roses and
P. W. F. Smith, Lancaster

Abstract: Certain issues concerned with graphical modelling with large numbers of variables are discussed. A rudimentary form of initial model selection and testing is proposed in the context of covariance selection models, similar in spirit to the screening procedure of Kreiner (1987), which avoids the explicit fitting of any graphical model. It is conjectured that a useful guide to assess the performance of the model is to compare its predictive power against that of principal components. This is illustrated by an example from a data set with 30 continuous variables.

Keywords : agorametry; covariance selection; graphical models; model selection; principal components.

1 Introduction

The seminal paper of Darroch, Lauritzen and Speed (1980) made graphical modelling a standard way to analyse the interaction structure from a sample of multi-dimensional random variables, and several examples with discrete, continuous and mixed variables have now been published : Kiiveri and Speed (1982), Edwards and Kreiner (1983), Lauritzen and Wermuth (1987), Whittaker (1987).

Their success is a function of several factors. The graph is a visually informative medium of communication. The graphical model has some deep properties, in particular, a separation theorem and the local property that the nearest neighbours of each variable constitute the set of best predictors for that variable. It has given the older techniques of path analysis and causal modelling a firm theoretical foundation and has unified the treatment of continuous and discrete variables. Model fitting algorithms based on the iterative proportional fitting procedure for discrete, continuous and mixed models are published and software to fit certain subsets of graphical models are available for standard PC-compatibles: Edwards (1987), Kreiner (1988).

However, as in regression analysis, model fitting and selection requires considerable effort and these examples have been restricted to a limited number of variables, generally less than ten or so. If the variables $X_1, X_2, \ldots, X_k$ increase in number then the number of possible edges goes up as $k(k-1)/2$ and new problems become manifest: in particular, those of information presentation and information digestion, and the computational demands on storage and execution time. Several different strategies and tactics are possible, for instance, restricting search to decomposable models is clearly desirable as these are the only ones to have direct estimates. We make no attempt to review the possibilities here; our aim is to present an application of principal components analysis that gives a rapid overall assessment of how well the graphical model with a given interaction graph fits the data, at the same time identifying variables which require further modelling. The next section proposes a rudimentary form of initial model selection, testing and model search, which avoids the explicit fitting of the model, and conjectures that a useful guide to assess the

Compstat 1988

performance of the model is to compare its predictive power against that of principal components.

2 Local Tests of Model Fit and Principal Components

For relatively small numbers of variables it is perfectly possible to fit any given graphical model by full maximum likelihood but as discussed above, with large numbers of vertices in the graph this is a formidable task.

Edge Exclusion Deviances: One possibility for exploratory work is to rely more heavily on the raw edge exclusion deviances: these are just the maximised log-likelihood ratio test statistic, or deviance for short, for testing if any one edge can be excluded from the *complete* graph. Under standard large sample theory these deviances have a chi-squared distribution, they are simple to compute from contingency tables and correlation matrices, and in the latter case there is a simple formulae :

$$-N \log(1 - r^2_{ij|rest}),$$

where $r_{ij|rest}$ is the observed partial correlation coefficient between variables X_i and X_j adjusted for the remaining $k-2$ variables. This can be computed for all pairs of variables in one pass: because the elements of the matrix obtained by inverting the correlation matrix, and then scaling it by the same transformation that takes a covariance matrix to a correlation matrix, are negatives of these partial correlations.

Small values of the deviance correspond to accepting the hypothesis that two variables are independent given the remaining variables, so that no edge is needed in their interaction graph. By choosing an appropriate cut-off point, the triangular array of edge exclusion deviances defines the incidence matrix of a graph, and thus determines those parameters in the graphical model that are to be set to zero.

It is recognised that there are dangers in using these raw edge exclusion deviances to select a graphical model; dangers which are directly analagous to those of a regression selection procedure entirely determined by the t-statistics computed from fitting the single regression containing all explanatory variables. A more careful backwards selection method would choose the least significant of this array, but then recompute the exclusion deviances taking the reduced model as the new base model. One remark, though, is that it is usually (though not universally) the case that edges detemined by this procedure form a minimal set of edges, in the sense that they cannot be excluded from any graphical model for the data.

Testing for the Local Markov Property: The incidence matrix of the interaction graph defines a graphical model by specifying which interaction parameters are set to zero. It remains to evaluate the other parameter estimates and test if it provides an acceptable fit to the data array. Explicit fitting is to be avoided, especially if the procedure for choosing the interaction graph gives no guarantee that the ensuing model is decomposable or provides a good fit. Rather than test the overall model as a whole, one way forward is to test some of the implications contained in the graph, notably the separation theorem. Though this too, is a computational task several orders of magnitude greater than the original, unless the choice of separating subsets is severely restricted.

So we suggest the strategy of testing for the local Markov property, where each vertex is examined in turn to see if it is adequately predicted by its nearest neighbours. Each examination corresponds to evaluating the residual sum of squares from fitting two regressions for each vertex: one with all other variables in the explanatory set and the other with only the nearest neighbours; thus computing the observed partial variance $s_{ii|a}$ where in one case, the subset a of $\{1, 2, \ldots, k\}$ contains all the other vertices, the 'rest', and in the other, just the 'nearest neighbours' of vertex i. As the variables are scaled to have unit variance this decomposition is

$$\begin{array}{ccccccc} 1 = & s_{ii|rest} & + & s_{ii|nn} - s_{ii|rest} & + & s_{ii} - s_{ii|nn} \\ & \text{unexplainable} & & \text{attributable to} & & \text{attributable to} \\ & \text{residual} & & \text{non-adjacent vertices} & & \text{nearest neighbours} \end{array}$$

The total attributable variation is, of course, just the squared multiple correlation coefficient. The F-test from the analysis of variance table for the variation attributable to the non-adjacent vertices is a test for the local Markov property. If the test is significant then further regression selection techniques can be used to determine which extra edges are needed, and examining each variable in turn provides a comprehensive screening of all missing edges in the graph.

There is an implicit 'filling-in' taking place in this procedure, Lauritzen and Spiegelhalter (1988). As an example consider analysing the interaction graph (a):

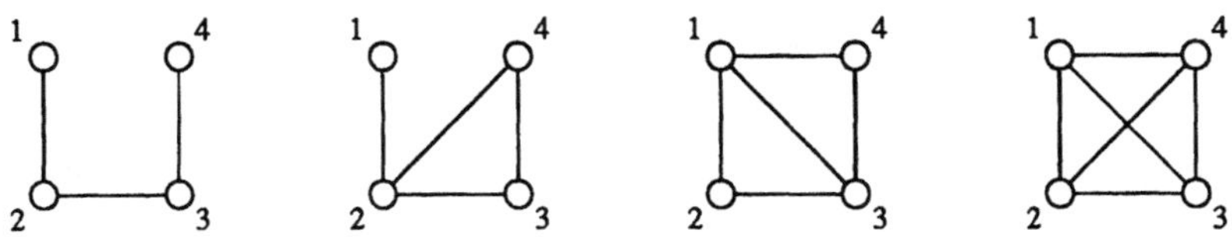

(a) (b) (c) (d)

The F-test at vertex 1 corresponds to testing graph (b) against the complete graph (d) to see if edges (1,3) and (1,4) require inclusion. The F-test at vertex 2 corresponds to testing graph (c) against (d) to see if edge (2,4) requires inclusion. Note how both comparisons augment the edges of (a) before carrying out the test. Other details of this procedure have to be decided, for as this example illustrates, each edge is tested more than once, and it may happen that edge (i, j) is required by vertex i, but not by vertex j.

Principal Components: A well known guide to determine how well the first few principal components explain the total variation is to check how near each variable in the loadings plot is to the perimeter of the unit circle, for instance, see Caillez and Pages (1976). We shall use this distance to give both a global and local assessment of the graphical model.

The underlying decomposition of variance is

$$\begin{array}{cccc} 1 = & s_{ii|pc} & + & s_{ii} - s_{ii|pc} \\ & \text{residual} & & \text{attributable variation} \end{array}$$

where $s_{ii|pc}$ is the residual or partial variance of i-th variable predicted from the first p principal components. The attributable variation can be easily calculated from

$$s_{ii} - s_{ii|pc} = \lambda_1 v_{i1}^2 + \lambda_2 v_{i2}^2 + \cdots + \lambda_p v_{ip}^2,$$

where λ_j is the j-th eigen value and v_{ij} the i-th element of the j-th eigen vector of the sample correlation matrix.

Plotting the explained variation due to the first p principal components, $s_{ii} - s_{ii|pc}$, against that due to prediction from the nearest neighbours, $s_{ii} - s_{ii|nn}$, gives an overall assessment of the graphical model, as well as a detailed variable by variable comparison of the two techniques. The measures are the same for both methods, and because of working with correlations, scale invariant. It indicates which variables are better represented by the first few components and which variables better predicted by their nearest neighbours. The plot is made more informative by indicating which variables show a significant attributable variation from non-adjacent vertices. As a side benefit the plot may also indicate whether additional components are needed for the spectral approximation of the correlation matrix.

3 An Example from Agorametry

The data in this example are measures of public opinion towards a cross-section of French political personalities, made on a sample of 247 individuals interviewed in 1987. The variables were constructed by scoring a 5 point scale (tres mauvaise opinion, mauvaise opinion, indifferent ou ne connait pas, bonne opinion, and tres bonne opinion) according to whether 'vous appreciez ou non, les personalites suivantes:' and there followed a list of 30 well known persons in French political life, such as Mitterand, Chirac, Rocard, Barre and Thatcher. Though a case could be made for arguing that these names form a random sample from a larger population, the consequent analysis is not pursued here. The scores were constructed from the marginal distributions of the responses and the data for analysis was reduced to the 30 × 30 correlation matrix. Typical values are -0.243, -0.242, 0.500, -0.194, 0.403, -0.118, -0.017, 0.241, -0.223, indicating that, as typical with the social sciences, the variables are only weakly dependent; on the other hand any test for mutual independence is overwhelmingly rejected.

Principal Components: Weak dependence is also a feature of the spectral decomposition: as many as eight of the 30 eigen values of the correlation matrix are greater than one, they are 6.287, 3.168, 2.197, 1.522, 1.495, 1.140, 1.118, and 1.067. The first two components account for only 31.5% of the total variation, the first three for 38.8%; and, in fact, the eight above only approach 60% of the total.

Though principal components are not often described within in a parametric model fitting framework, the implicit number of parameters entailed in using the first two components to represent the data is $2 + 29 + 28 = 59$ increasing to 86 in using three.

The Interaction Graph: The first step edge exclusion deviances, computed in the manner discussed above, but classified according to size are given in the table, where 1, 2, 3 and 4 denote a

	K	B	M	G	K	H	C	J	T	B	L	S	P	R	M	M	M	A	K	L	G	S	M	Z	L	T	G	R	L	B
Kra	x																													
Bar	1	x																												
Mon	.	3	x																											
Gat	.	1	.	x																										
Kou	.	.	.	.	x																									
Har	.	.	.	.	1	x																								
Chi	.	.	2	.	.	.	x																							
JP2	3	3	.	.	.	.	4	x																						
Tap	1	.	.	.	.	.	1	.	x																					
Ber	.	.	.	.	.	.	.	.	3	x																				
Lal	1	1	.	.	.	.	.	.	.	.	x																			
Sin	1	.	.	.	.	.	.	.	.	1	2	x																		
Pas	.	.	3	.	.	.	3	.	.	.	.	.	x																	
Rea	.	.	.	.	.	2	.	.	.	.	.	.	.	x																
Mit	.	.	.	.	.	2	.	2	.	.	.	1	.	.	x															
Mar	4	.	.	.	2	.	.	2	.	.	.	.	.	1	.	x														
Mai	.	.	.	.	.	.	.	.	.	.	.	.	.	.	.	1	x													
Aqu	.	.	.	.	2	.	.	.	.	.	.	.	.	.	.	.	.	x												
Koh	2	3	.	1	.	.	2	.	.	.	.	.	.	1	.	.	.	2	x											
Lah	.	2	.	.	.	.	.	3	1	.	.	.	.	.	.	.	.	.	.	x										
Gom	.	.	1	1	.	.	.	.	.	.	1	.	.	.	.	.	.	.	.	.	x									
Sab	.	.	.	.	.	.	1	.	.	.	.	.	.	.	.	.	.	.	.	4	.	x								
Mo2	.	.	.	.	.	.	.	1	.	2	.	.	.	.	.	.	.	.	.	2	.	.	x							
Zar	.	.	.	.	.	.	.	.	1	.	.	.	.	.	2	.	.	.	.	1	.	4	.	x						
Lag	1	.	.	.	.	.	.	.	.	2	.	.	.	.	.	.	.	.	.	.	1	.	1	.	x					
Tha	.	.	.	.	.	.	.	.	.	.	.	.	1	4	.	.	.	.	.	.	.	.	.	.	3	x				
Gor	1	.	.	.	.	.	.	2	.	.	.	.	.	.	.	.	.	.	.	.	.	.	.	.	.	1	x			
Roc	.	.	.	.	.	.	.	.	.	.	.	.	.	.	3	.	4	2	.	.	.	.	.	.	.	.	.	x		
LeP	.	.	.	.	.	1	.	.	.	.	.	.	1	.	.	.	.	.	.	.	.	.	.	.	.	.	.	.	x	
Be2	.	.	.	2	.	.	.	2	.	.	.	.	.	.	.	.	3	.	.	.	.	.	.	2	.	.	.	.	.	x

Table. Significant edge exclusion deviances.

deviance in the intervals 4-6, 6-10, 10-20 and (20+), respectively. The lower limit of 4 is chosen as it is near the 5% point of chi-squared with 1df. The resulting incidence matrix is fairly sparse with only 68 entries from a total of 435 possible, suggesting that there is plenty of scope for parsimonious graphical modelling.

Even so the resulting interaction graph is complicated, and remains so even if a higher value of the deviance cut-off is used to simplify the display. Suffice it to report that many directly connected vertices correspond to variables that are close in the principal component loadings plot, though interestingly, there are a few edges, for instance, one between Reagan and the anti-racist Harlem Desir, that are far apart in the loadings plot. In this instance using the cut-off value of 4 leads to including 68 parameters in the graphical model and so makes the results roughly comparable with using two principal components, as discussed above.

We turn now to the comparison and plot for each of the 30 variables the attributable variation calculated from the spectral decomposition of the correlation matrix, against the variation attributable to the variable's nearest neighbours.

Firstly, the diagram makes it clear that neither a model based on an interaction graph nor one on principal components will have a particularly high predictive power. Only very few variables suc-

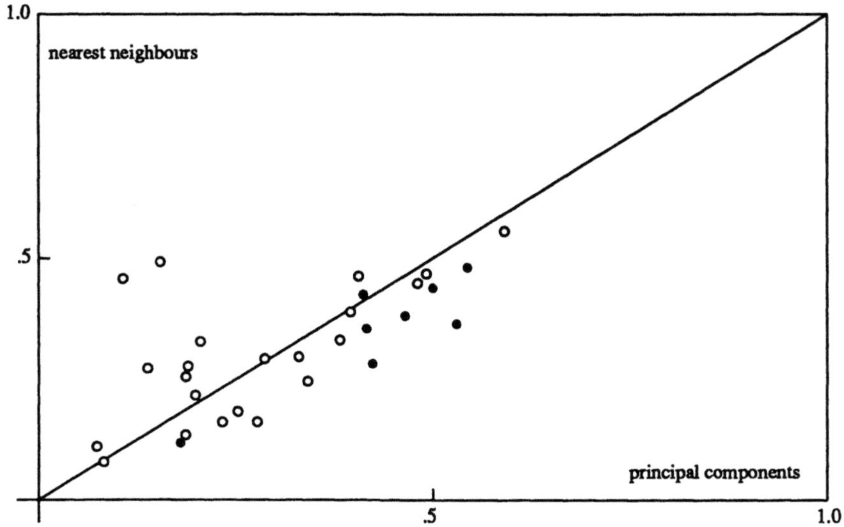

Figure. Explained variation for each variable: principal components against nearest neighbours.

ceed in reducing their original variance by more than 50%, and there are more values in the lower range 10%-20%. Secondly, it appears that more of the points lie below the 45° line, than above it, indicating that, on balance, principal components have the greater predictive power. None the less, a variable that is explained relatively well by one technique is also explained relatively well by the other, and conversely; but in this instance there are two exceptions, corresponding to variables that are well predicted by their nearest neighbours but not by the first two components. There are no exceptions in the other direction. This indicates a need for the third component; when it is included the two exceptional variables join the general cloud, which itself shifts marginally to the right. However to make the comparison fairer the cut-off point should be lowered to allow more parameters in the graphical model, which shifts the distribution of the cloud back to the 45° line. Finally, the filled in circles in the diagram indicate which variables have significant F-tests. Not surprisingly almost all lie below the 45° line. More surprisingly perhaps is that, again with the exception of one vertex, they are the more, rather than the less, predictable variables. These are the vertices that require further modelling, though we can tell immediately from the diagram that their position is not going to radically alter.

4 Discussion

The plot of the variation attributable to the nearest neighbours against that attributable to the principal components gives some guide to the overall efficacy of the graphical model under consideration, as well as the means to identify further directions for improvement. On the other hand it is not clear that this provides a fair way to give a global comparison of the techniques of graphical modelling and principal components. Firstly, using the partial variance of X_i given the components implicitly assumes that X_i is simultaneously predictor and predictand. Secondly, the nature of the 'maximal' model within the two techniques are rather different: including all components in the predictor reduces the partial variance given the components to zero, while including all variables in the set of nearest neighbours, does not. Similarly, the nature of the 'minimal' model is different. In the extreme case where the sample correlation is the identity matrix, then the attributable variance to any set of nearest neighbours is zero, while, depending on how the non-uniqueness of the eigen-vectors is resolved, the variance attributable to a component can vary between 0 and 1, and on average takes the value $1/k$.

Acknowledgement: to Jean Pierre Pages, CEA/IPSN/DPS/SEGP/LSEES for his support, and access to the data; to the Economic and Social Science Research Council, under grant number HOO232071, for their support of 'Graphical Modelling in Social Survey Analysis.'

References

Pages, J.P. et al. (Eds) (1987) Les Structures de l'Opinion en 1987. ssociation pour l'Etude des Structures de L'Opinion Agorametrie, Paris.

Cailliez, F. and J.P.Pages (1976) Introduction a l'Analyse des Donnees. SMASH, Paris.

Darroch, J., Lauritzen, S. and Speed, T. (1980). Markov fields and log linear interaction models for contingency tables. Ann. Stat. 8. 522-539.

Dempster, A.P. (1972). Covariance Selection. Biometrics, 28, 157-175.

Edwards, D.E. (1987) A guide to MIM. Research Report 87/1, Statistical Research Unit, University of Copenhagen.

Edwards, D. and Kreiner,S. (1983). The analysis of contingency tables by graphical models. Biometrika. 70, 3, 553-565.

Kiiveri H. and T.P. Speed (1982) Structural analysis of multivariate data: a review. In Leinhardt S.(ed) Sociological Methodology, Jossey Bass San Francisco.

Kreiner, S. (1987) Analysis of multidimensional contingency tables by exact conditional tests: techniques and strategies. Scand.J.Stat.14.

Kreiner, S. (1988) The EXAPC manual. Unpublished draft.

Lauritzen, S.L., and Spiegelhalter, D. (1988) Local computations with probabilities on graphical structures and their application to expert systems. J. Royal Statist.Soc. B. 50,2.

Wermuth, N. and Lauritzen S.L. (1983) Graphical and recursive models for contingency tables. Biometrika, 70,3,537-552.

Wermuth,N.and Lauritzen, S.L.(1987) Conditional independence graphs, graphical chain models, and data. Technical report 87-2, University of Mainz.

Whittaker, J. (1987) Graphical Modelling in Survey Analysis: Case Studies. Technical Report. Centre for Applied Statistics, University of Lancaster.

Some Graphical Displays for Square Tables

A. de Falguerolles and J. R. Mathieu, Toulouse

KEYWORDS :graphical displays, square table, log-linear models, singular-value decomposition.

Social mobility is one of the most studied topics in social science. In this context, the data commonly considered are square tables either simple or stratified. The availability of statistical softwares (GAUSS, GLIM, ...) enables to consider exploratory methods providing low dimensional graphical displays along with baseline log-linear models. Recently there have been attempts to combine these two strategies in order to serve various objectives: residual analysis, model selection, model description (BACCINI et al.(1987), CAUSSINUS and FALGUEROLLES(1987), HEIJDEN and LEEUW(1985), HEIJDEN(1987), WORSLEY(1987)). In this article, we present on real data some of the issues raised by this combined approach. We also use some results on separability (MATHIEU(1987)) to consider models of marginal homogeneity.

1. Introduction:

We consider here square mobility tables, that is tables where the same categories are used for rows and columns. The cells of the table give the counts of observations that share each combination of rows and columns. In some cases the table is stratified. The cells on the main diagonal of the table (and possibly some others) may deserve special attention : sometimes they are empty by design, or contain artificial values, or they are dominating the analysis because they have two high or too low frequencies. This distinction between non-diagonal and diagonal cells is an important feature for the statistical treatment of square tables.

Mobility matrices are often analysed by means of log-linear models such as Goodman quasi-independence or Caussinus quasi-symmetry. For an elementary and clear introduction to such models we refer to HOUT(1983). They are also widely analysed by singular-value decomposition : GOWER(1977),BURTSCHY(1984). For a further discussion we refer to HEIJDEN(1987).

These two approaches may seem unrelated. But, as pointed out in the introduction, there has been recently several attempts to use them in a combined fashion. The general idea is to analyse by singular-value decomposition either the residuals from tentative models or some parameters of particular models. We illustrate some of these issues on real data in the next two sections.

Compstat 1988

2. Mobility table, the standard case :

We consider the following mobility table from FEATHERMAN and HAUSER(1978 : 49) discussed at length by HOUT(1983) :

father's occupation by son's first occupation

UNM	LNM	UM	LM	F	fathers'occupation
1414	521	302	643	40	upper nonmanual
724	524	254	703	48	lower nonmanual
798	648	856	1676	108	upper manual
756	914	771	3325	237	lower manual
409	357	441	1611	1832	farm

(sons'occupations: UNM, LNM, UM, LM, F)

2.1) Log-linear approach :

Most baseline models can be fitted in GLIM whether or not the diagonal cells are taken into account. In the latter case, it suffices to give a null weight to the diagonal cells. Let ri (the row effect), cj (the column effect) and sij (the symmetrical interaction) denote three factors which after tabulation look as follows :

```
1 1 1 1 1      1 2 3 4 5      11  1  2  3  4
2 2 2 2 2      1 2 3 4 5       1 12  5  6  7
3 3 3 3 3      1 2 3 4 5       2  5 13  8  9
4 4 4 4 4      1 2 3 4 5       3  6  8 14 10
5 5 5 5 5      1 2 3 4 5       4  7  9 10 15
    ri             cj              sij
```

Moreover factors for other constrained diagonals models are easily constructed. Note that defining different levels on the diagonal of sij is equivalent, for some models, to imposing null weights to the diagonal cells.

The results for the illustrative data are summarised below. None of the models fit the data.

		deviance	chi-square	d.f.
independence	$fit ri+cj	6170	7165	16
quasi-independance	(same with weights)	683	719	11
symmetry	$fit sij	2805	2561	10
quasi-symmetry	$fit sij+ri+cj	27	27	6

2.2) Marginal homogeneity

Under marginal homogeneity the interaction effects are not symmetrical, but the row and column marginals are the same. Such model is not log-linear, but it can be tested using some separability results MATHIEU(1987).

The linear hypothesis of marginal homogeneity (MH) and the log linear hypothesis of quasi-symmetry (QS) are separable. In addition the hypothesis MH and QS is equivalent to the hypothesis of symmetry (S). Then either the deviance or the chi-square statistics for MH can be approximated by the difference between those statistics under S and under QS. Accordingly, the fitted values under MH can be approximated by fv(MH) $\approx$ f - fv(QS) + fv(S) and associated residual estimates are easily obtained. Note that the separability of MH and of independence (I) or of quasi-independence (QI) might also be used. In addition "direct" estimation is sometimes possible with GLIM (identity link, hence possible failures in the computation). For the data above the corresponding approximations to the test statistics all lead to reject marginal homogeneity (deviance= 2744, chi-square= 2451, degrees of freedom= 4).

2.3) Residual analysis :

The analysis of residuals from log-linear models in terms of correspondence analysis or of GABRIEL's biplot is considered in CAUSSINUS and FALGUEROLLES(1987), HEIJDEN(1987), HEIJDEN and LEEUW(1985), FALGUEROLLES and HEIJDEN(1987). There are different motivations for such practice: the associated scatter diagrams may suggest revisions of the current model or just visualize some interesting part of the data (the "model" filtering uninteresting or obvious features).

In some cases, adapted methods GOWER(1977) take advantage of the structure of the residuals : namely,the residuals from symmetry or from quasi-symmetry are skew-symmetric. Furthermore different types of residuals can be analysed (FALGUEROLLES and HEIJDEN(1987)). We consider here mainly the Pearson standardized residuals . Thus singular value decomposition (SVD) of the matrix of standardized residuals decomposes the chi-square statistic. Finally SVD and derived GABRIEL's biplot are easily obtained using GAUSS procedures.

Due to lack of space, we do not report in detail the analysis of the residuals from all models . Associated singular values are shown opposite. Their pattern reflects the improvement in incorporating structure by adapted models and the amount of complementary information left in the residuals.

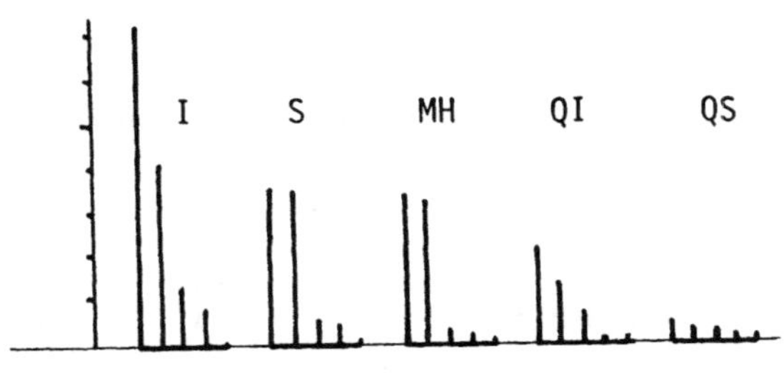

Analysis of the different approximations of the residuals from the model of marginal homogeneity (MH) (separability of MH and QS, of MH and Q, of MH and QI, and direct estimation) revealed very similar structures. Note that these approximations are defined as differences between the estimated frequencies under two models.

We recall that none of the models above fit. Hence graphical displays of their residuals are of interest. As an example, we give opposite the biplot of standardized residuals from QS. Visual inspection points out some asymmetry involving upward and downward mobility within the nonmanual strata -cells (1,2) and (2,1)- and within the lower strata -cells (2,4) and (4,2)-.

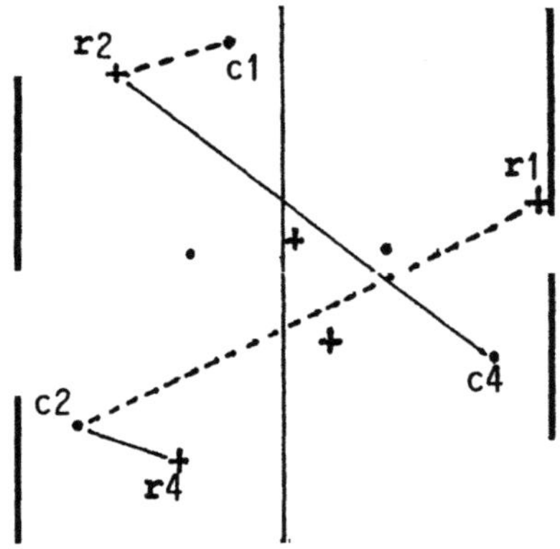

2.4) Model description

The symmetrical interaction for the quasi-symmetry model measures the similarity between two categories. Unfortunately the numerical values of the parameters depend on the constraints imposed for identification. However the constraints can be set in such a way that the parameters have row and column adding to one (the initial multiplicative formulation by CAUSSINUS(1966)). Arbitrarily posing diagonal similarities equal to one, the symmetrical interactions thus define a semi definite positive matrix. This property suggests the use of metric multidimensional scaling to obtain a graphical display of the categories. The re-evaluation of the symmetrical interaction parameters (marginal fitting on the exponentiated parameters) is straightforward. In the example, the following matrix is obtained :

	1	2	3	4	5
1	1.00000	0.94704	0.02514	0.01411	0.01343
2	0.94704	1.00000	0.02212	0.01707	0.01336
3	0.02514	0.02212	1.00000	0.47429	0.47868
4	0.01411	0.01707	0.47429	1.00000	0.49477
5	0.01343	0.01336	0.47868	0.49477	1.00000

Metric multidimensional scaling on this matrix constructs coordinates for the categories and thus visualize the intensity of the symmetrical interactions :

categories	axis 5	axis 4	axis 3	axis 2	axis 1	
1	0.	0.16	0.00	0.00	-0.76	upper nonmanual
2	0.	-0.16	-0.00	0.01	-0.76	lower nonmanual
3	0.	-0.00	-0.06	-0.59	0.49	upper manual
4	0.	0.00	-0.47	0.35	0.51	lower manual
5	0.	-0.00	0.53	0.24	0.51	farm

The interpretation of the different axes is obvious . Hence the associated scatter diagrams are not given.

3) Mobility table, the stratified case :

We consider now data on marital class endogamy in Britain between the mid-nineteenth and mid-twentieth centuries (PENN and DAWKINS (1983)). They consist in a series of five square tables. Each square table cross-classifies during a decennial period the occupational grouping of the father of the bride and of the father of the groom. The five decennial periods are 1856-1865 1875-1884 1900-1909 1920-1929 1955-1964. The occupational grouping has seven levels: from bourgeoisie to unskilled manual.

3.1) Log-linear approach :

Extensions of some of the models described in section two are easily fitted with GLIM. Let ri (the row effect), cj (the column effect), tk (the time effect) and sij (the symmetrical order one interaction between i and j) denote four factors. Null weights can be imposed to diagonal cells. These factors, the operators in model formulae and the directive $weight offer enough flexibility to cover a large scope of models.

The results of fitting various log-linear models to the endogamy data are given below :

	deviance	chi-square	d.f.
$fit (ri+cj)*tk	308	338	180
- with weights	190	189	145
$fit (ri+cj)*tk+ri*cj	134	118	144
- with weights	103	90	116
$fit sij+(ri+cj)*tk	146	135	159
- with weights	116	107	131
$fit sij*tk	101	87	105

It appears that the model of symmetry (sij*tk) or of quasi-symmetry (sij+(ri+cj)*tk and weights, CAUSSINUS and THELOT(1976)) offer interesting alternatives to the model of no order two interaction which was retained by PENN and DAWKINS(1983). Moreover for these models, graphical displays analogous to that introduced in section 2.4 can visualize the symmetrical interactions.

3.2) Other graphical displays

We follow here the general strategy set in section 2.3 of decomposing either the residuals from a model or the difference between the fitted values from two models. The resulting data have first to be organized into two-way tables. Associated SVD and biplots of these tables are then easily obtained.

Detailed outputs for these data, programs and interpretations can be obtained from the authors.

References :

Baccini, A. et al. (1987): Comparaison, sur un exemple, d'analyses des correspondances et de modélisations. Rev. de Statistique Appliquée, XXXV(3),21-34.

Burtschy, B. (1984): Analyse factorielle des matrices d'échanges. in Data Analysis and Informatics (E. Diday et al. eds) North Holland, 447-464.

Caussinus, H. (1966): Contribution à l'analyse statistique des tableaux de corrélation. Ann.Fac. Sci., Univ. de Toulouse, XXIX, 77-182.

Caussinus, H. et Falguerolles, A. de (1987): Tableaux carrés: modélisation et méthodes factorielles. Rev. de Statistique Appliquée, XXXV(3),35-52.

Caussinus, H. et Thelot, C. (1976) : Note complémentaire sur l'analyse statistique des migrations. Ann. de l'INSEE, N° 22-23, 135-145.

Falguerolles, A. de et Heijden, P.G.M. van der (1987) : Sur l'analyse factorielle des correspondances et quelques unes de ses variantes. Rev. de Statistique Appliquée, XXXV(3),7-12.

Gower, J.C. (1977): The analysis of asymmetry and orthogonality. In J. Barra at al. (Eds.), Recent developments in statistics. Amsterdam : North-holland, 109-123.

Heijden, P.G.M. van der (1987): Correspondence analysis of longitudinal categorical data. DSWO Press, Leiden

Heijden, P.G.M. van der and Leeuw, J. de(1985): Correspondence analysis used complementary to loglinear analysis. Psychometrika, vol. 50, 4, 429-447.

Hout, M. (1983): Mobility tables. Sage university paper 07-031. Sage Pubns. Beverly Hill and London.

Mathieu, J. R.(1987): On some linear and log-linear problems about the marginal distributions of two and three dimensional contingency tables. Statistics, XVIII(1), 49-63.

Penn, R.D. and Dawkins, D.C. (1983): Structural transformations in the british class structure: a log-linear analysis of marital endogamy in Rochdale 1856-1964. Sociology, vol.17, 506-523.

Worsley, K. J.(1987): Un exemple d'identification d'un modèle log-linéaire grâce à une analyse des correspondances. Rev. de Statistique Appliquée, XXXV(3), 13-20.

Data Plotting Methods for Checking Multivariate Normality and Related Ideas

T. Isogai, Osaka

Summary: The object of the paper is to find directions along which multivariate observations have the greatest multivariate skewness or kurtosis in an appropriate sense. Typical measures of multivariate skewness and kurtosis are Malkovich and Afifi's (1973) b_1^* and b_2^*, which are essentially nonlinear maximization problems. To avoid this nonlinearity we present an approach to reduce the above problems to easier ones, which are eigenvalue problems in linear algebra and closely related to some types of measures of multivariate skewness and kurtosis. By using the resultant directions we can project observations into the sample space and check normality of the data through probability plots and scatter plots. We also show that the proposed approach enables us to extend the usual principal component analysis to a higher order case.

Keywords: multivariate normality, measures of multivariate skewness and kurtosis, higher order principal component analysis

1. Introduction

Let $X' = (X_1, X_2, \ldots, X_p)$ be a p-dimensional random vector distributed according to a p-variate distribution function F with mean vector $\mu' = (\mu_1, \ldots, \mu_p)$ and covariance matrix $\Sigma = (\sigma_{ij})$, i, j = 1,..., p and also let $\psi(t) = \log E[\exp(t'X)]$ be the cumulant generating function of X, where E denotes the operation of expectation in the distribution F and t' =

Compstat 1988

$(t_1, \ldots, t_p)$ is a constant scalar vector.

The 3rd and 4th order multivariate cumulants of X are respectively defined by $\kappa_{ijl} = \partial^3/\partial t_i \partial t_j \partial t_l\ \psi(t)|_{t=0}$ and $\kappa_{ijlm} = \partial^4/\partial t_i \partial t_j \partial t_l \partial t_m\ \psi(t)|_{t=0}$ for i, j, l, m = 1,..., p, where ∂/t_i (i = 1,..., p) denotes the differential operator in the argument t_i.

Now let $a' = (a_1, \ldots, a_p)$ be a p-dimensional scalar vector. By using a linear conbination of the random vector X, i.e. a'X, Malkovich and Afifi's (1973) measures of multivariate skewness b_1^* and kurtosis b_2^* are respectively defined by

$$b_1^* = \max(\sum_i \sum_j \sum_l a_i a_j a_l \kappa_{ijl})^2,\ b_2^* = \max(\sum_i \sum_j \sum_l \sum_m a_i a_j a_l a_m \kappa_{ijlm})^2$$

where the maximization with respect to a vector a is carried out under the restriction that the variance of a'X (= var(a'X), say) is equal to one, i.e. $a'\Sigma a = 1$. The direction a which maximizes b_1^* or b_2^* gives us a reasonable solution to our problem. But it seems very difficult to calculate the maximum values of b_1^* and b_2^* as well as directions a's in a high dimensional case.

2. Measures of multivariate skewness and kurtosis

Let a, b, c and d be arbitraly p × 1 constant scalar vectors and put $\partial' = (\partial/t_1, \partial/t_2, \ldots, \partial/t_p)$. Let us denote the differentiation with the direction of a p × 1 vector $v' = (v_1, \ldots, v_p)$ by $v'\partial$ (= $v_1(\partial/t_1) + v_2(\partial/t_2) + \ldots + v_p(\partial/t_p)$). Isogai (1985) defined the 3rd order generalized multivariate cumulant K_{abc} along directions a, b and c by

$$K_{abc} = (a'\partial)(b'\partial)(c'\partial)\psi(t)|_{t=0} = \sum_i \sum_j \sum_l a_i b_j c_l \kappa_{ijl}$$

and similarly the 4th order generalized multivariate cumulant K_{abcd} along directions a, b, c and d by

$$K_{abcd} = (a'\partial)(b'\partial)(c'\partial)(d'\partial)\psi(t)|_{t=0} = \sum_i \sum_j \sum_l \sum_m a_i b_j c_l d_m \kappa_{ijlm}$$

where summations are carried out with indices i, j, l and m = 1,...,p.

By using these K_{abc} and K_{abcd}, Isogai (1985) defined various measures of multivariate skewness and kurtosis as follows: measures of multivariate skewness were defined by

$$K_{AAA}{}^2 = \max_a (K_{aaa})^2\ (= b_1^*),\ K_{AAB}{}^2 = \max_{a,b} (K_{aab})^2,$$

$$K_{ABC}{}^2 = \max_{a,b,c} (K_{abc})^2 \quad (= \text{Mardia's } \beta_{1,p})$$

and measures of multivariate kurtosis were defined by

$$K_{AAAA}{}^2 = \max_{a} (K_{aaaa})^2 \quad (= b_2^*),$$

$$K_{AAAB}{}^2 = \max_{a,b} (K_{aaab})^2, \quad K_{AABB}{}^2 = \max_{a,b} (K_{aabb})^2,$$

$$K_{AABC}{}^2 = \max_{a,b,c} (K_{aabc})^2, \quad K_{ABCD}{}^2 = \max_{a,b,c,d} (K_{abcd})^2 \quad (= tr(K_2))$$

where max means the maximization with respect to argument vectors a, b, c and d under the restriction that variances of a'X, b'X, c'X and d'X are all equal to one. We remark that $K_{ABC}{}^2$ is equivalent to Mardia's (1970) $\beta_{1,p}$ and $K_{ABCD}{}^2$ is equal to $tr(K_2)$ introduced by Isogai (1983).

Note that all of these measures are invariant under the affine transformation of the random vector X. They have the following properties. Put $\Sigma^{-1} = (\sigma^{ij})$, the inverse of Σ.

<u>Proposition 2.1.</u> [Isogai(1985)] Under $a'\Sigma a = 1$, $b'\Sigma b = 1$, $c'\Sigma c = 1$ and $d'\Sigma d = 1$, we have

(2.1) $$(K_{aaa})^2 \le a_j a_l a_{j'} a_{l'} \kappa_{ijl} \kappa_{i'j'l'} \sigma^{ii'}$$
$$\le a_l a_{l'} \kappa_{ijl} \kappa_{i'j'l'} \sigma^{ii'} \sigma^{jj'}$$
$$\le \kappa_{ijl} \kappa_{i'j'l'} \sigma^{ii'} \sigma^{jj'} \sigma^{ll'} = K_{ABC}{}^2$$

(2.2) $$(K_{aaaa})^2 \le a_j a_l a_m a_{j'} a_{l'} a_{m'} \kappa_{ijlm} \kappa_{i'j'l'm'} \sigma^{ii'}$$
$$\le a_l a_m a_{l'} a_{m'} \kappa_{ijlm} \kappa_{i'j'l'm'} \sigma^{ii'} \sigma^{jj'}$$
$$\le a_m a_{m'} \kappa_{ijlm} \kappa_{i'j'l'm'} \sigma^{ii'} \sigma^{jj'} \sigma^{ll'}$$
$$\le \kappa_{ijlm} \kappa_{i'j'l'm'} \sigma^{ii'} \sigma^{jj'} \sigma^{ll'} \sigma^{mm'} = K_{ABCD}{}^2$$

(2.3) $$K_{AAB}{}^2 = \max_a (a_j a_l a_{j'} a_{l'} \kappa_{ijl} \kappa_{i'j'l'} \sigma^{ii'})$$

$$K_{AAAB}{}^2 = \max_a (a_j a_l a_m a_{j'} a_{l'} a_{m'} \kappa_{ijlm} \kappa_{i'j'l'm'} \sigma^{ii'})$$

$$K_{AABC}{}^2 = \max_a (a_l a_m a_{l'} a_{m'} \kappa_{ijlm} \kappa_{i'j'l'm'} \sigma^{ii'} \sigma^{jj'})$$

(2.4) $$K_{AAA}{}^2 \le K_{AAB}{}^2 \le K_{ABC}{}^2 \quad \text{and}$$

$$K_{AAAA}{}^2 \le \left\{ \frac{K_{AAAB}{}^2}{K_{AABB}{}^2} \right\} \le K_{AABC}{}^2 \le K_{ABCD}{}^2$$

where for simplicity in the above inequalities we omit the symbol $\sum$, the notation of summation, and when the same index appears twice in different quantities, the summation is carried out with respect to the same index.

2.1. Plotting method based on inequalities of the most dominant measures of multivariate skewness K_{ABC}^2 and kurtosis K_{ABCD}^2

The relations (2.1) and (2.4) suggest that instead of dealing with K_{AAA}^2 (= b_1^*), we may deal with the third term of (2.1), which can be regarded as a function of the vector a and has K_{ABC}^2 as its maximum value. This leads us to the problem:

(2.5) $\max_a[\max_{b,c} (K_{abc})^2]$ under $\mathrm{var}(a'X) = a'\Sigma a = 1$.

This is related to the following eigenvalue problem:

(2.6) $(\sigma^{ii'}\kappa_{ijl}\sigma^{jj'}\kappa_{i'j'l'})a = \lambda\,(\sigma_{ll'})a$ under $a'\Sigma a = 1$.

For the measure K_{AAAA}^2 (= b_2^*), we may use the fourth term of (2.2),which leads us to the maximization problem:

(2.7) $\max_a[\max_{b,c,d} (K_{abcd})^2]$ under $a'\Sigma a = 1$.

The maximum value of (2.7) is K_{ABCD}^2. (2.7) can be rewritten as

(2.8) $(\sigma^{ii'}\kappa_{ijlm}\sigma^{jj'}\kappa_{i'j'l'm'}\sigma^{ll'})a = \lambda\,(\sigma_{mm'})a$

under $a'\Sigma a = 1$.

2.2. Approximate plotting method derived from measures of multivariate skewness K_{AAB}^2 and kurtosis K_{AABB}^2 and K_{AABC}^2

From (2.3) we know that K_{AAB}^2, K_{AABB}^2 and K_{AABC}^2 need the maximization with respect to bilinear forms of vectors a and b in their object functions respectively. Here we suggest that the maximization with respect to bilinear forms may be replaced by the sum of diagonal elements of each matrix (= the trace in the argument matrix).

Then $K_{AABB}^2 = \max_{a,b} (K_{aabb})^2$ is related to the following eigenvalue problems:

(2.9) $(\sigma^{ii'}\kappa_{ii'jj'})a = \lambda\,(\sigma_{jj'})a$

(2.10) $(\sigma^{ii'}\kappa_{ii'lm}\sigma^{jj'}\kappa_{jj'l'm'}\sigma^{ll'})a = \lambda\,(\sigma_{mm'})a$

under $\mathrm{var}(a'X) = a'\Sigma a = 1$. In Isogai (1983) it was shown that the trace of (2.9) is equal to Mardia's (1970) measure of multivariate kurtosis $\beta_{2,p}$.

Similarly, from (2.3), $K_{AAB}{}^2$ and $K_{AABC}{}^2$ are related to the following eigenvalue problems respectively:

(2.11) $(\kappa_{ijl}\sigma^{ll'}\kappa_{i'j'l'}\sigma^{i'j'})a = \lambda\ (\sigma_{ij})a$

(2.12) $(\kappa_{ijlm}\sigma^{ll'}\sigma^{mm'}\kappa_{i'j'l'm'}\sigma^{i'j'})a = \lambda\ (\sigma_{ij})a$

under $a'\Sigma\ a = 1$.

3. Higher order principal component analysis (P.C.A.)

We shall reconsider the usual P.C.A. by using the cumulant generating function $\psi(t)$. The usual P.C.A. is defined as a maximization problem with respect to var(a'X), i.e.

(3.1) $\max\ (a'\partial)^2\ \psi(t)|_{t=0} = \max\ (a'\Sigma a) = \max\ K_{aa}$

under $a'a = 1$. This definition is equivalent to

(3.2) $\max_a\ [\ \max_{\substack{b \\ b'\Sigma b = 1}}\ \{(a'\partial)(b'\partial)\psi(t)|_{t=0}\}^2\ ] = \max_a\ \{\ \max_{\substack{b \\ b'\Sigma b=1}}\ (K_{ab})^2\}$

under $a'a = 1$. The definition (3.2) is the problem of maximizing the squared covariance between a'X and b'X with var(b'X) = 1. By using the above two definitions (3.1) and (3.2) we shall define a higher order principal component analysis.

3.1. Higher order P.C.A. as an extension of the definition (3.2)

A 3rd order principal component analysis is defined as

(3.3) $\max_a\ \{\ \max_{\substack{b,c \\ b'\Sigma b=1 \\ c'\Sigma c=1}}\ (K_{abc})^2\}$ under $a'a = 1$.

(3.3) is equivalent to the following eigenvalue problem:

(3.4) $(\sigma^{ii'}\kappa_{ijl}\sigma^{jj'}\kappa_{i'j'l'})a = \lambda\ a$ under $a'a = 1$.

Similarly, a 4th order principal component analysis is defined as

(3.5) $\max_a\ \{\ \max_{\substack{b,c,d \\ b'\Sigma b=1 \\ c'\Sigma c=1 \\ d'\Sigma d=1}}\ (K_{abcd})^2\}$ under $a'a = 1$.

(3.5) is equivalent to the eigenvalue problem:

(3.6) $(\sigma^{ii'}\kappa_{ijlm}\sigma^{jj'}\kappa_{i'j'l'm'}\sigma^{ll'})a = \lambda\ a$ under $a'a = 1$.

3.2. Higher order P.C.A. as an extension of the definition (3.1)

The maximization problem of var(a'X) can be interpreted as the maximization problem of $(a'\partial)^2 \psi(t)|_{t=0}$ with respect to a vector a. Then 3rd and 4th order principal component analyses are respectively defined by

$$(3.7) \quad \max_{a} [\max_{\substack{b \\ b'\Sigma b=1}} \{(a'\partial)^2 (b'\partial)\psi(t)|_{t=0}\}^2] = \max_{a} [\max_{\substack{b \\ b'\Sigma b=1}} (K_{aab})^2]$$

and

$$(3.8) \quad \max_{a} [\max_{\substack{b,c \\ b'\Sigma b=1 \\ c'\Sigma c=1}} \{(a'\partial)^2 (b'\partial)(c'\partial)\psi(t)|_{t=0}\}^2]$$
$$= \max_{a} [\max_{\substack{b,c \\ b'\Sigma b=1 \\ c'\Sigma c=1}} (K_{aabc})^2]$$

under a'a = 1.

As a restricted case of (3.8), we may define a 4th order P.C.A. by

$$(3.9) \quad \max_{a} [\max_{\substack{b \\ b'\Sigma b=1}} (K_{aabb})^2] \quad \text{under } a'a = 1.$$

By using the idea of the section 2.2, we can relate (3.7), (3.8) and (3.9) to the following eigenvalue problems respectively:

$$(3.10) \quad (\kappa_{ijl}\sigma^{ll'}\kappa_{i'j'l'}\sigma^{i'j'})a = \lambda a,$$

$$(3.11) \quad (\kappa_{ijlm}\sigma^{ll'}\sigma^{mm'}\kappa_{i'j'l'm'}\sigma^{i'j'})a = \lambda a,$$

$$(3.12) \quad (\kappa_{ijll'}\sigma^{ll'})a = \lambda a$$

under a'a = 1.

References

Isogai, T. (1983). On measures of multivariate skewness and kurtosis. Mathematica Japonica 28, 251-261.

Isogai, T. (1985) Note on statistics from the 3rd and 4th order multivariate cumulants. Studies in Humanities and Social Sciences 33, Osaka Univ., 67-72.

Malkovich, J.F. and Afifi, A.A. (1973). On tests for multivariate normality. J.Am.Statist.Assoc., 68, 176-179.

Mardia, K.V. (1970). Measures of multivariate skewness and kurtosis. Biometrika 57, 519-530.

Computer-Aided Illustration of Regression Diagnostics

T. Nummi, M. Nurhonen and S. Puntanen, Tampere

Abstract: In this paper we describe a microcomputer system developed by the authors for illustration of various concepts related to regression analysis. Particular attention is paid to the diagnostics for analysing influential observations and outliers. Naturally graphical methods play a central role in this illustration. The system is mainly planned for a student taking a course in regression analysis or a person who is applying regression analysis and wants to know the meaning of diagnostics in practice. The programming language is the APL and thus the user familiar with the APL can easily extend the system by his own functions. The system is implemented on Apple Macintosh personal microcomputer and it is a part of a larger system, called KONSTA 88, which is planned for illustrating statistical concepts.

Key words: Influential observation, leverage, microcomputer software, outlier, perturbation, residual, teaching of statistics.

1. INTRODUCTION

The recent growth of the computational capacity has activated lots of matrix formulas, related to regression diagnostics, which earlier were merely of theoretical interest. The availability of possible diagnostics, such as various residuals, leverages, Cook's distance etc., causes no problems for most users. As a matter of fact, the number of diagnostics suggested in the literature is almost confusingly high. However, their interpretation and practical meaning may not be that easy or obvious.

In this paper we describe a microcomputer system developed by the authors for illustration of various concepts related to regression analysis. Particular attention is paid to the diagnostics for analysing influential observations and outliers. Naturally graphical methods play a central role in this illustration. The system is mainly planned for a student taking a course in regression analysis or a person who is applying regression analysis and wants to know the meaning of diagnostics in practice. The system is implemented on Apple Macintosh personal microcomputer and the programming language is the APL. The user familiar with the APL can easily extend the system by his own functions.

The main motivation in developing the system was that the pure numerical values of the diagnostics do not tell the whole story. Namely the situation may often be so complicated that an interactive graphical manipulation of the data set is needed. On the other hand, one has to learn that the quantitative measures may tell something which is hidden from the eye.

Let us imagine what happens in a student's mind when he sees a scatterplot of the type of Figure1a. One natural idea to come to a student's mind could very well be

Compstat 1988

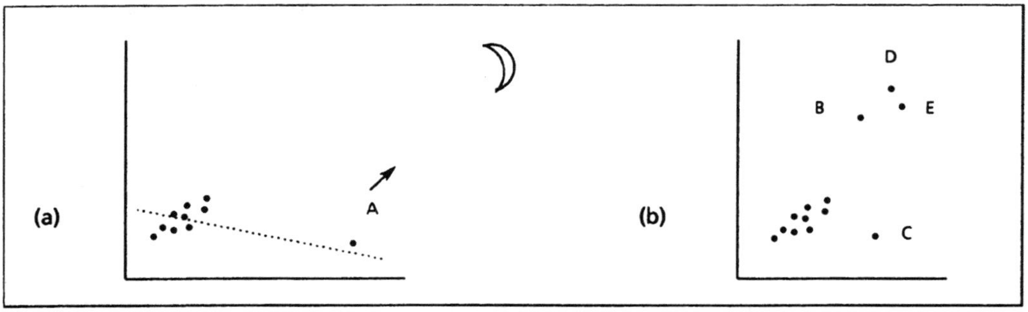

FIGURE 1. Illustration of moving and deleting data points.

something like: "I really would like to move the data point A and see what happens then to the regression line." The student might indeed be curious to know how far and to which direction he has to move the data point to obtain certain regression line or correlation coefficient. Similarly, one might wonder about the consequences of deleting some of the points B, C, D or E in Figure 1b. Furthermore, we might imagine two students arguing with each other about the place where point A should be moved so that a statistic would attain a given value. But what are we involved with now? Isn't it quite near to a kind of a "computer game"? Yes, we think so, and in fact this is one of our main ideas: this program could also be a *toy* for the user of regression analysis. We feel that this toy (like other "good" toys) might be educational.

Our program includes for example a game, where the player sees a scatterplot on the screen and he has to guess some statistics like the correlation coefficient or he has to draw the regression line. Then he is asked to add one data point such that the statistics would change to certain values, or the program adds one point and asks for the changes. On each step the player will be fined or rewarded according to his answers. For a novice user of regression analysis the practice of eyes is an essential matter; cf. Anscombe (1973) and Mosteller *et al.* (1981).

As Cook and Weisberg (1982, p. 101) state, the basic idea in influence analysis is quite simple: we introduce small perturbations in the problem formulation, and then monitor how the perturbations change the outcome of the analysis. The leading idea behind our system is just to demonstrate this principle so that the user could concretely and interactively participate in this process.

One typical application of our system is that the user moves an observation on the screen and simultaneously sees the (graphically presented) effect on the diagnostics he is interested in. Similarly the user can see e.g. how the estimated regression line is moving or the residual variance is changing when one observation or a group of observations is moved. The points which stand out from the group on their diagnostic measures can be "flagged" if the user wants; the user can then perturb the data and ask for a new flagging. One possibility is to have a scatterplot on one window and the standard regression output summary of these data on another window and then to perturb the plot and to observe the consequences to the output. It is also possible to generate data sets with desired values of diagnostics. The user can also violate the standard assumptions of the linear model in various ways (e.g. non-normal, heteroscedastic, correlating errors), generate observations from these violated models and then observe the consequences.

One important aspect is to clarify the differences and similarities between an outlier and an influential case. Other central concepts to be illustrated are e.g. multiple outliers, jointly influential observations, masking, leverage values, transformation to normality, collinearity etc.

2. EXAMPLES

In this section we consider some examples of the use of KONSTA 88 in illustrating concepts related to regression analysis. It is clear that it is not at all an easy task to show on paper how the program works because the leading idea behind the program is that it should work dynamically, interactively and visually. In the following examples we consider the simple regression model with one explanatory variable, i.e., $y = \beta_0 + \beta_1 x + \varepsilon$, where $var(\varepsilon) = \sigma^2$.

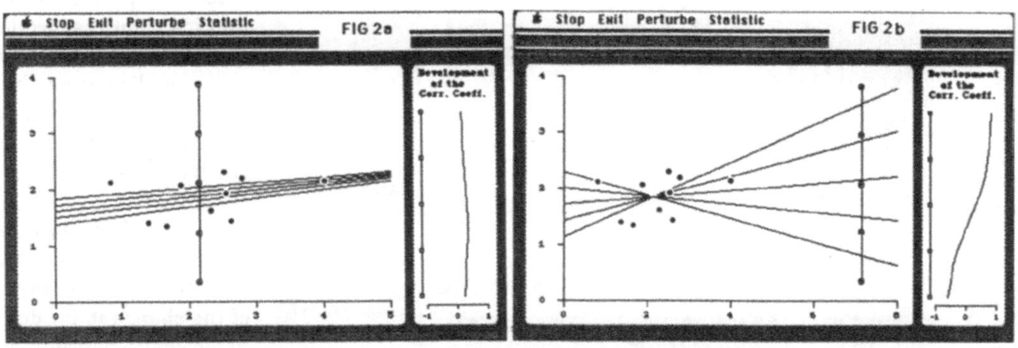

FIGURE 2. ILLUSTRATION OF THE MOVEMENT OF THE POINT ALONG A LINE. This figure shows (or at least gives some idea on) how the user can move a data point and observe the consequences on the screen. In Figures 2a and 2b the user has defined a line along which the point will move. The user can also define the statistics whose development, while moving a data point, he wants to see plotted. The path of the data point can be marked also completely free by hand or the user can define a circle along which the point will move.

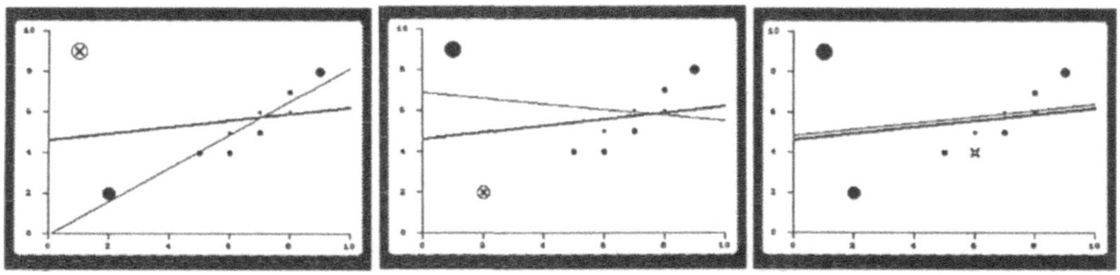

FIGURE 3. ILLUSTRATION OF THE CONSEQUENCES OF DELETING A DATA POINT. **Each point is marked with a circle whose area is proportional to the corresponding Cook's distance. The points will be deleted one by one and the resulting regression line is drawn. The deleted point is marked by a cross.**

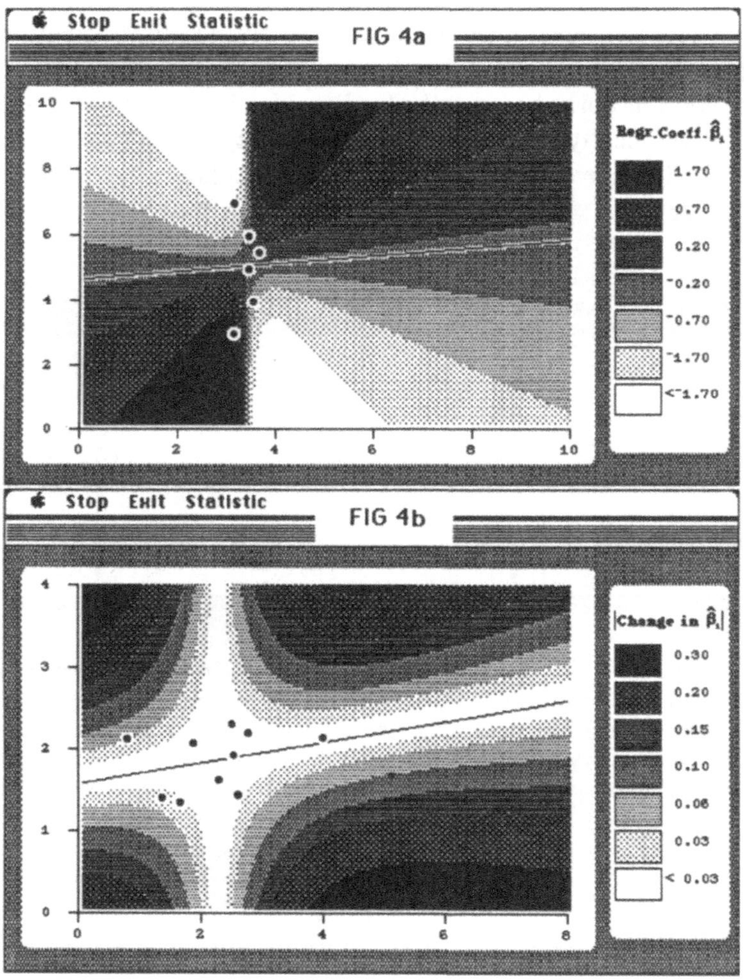

FIGURE 4. SHADED PLOT ILLUSTRATING THE EFFECT OF AN ADDITIONAL POINT ON $\hat{\beta}_1$ (SLOPE).

In Figure 4a dark shades correspond to large values of $\hat{\beta}_1$ indicating that if a point is added in that place then the corresponding new value of $\hat{\beta}_1$ is high. Light shades correspond to small new values of $\hat{\beta}_1$.

In Figure 4b we have a different scatterplot and here the absolute value of the change in $\hat{\beta}_1$, due to the additional point, is calculated.

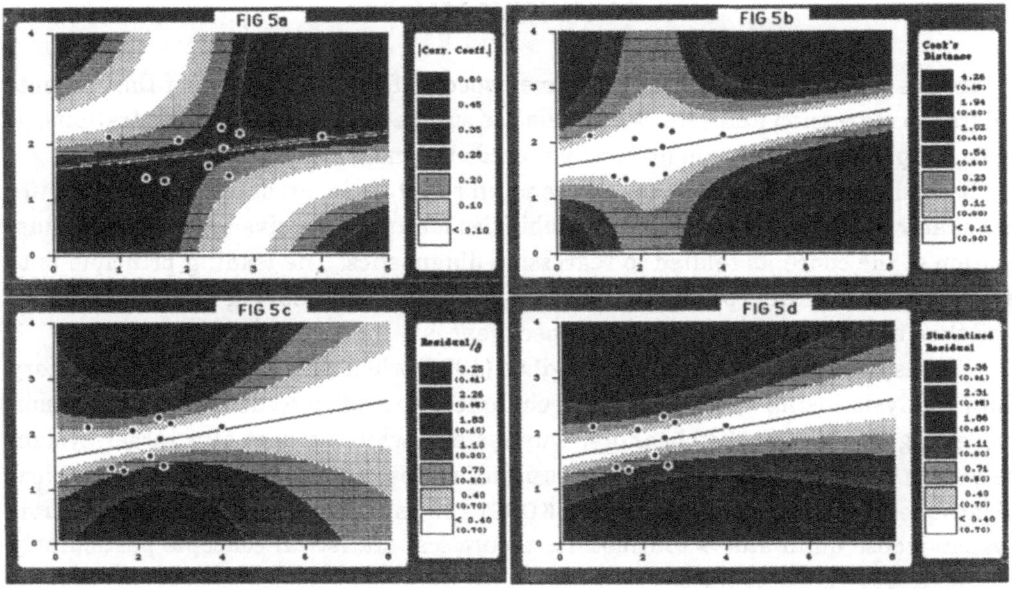

FIGURE 5. SHADED PLOTS ILLUSTRATING THE EFFECT OF AN ADDITIONAL POINT ON SOME STATISTICS. **Dark shades in Figure 5a correspond to large absolute values of correlation coefficient r indicating that if a point is added in that place then the corresponding new value of $|r|$ is high. In Figures 5b, 5c and 5d the dark area indicates that if a point is added in that place then that point's Cook's distance, residual (divided by $\hat{\sigma}$), or externally Studentized residual, respectively, is large. We can clearly observe e.g. that the leverage value affects less the externally Studentized residual than the ordinary residual.**

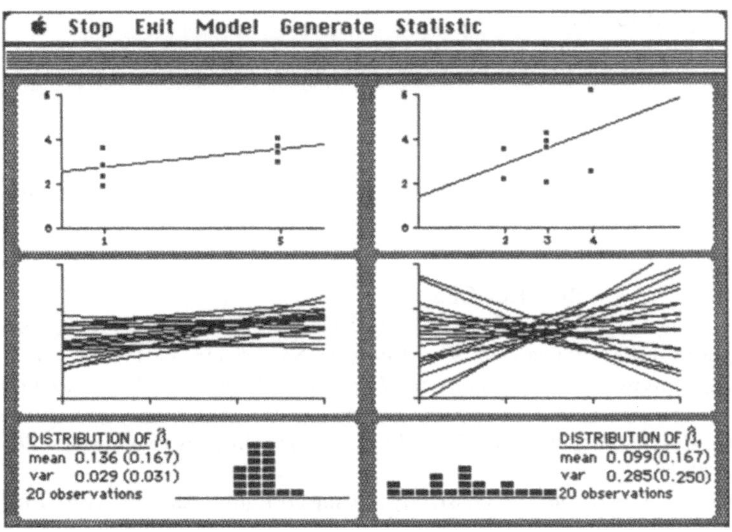

FIGURE 6. **Here we generate observations from two identical models, but under different values of x.
The program repeatedly generates observations from both models. After each generation the estimated regression line will be drawn (upper window), it will be "added" to the set of earlier lines (middle window) and also a histogram of observed values of $\hat{\beta}_1$ (or some other statistics) will be drawn.**

3. CONCLUDING REMARKS

In this paper we have described some aspects of the illustration of the concepts related to regression analysis. We emphasize simple possibilities for illustrations; at least simple methods should naturally precede the more complicated ones.

It should be emphasized that our computer system is not supposed to be a pure calculator: its main goal is to offer possibilities for an interactive and dynamic illustration of the concepts related to regression diagnostics. The leading principle is to allow the user himself perturb the data (e.g. by deleting, moving or adding points) and then immediately recognize the consequences.

We also agree with Denby and Pregibon (1987) when they argue that exploratory data analysis, using rather simple techniques, should precede formal regression modelling and diagnostic checking. Our system, which is part of a larger system called KONSTA 88, offers flexible possibilities for applying low-tech graphical techniques for looking at the data. KONSTA 88 is an interactive microcomputer system whose main aim is to illustrate theoretical statistical concepts particularly on the first statistics courses.

Furthermore, we would like to emphasize the possibilities to use the program as a toy: certainly the regression fans have earned their own toys. The program has several features which might be rather useless, in the serious classical sense, for the ordinary user of regression analysis. However, we think that they may give new ideas for the user's imagination, ideas which otherwise might not occur to him so easily.

Anyway, though life is short (cf. Cook and Weisberg 1982, p. 8) it seems fair to devote some part of it to regression diagnostics.

ACKNOWLEDGEMENT

This paper was prepared when the first author was an acting research assistant in the Academy of Finland.

REFERENCES

Anscombe, F.J. (1973). Graphs in statistical analysis. *The American Statistician*, 27, 17-21.

Cook, R. Dennis and Weisberg, Sanford (1982). *Residuals and Influence in Regression*. Chapman and Hall, New York.

Denby, Lorraine and Pregibon, Daryl (1987). An example of the use of graphics in regression. *The American Statistician*, 41, 33-38.

Mosteller, F.; Siegel, A.F.; Trapido, E. and Youtz, C. (1981). Eye fitting straight lines. *The American Statistician*, 35, 150-152.

Computer Guided Diagnostics

D. A. Belsley, A. Venetoulias and R. E. Welsch*, Cambridge

INTRODUCTION

The battery of diagnostic techniques developed during the past ten years to find influential observations, assess collinearity, check for data transformations, etc. can overwhelm an expert data analyst, let alone a novice. We are working on software systems that guide the analyst through these procedures and aid in interpreting the results. The tools we use are the S statistical analysis system from Bell Laboratories and Hypercard from Apple Computer. We discuss examples involving collinearity (New S) and solicitation of prior data information for use in diagnostic procedures (Hypercard).

THE LINEAR REGRESSION SYSTEM IN NEW S

The linear regression module comprises five interdependent basic modules: The *Identification of Relevant Information* module queries the user about the goal of the regression (prediction, explanation, measurement, ...), the user profile (to determine the degree of system intervention), and the data. The *Transformations* module verifies the functional form of the model. Details of its implementation are currently under investigation. The *Assumption* module runs tests on the basic regression assumptions for the selected model. Submodules handle each of the three main issues: normality, homoskedasticity, and independence. The *Data Diagnostics* module comprises four basic submodules. The *Collinearity* submodule, described below, reports on data conditioning. A second submodule performs influential-data diagnostics and reports its findings. This module contains many of the techniques found in Belsley, Kuh and Welsch (1980), Chatterjee and Hadi (1988) and Atkinson (1985). A third submodule examines the data for peculiarities as described in Wetherill (1986). The fourth submodule uses the results of the previous three and, with the user's aid, determines which variables to include in the model. Belsley (1988a) describes some philosophical foundations for model building underlying this effort. A final module, *Regression-Output*, performs the resulting regression and produces standard output. All other information gathered during the guidance process (sometimes automatically) is also available here for user inquiry.

Description of the Collinearity Submodule

The *Collinearity* module is a submodule of the *Data Diagnostics* component. It detects, and perhaps rectifies, existing potentially harmful collinearities in the data. This module can be

* This work was supported in part by National Science Foundation grant IST-8420614 and Army Research Office Grant DAAG29-84-K-0207.

invoked either explicitly by the user or automatically. Thus, if the user needs information on collinearities in the design matrix, his request calls the *Collinearity* submodule. However, the diagnostic analysis is performed anyway as a routine background task, and, if concern arises, the user is warned and given options as described below.

The *Collinearity* module performs three basic tasks: 1) recognize whether linear dependencies exist and assess their degree of importance; 2) identify the variables involved; and 3) attempt action to improve the situation. The first task is accomplished by the Multiple-Hard-Cuts-With-Soft-Edges algorithm of Belsley and Oldford described in Oldford and Peters (1984). The information gathered here identifies the number of dependencies, the number of competing dependencies, and the singular values associated with each. A descriptive evaluation of the situation formulates advice for the user and quantifies the degree of confidence in this advice. The second task, identifying the variables involved in each dependency, is accomplished through a decomposition of the variances of the regression coefficients as described in Belsley, Kuh, and Welsch (1980, pp 105-111). The third task, rectifying a potentially harmful situation, is not yet implemented since appropriate actions depend also on the overall state of the analysis.

An Example

We illustrate the use of the collinearity module with the hospital data (five independent variables plus intercept) from Myers (1986, page 82). The output is shown in Display 1 (these screen prints are not of the quality that we ultimately intend). After identifying any linear dependencies, a brief evaluation is reported. Here, the user is advised that "At least 1 near dependency exists". This advice is quantified by the degree of confidence the system has in its own advice and by the strength of the suspected collinearity.

At the time the user is warned, it has only been determined that dependencies exist; no attempt has been made to identify the variables involved. The user is then presented with four options. If option 1 is chosen (ignore the warnings), the system goes no further with the collinearity analysis but proceeds with the rest of the regression analysis. If, however, option 2 is chosen, control is returned to the user to decide what to do. If option 4 (autocorrection) is chosen, all accumulated knowledge is used to deal with the existing collinearities. This part of the module depends on other modules (*Identification of Relevant Information, the Influential Diagnostics*, and the *Transformation modules*) and is not yet implemented.

Display 1 shows the most interesting case, where the user chooses option 3 instructing the system to investigate the collinearities. The system then obtains all the relevant information on the design matrix, computes the variance decomposition, identifies the variables involved , resolves competing and dominating collinearities, separates variables affected by the collinearities from those which are not, and provides the user with the detailed report shown in Display 1. Now the user can either explore a prudent way to continue or request the system to do so. In our example, the system has correctly identified variables 2 and 4 as involved in a collinear relation associated

```
>
> col.collinearity(scaledHospital)

*********************      WARNING      *******************
There are indications of linear dependencies in the data!

                COLLINEARITY DIAGNOSTICS:  BRIEF DESCRIPTION

Advice:                         At least 1 near dependency exists
Strength of Advice:             Extremely strong
Strength of Collinearities:     Very strong

                COLLINEARITY DIAGNOSTICS:  USER OPTIONS
You can choose to:      Ignore these indications
                        Obtain a short evaluation
                        Investigate these indications
                        Request some corrective action to be taken automatically
1: Ignore
2: Short Evaluation
3: Investigate
4: Rectify automatically
Selection: 3

                COLLINEARITY DIAGNOSTICS:  DETAILED REPORT

Advice:                         At least 1 near dependency exists
Strength of Advice:             Extremely strong
Strength of Collinearities:     Very strong
Suspected Collinearities:       1
Competing Collinearities:       Unknown
Condition Number:               427.32557629699


Variance Decomposition Proportions:
(each row corresponds to the singular value shown on the left)

      Coef 1 Coef 2 Coef 3 Coef 4 Coef 5 Coef 6
2.281  0.001  0.000  0.002  0.000  0.001  0.000
 .816  0.014  0.000  0.011  0.000  0.002  0.006
 .281  0.026  0.000  0.379  0.000  0.012  0.018
 .212  0.009  0.000  0.464  0.000  0.294  0.017
 .091  0.805  0.000  0.142  0.001  0.254  0.757
 .005  0.146  0.999  0.003  0.999  0.438  0.200


Variables Involved in Collinearities:

 Singular Values Involved Variables Unaffected Variables
 ".005"          "2, 4"             "1, 3, 6"

NOTE:  If a variable appears to be involved in a dependency by itself,
       there may exist dominating dependencies which obscure the
       involvement of other variables.


*** collinear.takeAction ***

The module which tries to take some corrective action on the collinearity
problems has not been implemented yet.  Its actions must depend on the user's
expertise and desire, the stage of the analysis and the magnitude of the
encountered problems.  The first two of these factors are still under
investigation.

*** collinear.takeAction ***

>
>
```

DISPLAY 1

with the singular value 0.005. Since variable 1 is the intercept, variables 2 and 4 represent the first and third regressors, precisely the result derived in Myers (1986, page 223). Further, this is the only collinear relation reported by the system. The dependency associated with the singular value 0.091 is insignificant by comparison, a result again in agreement with Myers.

HYPERCARD

The tableau shown in Figure 1 is a Hypercard Stack entry for a Stack we have constructed, called Data Header. A data series cannot be introduced into a regression analysis until such a Header Card has first been created for it and a minimally completed. A moment's examination reveals that there are five informational groupings requiring user choice: data type, characteristics, measurement basis, quality, and importance to analysis. The format of the groupings makes it transparent to the user how to proceed, even without ever having previously encountered the header. The format also conveys information to the user about what the statistical system is capable of doing and what distinctions it considers important.

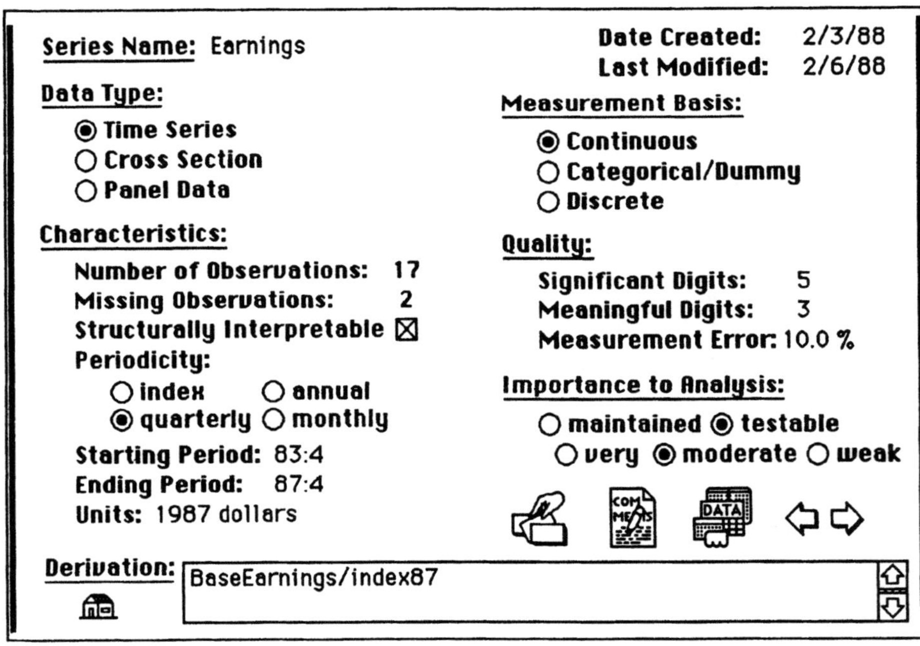

FIGURE 1

Data Type. The Data Type grouping is set merely by clicking with the mouse. A major advantage of presenting choices in this manner is that the user is able to see at once the extent of the choices available and the fact that they are to be treated separately.

Characteristics. The second grouping receives the Characteristics information for the data. Clicking the check box beside Structurally Interpretable lets the user indicate whether the data are structurally interpretable or not. While terms like time series and cross section are familiar and would cause no concern, structurally interpretable is not. So, in the future, there will be a position for each entry for clicking to produce an appropriate page in an accompanying help stack defining and/or describing the concept. The user must also enter the Units of the data. This serves to document the data, and the information will eventually be used to determine the proper wording of reports and to check for units compatibility in derived variables. An important element of "automatic guidance" is exemplified here: the information on the data header is dynamic, automatically altering to remain relevant as various buttons are clicked. Thus, Figure 1 shows the card configuration for a time series. Should, however, the user employ cross-sectional data, the time-related entries are no longer relevant and should no longer encumber the user's vision. So, once Cross Section is selected in Data Type, the time-related periodicity entries disappear and are replaced by cross-section entries. A similar change occurs when the data are typed as panel data.

Measurement Basis. The Measurement Basis category lets the user record for his own archiving and inform the computer of the nature of the data measurement values. Future versions may allow for other types, such as truncated or censored data. This selection also exemplifies aspects of "painless guidance." Different statistical computations should be done in different ways depending upon the nature of the data, and the interpretation of the results should also differ.

The next two categories, Quality and Importance to Analysis, are, to our knowledge, unique among statistical environments and certainly exemplify a major component of our guidance philosophy. The user must be made to think about the quality of the data and about the relation of those data to the analysis at hand. This information is context dependent and not strictly data-series dependent, so the same data series could be very important in one analytical context and only marginally so in another. Thus these data headers are not viewed as pertaining to the series alone, but rather to the specific data series in a specific statistical context. If the user exports the same data to another context, he will adjoin to it a new data header relevant to that context. Only then can guidance be made relevant and consistent.

Quality. In the Quality grouping, the user enters information on the number of significant digits for the given data series (this will eventually be determined by the machine), and in the next entry the number of them he considers meaningful.

Importance to Analysis. The variate's Importance to the Analysis constitutes the final grouping, actually two subgroups. In the first, the user must determine whether the variate is considered Maintained or Testable. This important distinction is rarely made by the typical statistical practitioner. For example a t statistic is relevant only to a test of significance and is therefore of interest only if the parameter's significance is in fact testable. This is often not the case, and any report will differ substantially when this is so. Thus, consider a consumption-function including, among other variates, disposable income. In the real-life context leading to the specification of this function, the presence of the income term is not subject to test; it is maintained on theoretical grounds. A t test of its significance, then, is logically irrelevant and should not be viewed as for a testable variate. Should this t show insignificance, it would not be proper to suggest that the income variate be removed from the model. But it would be appropriate to report that an important effect in the model is not being supported by the data. Guidance suggestions, then, would lean toward trying to find out why this unsuspected outcome occurred (data errors, misspecification of functional form, etc.) rather than removing the variate from the analysis.

Derivation. The final heading of the Data Header card is Derivation, a text field allowing any formula used in deriving the given series to be displayed. The guidance facility will also use this information to determine how to tailor reports and suggestions and to help determine the way to carry out certain calculations. Thus, logarithmic transforms of structurally interpretable variates require a different normalization in a conditioning analysis from those whose levels are structurally interpretable. See Belsley (1988b).

FINAL REMARKS

At this point, there is no statistical system on the Macintosh that communicates well with Hypercard. We know of several that are considering modification to make rapid communication with Hypercard possible. When this happens, we will be able to provide guidance directly in the Macintosh environment. However, we have found Hypercard very useful for prototyping ideas that are then incorporated into the S-based guidance facility.

BIBLIOGRAPHY

Atkinson, A. C. (1985), *Plots, Transformations, and Regression: An Introduction to Graphical Methods of Diagnostic Regression Analysis*, Clarendon Press: Oxford.

Belsley, D. A. (1988a), "Modelling and Forecasting Reliability," forthcoming in "The Future of Forecasting," a special issue of the *International Journal of Forecasting*, edited by Everette S. Gardner and Spyros Makridakis.

Belsley, D. A. (1988b), "Conditioning in Models with Logs," forthcoming in the *Journal of Econometrics*.

Belsley, D. A., E. Kuh and R. E. Welsch (1980), *Regression Diagnostics*, John Wiley & Sons: New York.

Chatterjee, S. and A. S. Hadi (1988), *Sensitivity Analysis in Linear Regression.*, John Wiley & Sons: New York.

Myers, R. H. (1986), *Classical and Modern Regression Analysis with Applications*, Duxbury: Boston.

Oldford, W. and S. Peters (1984), "Building a Statistical Knowledge Based System with Mini-Mycin", *Technical Report No. 42*, Center for Computational Research in Economics and Management Science, MIT.

Wetherill, B. G. (1986), *Regression Analysis with Applications*, Chapman and Hall: London.

Expert Systems

How Should the Statistical Expert System and its User See Each Other?

J. A. Nelder, London

0. Introduction

This paper draws on experience that my colleagues and I have had in constructing GLIMPSE, a knowledge-based front-end for GLIM. GLIM is a statistical package that facilitates the specification, fitting, and checking of the class of generalized linear models (McCullagh and Nelder, 1983). It has its own interpretive language, sufficiently powerful to allow the user to program his own non-standard analyses if a model falls outside the built-in set. GLIM gives the user little on-line syntactic help (how to do things), and almost no semantic help (what to do), except to comment on unsuitable models which produce, for example, negative fitted values when these must be positive. Front ends are designed to remedy these deficiencies, by providing help of both kinds. The result, in the case of GLIMPSE, is a system with three-way communication taking place between the user, the front-end and GLIM itself, which serves as the algorithmic engine.

The front-end must give statistical help to the user the form of guidance in, e.g., transforming data, selection of one or more parsimonious models, and the checking of the models selected. This guidance is embodied in a set of rules, which in GLIMPSE we call the abstract statistician. However, a useful front-end must contain much more than statistical expertise in the form of rules; it must contain an interface through which the user and the abstract statistician convey information to each other. The design of this interface is the main topic of this paper.

Compstat 1988

1. Some basic principles

The construction of GLIMPSE assumes that the user may wish at times to go his own way, e.g. to try out a new idea that has occurred to him; thus the interface must make this possible. There are two possibilities: (i) the user gives direct commands to GLIM using the GLIM language, or (ii) the user invokes *tasks* of the same form as that used by the abstract statistician. Both are allowed in GLIMPSE, but the implications are different.

Direct use of GLIM causes the front-end to dump its current state (including that of GLIM), and to restore it after the user has finished. Thus in this mode the front end does not know what the user has done, and cannot make use of his results. The user, however, may learn something from this activity, then use what he has learnt within the front-end later.

In the second mode the abstract statistician notices the tasks that the user has invoked, and if its own advice later requires the evaluation of any of these tasks, the information previously found by the user will be extracted from the accumulated data and the task will not be re-done.

There is a high-level control language for the expression of tasks, considerably more general than that of GLIM, and also more uniform syntactically. An important property of GLIMPSE is that of *transparency*, which means that the user has access to all the tools that the abstract statistician uses. This is achieved through the definition of the task language, which is used by the abstract statistician in structures that comprise the statistical expertise. For the future, the existence of this common language has exciting possibilities which we discuss in section 4.

A second basic principle is that the abstract statistician always gives *advice,* rather than *commands.* This makes GLIMPSE a libertarian system, rather than an authoritarian one. The basic reason for adopting the libertarian appraoch has been our realization that the user will almost certainly have specific information about his data that the abstract statistician will be unaware of, and that he must be able to use that information to the best advantage. A simple example might involve a (y,x) relationship that, while being substantially linear over the range of x involved, is sufficiently curved for the abstract statisticain to detect a quadratic component. The user may have good reason, however, to know that use of the linear approximation will be adequate for his purpose. In GLIMPSE the front-end may advise the user to incorporate a quadratic term, but the user need not take the advice. We believe that libertarianism is an essential principle if the user's background knowledge is to be incorporated effectively in an analysis.Of course there are dangers; thus GLIMPSE cannot prevent the user from assuming that he knows best and fitting a model which the data may contradict. Editors and referees will have to be clear that the fact that a user says he has used GLIMPSE does not necessarily imply that he has taken notice of the expertise contained in it.

2. Asking the user for information

In order to apply its rules the abstract statistican must obtain information of various kinds from the user. The expert-system shell APES (Hammond and Sergot, 1984) which GLIMPSE uses, includes a query-the-user feature whereby, whenever a piece of information is missing in its database, it generates a question to the user to obtain it. Thus the front-end is using the information in the user's head as an extension of its own stock of information. Such a piece of information may be something basic, such as the number of units (rows) in the data matrix underlying the current analysis. Such information the user must have. Something less basic is a value for the baseline mean deviance, to be used in the denominator of F-values produced at the model-selection stage. Here the user may not know a value, in which case the front end must use a decision rule to enable one to be calculated.

The process of questioning the user has to allow for various kinds of response if the front-end is to be seen as helpful. The following kinds of response are being built into GLIMPSE.

2.1. What replies are possible ?

The user must be able to discover, for instance, whether he can reply 'don't know' to a yes-no question. More generally he must be able to discover what other types of indirect response are available, whereby he may respond to a question with another question of his own.

2.2. What is the background to this question ?

The transparency of GLIM requires that the user should be able to understand the strategy which the front-end is using, not only statically (by reading background information before beginning, say, model selection), but also dynamically during the process itself. This implies the ability to ask to see appropriate parts of the background files as a response to a question.

2.3. Why is this question being asked ?

This is a more specific response, which asks to see the actual rules currently being invoked by the front end when the question arose. This is part of a general explanation facility available in GLIMPSE, which in turn builds on the explanation provided by APES. The rules in their Prolog form are often not immediately intelligible to the user, so that APES allows a plain-language template to be specified for a rule, to be used in the explanation.

To provide good, generally intelligible, explanations is no easy task. The high-level rules currently being invoked may not mean much to the user, so decisions have to be made on whether to use top-down or bottom-up display of rules. The depth of explanation also needs to be controllable. We attempt to provide 3 depths of explanation in GLIMPSE. When a tree branches the user can opt to follow down any of the branches, and also to cut off explanations when he has seen enough.

2.4. *What would the front-end suggest ?*

At various points the front end presents the user with information, say in the form of a scatter-plot, and asks, e.g.'is this plot linear?' The point of this question is to make use of the user's skill in assessing such properties, using again any special knowledge of the data that he may have. The user is entitled to reply 'what do you think ?'. Exercise of this option leads to the system executing a decision rule and presenting the user with the result. The libertarian principle implies that the user need not accept the advice.

2.5. *What does this word mean ?*

If the system uses GLIM jargon in a question, the user may ask for a definition of an unfamiliar word. GLIMPSE provides a lexicon of such definitions, together with a list of other words in the lexicon relevant to any given word. For instance the definition of 'Pearson residual' gives the following list of relevant words:

residual
deviance residual
standardized residual
fitted value
variance function
Pearson chi-squared statistic

Such a lexicon constitutes a simple browser, which can be used by the user (at any time) to learn something of the principal ideas underlying generalized linear models. If taken further the lexicon could be expanded to become a self-teaching aid to wider areas of statistics. This, however, would be a major project in itself.

2.6. *Backtracking*

Most users will eventually find themselves in the position of realizing that they have given a wrong answer to an earlier question, the result of which could nullify any subsequent progress. Hence the front end must provide a method of back-tracking through the dialogue between user and front end, correcting the faulty answer and then moving forward again. Without this facility the user will be quite unsure how to cope with the consequences of a faulty answer.

3. What sort of expertise can the user expect ?

The lazy-minded user might hope to press one button, which would cause GLIMPSE, say, to produce *the* analysis of his data. Any user who expects this of GLIMPSE, or indeed of any current statistical expert system, is going to be disappointed. There are several reasons for the current limitations on the kind and amount of expertise that can be provided.

One problem arises from the potentially enormous scope of a single statistical method in different fields of applications, and this applies even to the relatively limited class of models provided by GLIM. GLIM has been successfully and correctly used in fields of application that were never thought of by the originators of the package. This is one reason for making the rules in the front-end application-free. There is then some hope that it may be useful in an entirely new field of application. The corresponding disadvantages are obvious; for example GLIMPSE contains expertise for doing probit analysis on data from a bio-assay, yet does not know what a poison is, let alone about a specific group of insecticides. With correct design, however, a front-end should be able to be expanded for particular applications by adding application-dependent rules.

Another important aspect of the front-end rules is that they should include rules for discovering if the user's data is of a kind that makes it unsuitable for analysis by any of the class of models which the back end supports. For instance GLIM is not a suitable tool for fitting ARIMA models to time series. and interrogation of the user by the front end should include warnings and questions designed to discover misfits of this kind.

Some front-ends help with the formulation of a problem (the input) but then do nothing beyond computing the necessary numbers and presenting the answers to the user. A statistical front-end must go further than this and give help with the interpretation of the output as well as formulation of the input. Such interpretation can in its turn seek to make use of the user's own expertise (for example, in assessing the linearity of a graph) to augment its own rules.

4. Can user-system interaction evolve ?

At any time the expertise embodied in the rules of a front-end reflects the knowledge of the builders, with this knowledge being encoded as a set of rules. Now knowledge is itself continually growing, and this is certainly the case with statistical analysis. For example, the techniques of model checking have developed rapidly over the last 15-20 years (O'Brien, 1988). The result has been that a loop has been put into the analytic process, whereby model checking may expose deficiencies in the model chosen, leading to revision of the model class, followed by re-analysis, etc. Further new methods are certain to be formulated. The question then arises: how can a front-end be updated ?

The standard method requires that the rules, expressed, perhaps, in Prolog, be amended to allow for new knowledge. We know the problems when a large program in, say, Fortran is to be amended. Very detailed knowledge of the program structure is required if the full implications of any changes are to be understood. If they are not understood then all kinds of undesirable side-effects may be introduced by the changes. Although there is currently little experience of updating large Prolog programs, we may expect similar difficulties to arise.

There are two routes which we might follow at this point; these lead to what I call adaptive front ends and evolving front ends.

4.1 Adaptive front ends.

An adaptive front end is one that can modify its rules as the result of interaction with its users. In order to do this it must have a tutorial mode in which an expert is questioned by the system about the methods he would use to analyse certain sets of data (the *training set*), and from his replies a consistent set of rules is constructed. The rules are a function of both the expert's knowledge and the training set. Also the system must itself contain a basic framework which guides the questions the expert is asked when his rules are being elucidated. The use of different experts will, in general, produce different rule sets, and it would be a major task to merge such sets into a consistent superset of rules. STUDENT (Gale, 1985) is a project to produce an adaptive front end; it is an ambitious project whose outcome is not yet clear. An interesting consequence of this approach is that the user might later opt to analyse his data by the rules derived from Prof. X, or Prof. Y, or....etc. It may not be easy, however, for the user to update the rule-set of interest, except by asking Prof. X to add the current data set to his training set. A possible way out of this difficulty is to use an evolving rather than an adaptive system.

4.2 Evolving front ends

Consider a system of the GLIMPSE type with a task language used by both user and front end, and with each task translated into a macro for the back end. The front end defines a strategy for, say, model selection, which combines a sequence of taks with queries to the user for information. The user can develop, for his own use, his own sequence of tasks for an alternative approach to model selection; however, he will be limited to using the tasks already defined by the system. With a few extensions to the existing control language, however, the user could go much further. For suppose that the control language supported

- (i) the definition. storage, and invocation of procedures;
- (ii) a branching if-then-else construction;
- (iii) the definition of questions to the user, and the handling of responses;
- (iv) the definition of the syntax and semantics of new tasks.

It then becomes possible to define, and make available to others, new strategies for an activity like model selection. Although the extensions listed above are not trivial, they are well within the scope of existing computing techniques. Numbers (i) and (ii) are straightforward language extensions, while (iii) involves bringing a lower-level facility up to the higher level of the task language. In (iv) syntactic definition, given a good existing parser, is largely a matter of updating tables, while the semantics involves supplying the appropriate GLIM macro with a link via the translator to the front end.

Once new strategies can be defined they can (i) be exchanged between users, and (ii) be compared with each other and with the existing standard strategy. The development of standards of comparison is not an easy task, but compared the different strategies will be, and users will select from them. The stage is now set for the front end to evolve; the evolution will not be exactly Darwinian, because the strategies will not develop by random mutations of existing strategies. However, the population of users closely parallels the environment, in which selection will operate. It may be that the outcome of this process will be more rapid evolution than in a Darwinian system, precisely because the variants on which selection is practised have not been generated at random, but according to some rational basis for improvement. It is important to note that such evolution does not depend on the originators' sole efforts, but incorporates a much wider source of expertise.

5. Conclusion

I hope to have shown that the form of the interaction between the user and a front end is worthy of study, independently of the expertise that the front end contains; also that collaboration between users and the system has the potential to lead to evolving systems in which expertise is continually improved and expanded.

6. Acknowledgements

GLIMPSE is a co-operative project between the Computing and Mathematics Departments of Imperial College and the Numerical Algorithms Group (NAG) of Oxford. I am especially indebted to Carl O'Brien (NAG) and David Wolstenholme (Computing Department) for work in developing the system, and for many useful discussions. In particular D.W. developed the task language, and wrote the parser for it.

7. References

Gale, W.A. (1985) Student phase 1 - a report on work in progress. *Artificial Intelligence and Statistics.* Addison-Wesley, Reading, Mass.

Hammond, P. and Sergot, M.J. (1984) *APES: Augmented Prolog for Expert Systems, reference manual.* Logic Based Systems, Surrey.

McCullagh, P. and Nelder, J.A. (1983) *Generalized linear models.* Chapman and Hall, London

O'Brien, C.M. (1988) Assessment of GLM mis-specification. Paper presented at *Third International Workshop on Statistical Modelling,* Vienna.

Payne, C.D. (ed.) (1985) *The GLIM system, release 3.77.* Numerical Algorithms Group, Oxford.

Towards a Probabilistic Analysis of MYCIN-like Expert Systems (Working Paper)

P. Hájek, Prague

ABSTRACT: A formal apparatus is developed here for the study of the compatibility of MYCIN-like expert systems with axioms of probability theory. Some partial results are presented; but the main stress is on open problems.

0. Introduction. This is a working paper, a free continuation of [HV] and [H84].The knowledge of the framework of [HV], even if helpful, is not presupposed. The main question may be formulated as follows: what can MYCIN-like (extensional) systems be good for from the probabilistic point of view?

To avoid prejudiced and/or simple minded answers we elaborate a precise formal apparatus and formulate precise mathematical questions. Till now, we have only rather partial answers; but I do not exclude the possibility that to find the answers is not difficult. Comparative approach (see[HV]) is stressed; thus papers discussing probabilistic properties of MYCIN-like systems (Duda et al., Heckerman, Pednault, Johnson, Hájek and Havránek) even if relevant and interesting in the present context, do not annswer our main questions.

1. Preliminaries. To simplify matters, we shall make three restricting assumptions: *First,* we shall investigate only knowledge patters with no intermediate propositions. Thus propositions are divided into questions and goals; each rule has its antecedent which is an elementary cojunction of some questions and the succedent is a goal. (*Ques* and *Goal* denote the set of all questions and goals respectively; $EC(Ques)$ is the set of all elementary conjunctions of questions, including the empty conjunction. *Second,* we let weights run over $[0,1]$; it is easy to show that in this case the PROSPECTOR's operation becomes

$$x \oplus y = \frac{xy}{xy + (1-x)(1-y)}.$$

Third, we allow the user to answer the questions only using 1, 0 and 0.5 (yes, no, unknown). This makes possible to identify questionnaires (mappings $q : Ques \rightarrow \{1, \frac{1}{2}, 0\}$) with elementary conjunctions of questions, e.g. if $q(1) = 1, q(2) = 2$ and $q(i) = \frac{1}{2}$ for other questions then q is identified with $1 \,\&\, \neg 2$. (Cf. [H84]).

Note that there is no reason to exclude the *empty* antecedent; $\Rightarrow H\,(w)$ means that apriori weight of H is w. If Θ is a knowledge pattern and κ a weighting then we write $\underline{\Theta} = (\Theta, \kappa)$. For each goal H and each elementary conjunction E of some questions we have the *global weigth* of H given E defined as follows:

$$W_{\underline{\Theta}}(H|E) = \bigoplus_{E' \subseteq E} \kappa(E' \Rightarrow H).$$

(note that our setting makes the combining functions $CTR, CONJ, NEG$ superfluous.)

Compstat 1988

2. Main notions. Assume *Ques*, *Goal* given. A *belief system* is a mapping β of a subset of $EC(Ques) \times Goal$ into $[0,1]$. If $E \in EC(Ques), H \in Goal$ and β is defined for the pair (E,H) we denote the value by $\beta(H/E)$ (resembling conditional probability). β is *total* if $dom(\beta)$ is the whole of $EC(Ques) \times Goal$; otherwise it is *partial*.

Two belief systems β, β' are *numerically compatible (n.c.)* if for each for each $(E,H) \in dom(\beta) \cap dom(\beta')$ we have $\beta(H/E) = \beta'(H/E)$. They are *comparatively compatible (c.c.)* if for each $(E,H), (E',H') \in dom(\beta) \cap dom(\beta')$ we have

$$\beta(H/E) \leq \beta(H'/E') \text{ iff } \beta'(H/E) \leq \beta'(H'/E').$$

(We can still generalise to *locally c.c* systems postulating the last equivalence only for $E = E'$).

Let $\underline{\Theta} = (\Theta, \kappa)$ be a knowledge base. The belief system *generated* by $\underline{\Theta}$ is the above $W_{\underline{\Theta}}$. *Caution:* if $(E \Rightarrow H)(w)$ belongs to $\underline{\Theta}$ (i.e. $(E \Rightarrow H) \in \Theta$ and $\kappa(E \Rightarrow H) = w$) and $\beta = W_{\underline{\Theta}}$ is the belief system generated by $\underline{\Theta}$ then we need not have $\beta(H/E) = w$; see the formula above.The weight of the rule is understood as the *modification* of the joint effect of simpler rules leading to H. Let β' be an arbitrary belief system; β' is *n.c. (c.c.)* with $\underline{\Theta}$ iff β' is n.c. (c.c.) with $W_{\underline{\Theta}}$.

A *joint distribution* on $Ques \cup Goal$ is a mapping D associating with each elementary conjunction of elements of $Ques \cup Goal$ of maximal length a real from $[0,1]$ such that $\sum_{K \in dom(D)} D(K) = 1$. The probability $P_D(K)$ of any elementary conjunction of elements of $Ques \cup Goal$ is

$$P_D(K) = \sum_{K' \in dom(D) \& K \subset K'} D(K').$$

The conditional probability $P_D(H/E)$ for $E \in EC(Ques)$ and $H \in Goal$ is defined by

$$P_D(H/E) = \frac{P_D(H\&E)}{P_D(E)}.$$

The mapping $\beta(H/E) = P_D(H/E)$ is a total belief system *generated* by the distribution D. A belief system β is *numerically (comparatively) probabilistically sound* (n.p.s. or c.p.s. respectively) if there is a distribution D such that β is n.c. (c.c.) with D. β is *weakly sound* (cf. [H84] if for each E, H such that $\beta(H/E)$ is 0 or 1 and each $E' \supseteq E$ we have $\beta(H/E) = \beta(H/E')$ if defined.

3. Formulation of problems.

Our starting point is the following

3.1 THEOREM: (See [H84].) If β is weakly sound then there is a knowlegde base $\underline{\Theta} = (\Theta, \kappa))$ such that $dom\beta = dom\kappa$ and β is n.c. with $\underline{\Theta}$.

Note that if $\beta(H/E)$ is defined then it is enough to put

$$\kappa(E \Rightarrow H) = \bigoplus_{\substack{E' \subseteq E \\ \beta(H/E') \text{ defined}}} (-)^{|E-E'|} \beta(H/E)$$

where $(-)^{2n-1} p = 1 - p$ and $(-)^{2n} p = p$.

For example, if $\beta(H/E_1) = w_1, \beta(H/E_2) = w_2$ and $\beta(H/E_1 \& E_2) = w_3$ then $\underline{\Theta}$ is the following knowledge base:

$$E_1 \Rightarrow H\,(w_1), \quad E_2 \Rightarrow H\,(w_2), \quad E_1 \& E_2 \Rightarrow H\,(w_3 \oplus (-)w_1 \oplus (-)w_2).$$

Note that evidently each n.p.s. belief system is weakly sound.

3.2 COROLLARY: For each n.p.s. (c.p.s.) belief system β there is a knowledge base $\underline{\Theta} = (\Theta, \kappa)$ such that $dom\beta = dom\kappa$ and β is n.c. (c.c.) with $\underline{\Theta}$.

The condition on domains implies that the cardinality of the set of rules is the same as the cardinality of $dom\beta$. This looks nice *but* the problem is whether the total belief system $W_{\underline{\Theta}}$ generated by $\underline{\Theta}$ is probabilistically sound (see below). Clearly, since β is probabilistically sound we may first find a distribution D such that β is compatible with the belief system P_D and then apply the corollary to P_D; thus we have the following

3.3 COROLLARY: Each n.p.s. (c.p.s.) belief system β is n.c. (p.c.) with a total belief system β' which is generated by a knowledge base $\underline{\Theta}$ and is n.p.s. (c.p.s).

This looks again nice *but* now we have no control on the cardinality of the set of rules in $\underline{\Theta}$. Thus we are lead to the following two problems that appear to be of great significance for the correct evaluation of MYCIN-like systems from the probabilistic point of view:

3.4 PROBLEM A: Let β be a belief system with rational values. How to decide simply whether β is comparatively probabilistically sound? (Note that the question *is* decidable thanks to Tarski's result on the decidability of elementary algebra.)

3.5 PROBLEM B: Assume that β is c.p.s. How to find a *possibly simple* knowledge base $\underline{\Theta}$ such that β is c.c. with $\underline{\Theta}$ and the belief system $W_{\underline{\Theta}}$ generated by $\underline{\Theta}$ is c.p.s?

Problem A is relevant also for the purely probabilistic (intensional) approach to rule-based systems. Note that at the first glance it is possible that the computational complexity of either problem is high.

We shall collect some additional information concerning Problem A.

3.6 DEFINITION AND LEMMA: Let a belief system β be given. Write $(H/E) \preceq (H'/E')$ for $\beta(H/E) \leq \beta(H'/E')$ and $(H \preceq H')/E$ for $\beta(H/E) \leq \beta(H'/E)$. If β is c.p.s. then we have the following:

(R1) $$(H/E\&E') \preceq (H/E) \text{ implies } (H/E\&\neg E') \succeq (H/E)$$

(R2) $$(H \preceq H')/(E\&E') \text{ and } (H \prec H')/(E\& \not E') \text{ implies } (H \prec H')/E$$

whenever the corresponding beliefs are defined.

3.7 EXAMPLES: (1) A simple belief system which is not c.p.s.:

$$\beta(H/E) = r, \beta(H/ \not E) = r, \beta(H/true) = p, p \neq r$$

(2) A belief system which is n.p.s. but no $\underline{\Theta} = (\Theta, \kappa)$ such that $dom\beta = dom\kappa$ and comparatively compatible with β generates a probabilistically sound belief system $W_{\underline{\Theta}}$ (this example is inspired by an example due to Jeff Paris).

$$\beta(H/E_1) = q \quad \beta(H/E_2) = q \quad \beta(H/E_1\&\neg E_2) = p$$
$$\beta(H'/E_1) = q \quad \beta(H'/E_2) = q \quad \beta(H'/E_1\&\neg E_2) = r$$

where $1 > p > q > r > 0$.

(If $\underline{\Theta}$ were as said then for $\beta' = W_{\underline{\Theta}}$ we would have $(H = H')/E_1, (H \succeq H')/(E_1\&\neg E_2)$ and $(H = H')/(E_1\&E_2)$.)

3.8 PROBLEM: Is each belief system satisfying (R2) above locally comparatively probabilistically sound? Cf [Scott], [Luce].

4. Remarks on Problem B.

We present a probabilistic inrerpretation of a knowledge base generating a belief system compatible with a given probability.

4.1 THEOREM: Let P be a probability on a set $Prop = \{1, 2, \ldots n\}$ of propositions and assume for simplicity that for each $E \in EC(Prop)$, $P(E) \neq 0$. There is a unique system $\Lambda = \{\lambda^E : E \in EC(Prop)\}$ such that $\lambda^\emptyset = 0$, and for each $E \in EC(Prop)$, $\log P(E) = \sum_{E' \subseteq E} \lambda^{E'}$.

Clearly, $\lambda^E = \sum_{E' \subseteq E} (-1)^{|E-E'|} \log P(E')$. Call the system Λ the *hereditary log linear representation* of P. Let $Prop = Ques \cup Goal, E \in EC(Ques), H \in Goal$.

4.2 COROLLARY: Let $O(H/E) = \frac{P(H/E)}{P(\neg H/E)}$ (odds). Then

$$\log O(H/E) = \sum_{E' \subseteq E} \lambda^{E'\&H} - \lambda^{E'\&\neg H} = \sum_{E' \subseteq E} \lambda_*^{E\&H}$$

where $\lambda_*^{E\&H}$ is $\lambda^{E\&H} - \lambda^{E\&\neg H}$.

Note that $f(x) = e^x/(1-e^x)$ maps logarithms of odds to probabilities; f is an isomorphism of the additive group of reals onto the interval $(0,1)$ with the operation $\oplus$ (see [HV]). Thus if we define, for each $E \in EC(Ques)$ and each $H \in Goal$, $\kappa(E \Rightarrow H) = f(\lambda_*^{E\&H})$, we get a knowledge base generating P.

4.3 COROLLARY: Under the above notation,

$$P(E/H) = f(\sum_{E' \subseteq E} \lambda_*^{E\&H}) = \bigoplus_{E' \subseteq E} \kappa(E \Rightarrow H).$$

4.4 REMARK: If we start with a probabilistically sound belief system we would like, in accordance with Problem B, to find a probability P compatible with β such that, for *big* E, $\lambda_*^{E\&H} = 0$ (so that the rule $E \Rightarrow H$ need not occur in the constructed knowledge base.) *What* E is big may depend on H. Havránek suggested to employ here the apparatus of graphical models, see e.g. [Laur-Spieg] and Asm-Edw]. This is being elaborated and the result is supposed to form the content of a future joint paper with Havránek. Thanks are due to Havránek for having called my attention to references [Luce], [Asm-Edw] and for many helpful discussions.

References.

[Asm-Edw] S. Asmussen, D.Edwards: Collapsibility and response variables in contingency tables, Biometrika 70 (1983) 567-578

[Duda] R. Duda, P. Hart, N. Nilsson: Subjective bayesian methods for rule-based expert systems, SRI International, Technical note 124, Stanford 1976

[H85] P. Hájek: Combining functions for ceratinty factors in consulting systems, Int. Journ. Man-Machine Studies 22 (1985), 59-67

[H84] P. Hàjek: Combining functions in consulting systems and dependence of premisses - a remark, (Plander, ed.) Aritificial Intelligence and Information - Control Systems of Robots, North-Holland 1984, 163-166

[H-Havr] P. Há jek, T. Havránek: A note on the independence assumption underlying subjective bayesian updating, (Plander, ed.) Artificial Intelligence and Information - Control Systems of Robots, North-Holland 1987

[HV] P. Hájek, J. Valdes: Algebraic foundations of uncertainty processing in rule-based expert systems, submitted (also: Preprint of the Matematical Institute, ČSAV, Prague 1987)

[Heck] D. Heckerman: Probabilistic interpretations of MYCIN's certainty factors, (Kanal and Lemmer, eds.) Uncertainty in Aritificial Intelligence, North-Holland 1986, 167-196

[Johns] R. W. Johnsson: Independence and Bayesian updating methods, *ibid.* 197-201

[Laur-Spieg] S. L. Lauritsen, D. J. Spiegelhalter: Local computations with probabilities on graphical structures and their application to expert systems, J. R. Statist. Soc. B 50 (1988) No. 2

[Luce] D. Luce: On the numerical representation of qualitative coinditional probabilities, Annals Math. Stat. 39 (1968) 481-491

[Pedn] E. P. D. Pednault, S. W. Zucker, L. V. Muresan: On the independence assumption underlying subjective Bayesian updating, Artificial Intelligence 16 (1981) 215-222

[Scott] D. Scott: Measurement structures and linear inequalities, J. Math. Psychology 1 (1964) 232-240

An Expert System Accepting Knowledge in a Form of Statistical Data

R. Jiroušek and O. Kříž, Prague

A NEW PROBABILISTIC APPROACH

Bottle-neck effect in application of most expert system shells nowadays lies in knowledge acquisition. Experts are forced to express their knowledge in a formalized way acceptable by computers. Most of this burden lies on shoulders of knowledge engineers and besides being tedious it drags into the process an unwanted subjective factor. Therefore, recently some authors started to look for methods of automatical knowledge acquisition from the statistical data.

Being based on new ideas, our approach - INtensional Expert System INES - solves the problem in the most natural way. The original idea belongs to Perez, who inspired by his preceding work [1] suggested in [2] a model of an expert system strictly based on probabilistic foundations. Thus, in INES partial knowledge from a specific area is expressed in a form of (conditional or unconditional) probability distributions of minor dimensions. The dimensions of these distributions are usually 2, 3 or 4 and (therefore we call them oligodimensional in the sequel) it is easy to obtain their statistical estimates from data files. But naturally, these distributions cannot be directly used for decision-making. The system INES, like some other probabilistic methods proposed for expert systems (c. f. Lauritzen and Spiegelhalter's approach [8] or Shachter's one [7]), decides according to some multidimensional probability distribution embodying all the given pieces of partial knowledge, i. e. all oligodimensional distributions acquired. Shachter as well as Lauritzen and Spiegelhalter suppose that the acquired system of

Compstat 1988

oligodimensional distributions contains distributions describing all substantial dependences (i. e. if two variables are not present simultaneously in any from the given distributions then these variables are supposed to be conditionally independent) and therefore there is a unique joint distribution consistent with the given system of oligodimensional distributions. There is no such assumption in the original Perez's model. Within this model, one has to consider the whole class of joint distributions consistent with the given system of marginal distributions. There is no reason why to prefer one of them to others unless the final goal - the decision-making problem is taken into consideration. In this case, having accepted the minimax strategy there should be chosen that distribution which minimizes the maximal potential increase of probability of error (or risk, in general).

KNOWLEDGE-BASE LEADING TO AMBIGUOUS DECISION

The following example illustrates why we do not feel the assumption "unmentioned = conditionally independent" as natural as it is supposed in [8].

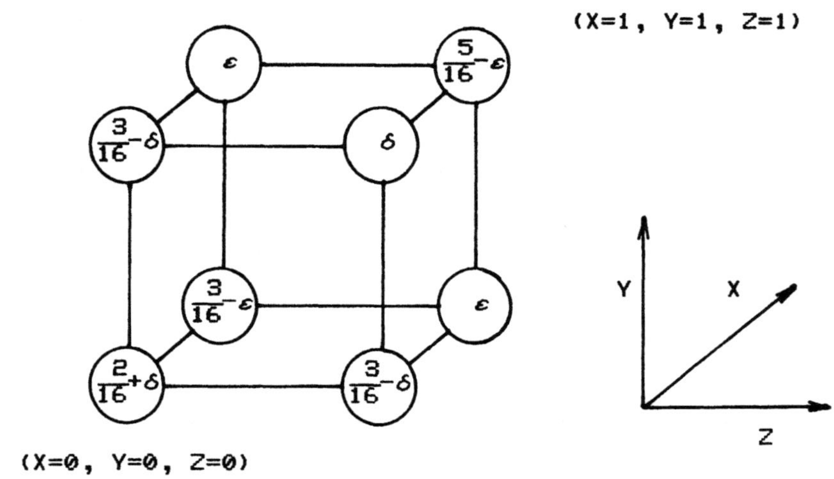

Fig. 1

Let us consider three binary variables X, Y, Z describing some hypothetical system with eight possible states. Suppose that all acquired knowledge is embodied in the following distributions:

$$P(X=0) = P(X=1) = 1/2$$

$$P(Y=0|X=0) = P(Z=0|X=0) = P(Y=1|X=1) = P(Z=1|X=1) = 5/8$$

$$P(Y=1|X=0) = P(Z=1|X=0) = P(Y=0|X=1) = P(Z=0|X=1) = 3/8$$

One can easily see that for every $\varepsilon \in \langle 0,3/16\rangle$ and $\delta \in \langle 0,3/16\rangle$ there exists a 3-dimensional distribution P_{XYZ} (described in Fig. 1.) consistent with the above system of distributions. When considering the problem of optimal estimation of the variable Z when values of X and Y are known, we find out that nine (from 16 possible) different decision functions are optimal (with respect to the probability of error) for different distributions in dependence on ε and δ. This is illustrated for some specific cases in the following table.

	$\varepsilon=0$ $\delta=0$	$\varepsilon=0$ $\delta=\frac{1}{16}$	$\varepsilon=0$ $\delta=\frac{3}{16}$	$\varepsilon=\frac{2}{16}$ $\delta=0$	$\varepsilon=\frac{2}{16}$ $\delta=\frac{1}{16}$	$\varepsilon=\frac{2}{16}$ $\delta=\frac{3}{16}$	$\varepsilon=\frac{3}{16}$ $\delta=0$	$\varepsilon=\frac{3}{16}$ $\delta=\frac{1}{16}$	$\varepsilon=\frac{3}{16}$ $\delta=\frac{3}{16}$
X=0,Y=0	Z=1	Z=0	Z=0	Z=1	Z=0	Z=0	Z=1	Z=0	Z=0
X=1,Y=0	Z=0	Z=0	Z=0	Z=1	Z=1	Z=1	Z=1	Z=1	Z=1
X=0,Y=1	Z=0	Z=0	Z=1	Z=0	Z=0	Z=1	Z=0	Z=0	Z=1
X=1,Y=1	Z=1	Z=1	Z=1	Z=1	Z=1	Z=1	Z=0	Z=0	Z=0

OPTIMAL APPROXIMATION OF JOINT DISTRIBUTION

From the above example it can be seen that a careless choice of a joint distribution can lead to catastrophic conclusions. Perez showed in [1] that the increase of decision error $\Delta e(P,\bar{P})$ in the case when a decision function is adapted to probability distribution $\bar{P}$ instead to actual P can be upper-bounded with

$$\Delta e(P,P) \le C \cdot H(P,\bar{P}) = C \cdot \int \log \frac{dP}{d\bar{P}}\, dP,$$

where $H(P,\bar{P})$ is the relative Shannon's entropy of $\bar{P}$ with respect to P (or Kullback-Leibler divergence) and C is a constant. Therefore, our first goal is to choose that joint distribution with the smallest maximal Kullback-Leibler divergence from all other distributions coming into consideration.

Up to now, we have discussed theoretical problems only. But, we have to take into consideration the pragmatic point of view, too. Therefore, we are compelled to deal with only such distributions which can be fully represented within a computer program. For this reason, only a subclass of all pertinent multidimensional distributions is considered in INES. Namely only those approximations $\bar{P}$ that can be expressed in the following form

$$\bar{P} = P_{i_1} \ldots P_{i_K} \,/\, P_{j_1} \ldots P_{j_{K-1}}$$

where P_{i_k} $(k=1,\ldots,K)$ is either a distribution from the given input system of oligodimensional distributions or a distribution reconstructed from several of them with help of IPFP procedure (c.f. [3]). Distributions P_{j_k} $(k=1,\ldots,K-1)$ are those marginals of $P_{i_{k+1}}$ that

$$P_{i_1} \ldots P_{i_k} \,/\, P_{j_1} \ldots P_{j_{k-1}}$$

is a probability distribution for every $k = 2,\ldots,K$.

In [5] it is shown that the minimization of Kullback-Leibler divergence is equivalent to maximization of dependence tightness $I(\bar{P})$ of the approximating distribution $\bar{P}$. The dependence tightness for a distribution in the supposed form is defined by the formula

$$I(\bar{P}) = \sum_{k=1}^{K} I(P_{i_k}) - \sum_{k=1}^{K-1} I(P_{j_k})$$

where

$$I(P_i) = \int \log \frac{dP_i}{dP_i^*} \, dP_i .$$

P_i^* denotes the distribution defined for the same variables as P_i and having the same one-dimensional marginals taken independently (i. e. P_i^* is a product of one-dimensional marginals of P_i). In [5] there is also described an suboptimal algorithm for construction of such optimal approximation $\bar{P}$. The distribution $\bar{P}$ can be directly used for computation of conditional probabilities at decision-making. In present paper we shall avoid dealing with computational problems arising when the number of random variables N described by $\bar{P}$ is high (for details see [5]).

POSITION OF AN EXPERT WITHIN THE NEW MODEL

The described oligodimensional distributions are not the only way of feeding knowledge into INES. It is also possible to accept information from experts in a rule-like way where "weights" have the strict interpretation of conditional probabilities. All sorts of information are then transformed into modified oligodimensional distributions [6] and submitted to INES to generate global joint distribution.

But the active role of an expert is desirable even in case when only statistical estimates of oligodimensional distributions are available. The structure of resulting joint distribution can be significantly influenced by an expert indicating which distributions should be taken as input for the integration process. In this way, the expert can choose which dependencies are important from the decision point of view and which he would omit. If the expert is really an expert it may lead to substantial increase in discernment power of the system as it is illustrated in the sequel.

This example is taken from prediction of occurrence of rheumatic diseases based on 11 anamnestic factors for more than 1000 persons. It is easy to see that the higher the dimensionality of "relevant" distributions, the higher is the proportion of right diagnoses and the less is the number of undecided cases. The decision rule was namely adjusted to contain maximally 15 % of wrong diagnoses. It should be mentioned on the other hand that robustness is inversly proportional to dimensionality of input distributions. The effect of competent preselection is visible in Fig. 2. where rather different results were achieved when 9 four-dimensional distributions were selected in different ways.

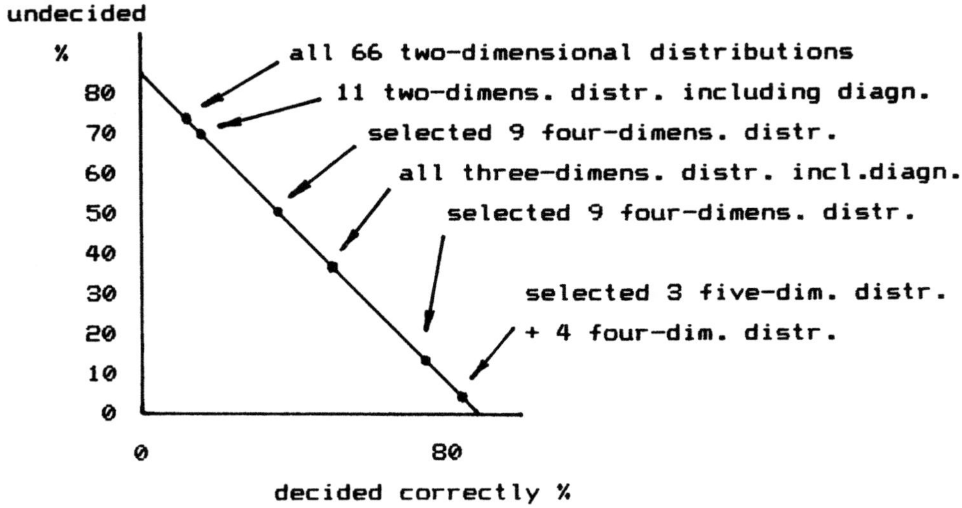

Fig. 2. Decision efficiency adjusted to 15 % error

REFERENCES

[1] Perez, A.: ε-admissible simplification of the dependence structure of a set of random variables. *Kybernetika 13(1977)*, 439-449.

[2] Perez, A.: Probability approach in integrating partial knowledge for medical decision-making (in Czech). In: Trans. *BMI'83 Conf. on Biomedical Engineering*, Marianske Lazne (1983), 221-226.

[3] Perez, A., and Jirousek, R.: Constructing an intensional expert system INES. In: J. H. van Bemmel, F. Gremy and J. Zvarova (Eds.), *Medical Decision-making: Diagnostic Strategies and Expert Systems*. North-Holland (1985), 307-315.

[4] Jirousek, R., and Perez, A.: Graph-aided knowledge integration in expert system INES. In: *Proc. Int. Conf. on Inform. Processing in Knowledge-based Systems*, Paris (1986), 260-264.

[5] Jirousek, R., and Perez, A.: A partial solution of the marginal problem. In: *Trans. of the 10th Prague Conf. on Inf. Theory e.t.c.* (1986), Academia, Prague (in print).

[6] Kriz, O.: Knowledge preprocessing for intensional type expert systems. In: *Trans. of the 10th Prague Conf. on Inf. Theory e.t.c.* (1986), Academia, Prague (in print).

[7] Shachter, R. D.: Intelligent probabilistic inference. In: *Uncertainty in Artificial Intelligence*. North Holland, (1986), 371-382.

[8] Lauritzen, S. L., and Spiegelhalter, D. J.: Local computations with probabilities on graphical structures and their applications to expert systems. *J. R. Statist. Soc. B*, 50 (1988) (in print).

Building a Statistical Expert System with Knowledge Bases of Different Levels of Abstraction

K. M. Wittkowski, Tübingen

INTRODUCTION

Statistical analysis is predetermined by the way a (prospective) experiment is planned or data are collected in a (retrospective) study. The a-priori knowledge of observable, theoretical and hypothetical relations (WITTKOWSKI 1987) determines the semantically meaningful database activities and statistical analyses. For instance, relations between variables and types of observational units may be used to determine whether or not the meaning of a value depends on values of other variables. The models underlying the statistical methods are determined by theoretical knowledge on the sampling strategy of factors, scales, and constraints (WITTKOWSKI 1985). For confirmatory analyses, the primary goal (hypothesis) needs to be specified at the time the sample size is computed.

METHODS

Based on this classification of relevant knowledge, a statistical expert system can be logically divided into six layers according to different levels of abstraction in the knowledge: EXECUTION, ACCESS, SEMANTICS, STRATEGY, DOMAIN, DIALOG (ELLIMAN, WITTKOWSKI 1987). Each layer contains knowledge of a different area of expertise implemented using different techniques of knowledge representation.

The bottom EXECUTION layer contains data and programs, i.e. observed knowledge from observational units and algorithmic

Compstat 1988

	U S E R	
dialog KB	INTERACTION	user KB
domain KB (old)	DOMAIN	domain KB (new)
model KB	STRATEGY	statistics KB
design KB	SEMANTICS	methods KB
data KB	ACCESS	programs KB
data	EXECUTION	*programs*

Fig. 1 Layers of a statistical expert system (KB = knowledge base)

knowledge implemented in the code of the programs. Because we do not want to invent the wheel a second time, we use commercially available data and program management systems.

The ACCESS layer contains knowledge of how to access programs and data. In the data and program knowledge bases we have the tree of links between original and derived attributes (c.f. e.g. BLUM 1982) and the parameters and data structures required by the programs, resp.. Because there is no uniform language for calling statistical programs, the latter knowledge base consists of small transformation procedures that generate the appropriate command sequences.

The abstractions then move away to experimental designs and statistical methods. The SEMANTICS layer contains knowledge of the (lattice) structure of observational units, i.e. whether or not these units are nested or crossed both in the study design and (implicitely) in the statistical methods. The lattice structure is represented using the method of WITTKOWSKI (1985).

Functional models and statistical concepts are represented in the STRATEGY layer. Functional models link the design to the realitiy. SI-units, level and type of scales, ranges, the distinction of factors, strata and observed variables, and the sampling strategy are represented in an object oriented approach (c.f. OLDFORD, PETERS 1986). The knowledge on statistical concepts is represented in form of production rules.

At the DOMAIN level, domain knowledge (e.g. in medicine or agriculture) is either entered or generated. Because domain knowledge is relatively independent on the other knowledge bases, this level will not be discussed in the present paper.

In the top DIALOG layer knowledge of the user is handled and all activities concerning a given experiment are monitored. Some approaches to implement this layer have been discussed in THISTED (1986).

RESULTS

In the development of PANOS-ES, we have started with a three-layer knowledge based front end (ACCESS, SEMANTICS, STRATEGY) to commercially available statistical analysis systems (EXECUTION layer). Within each layer different activities of human experts are to some degree overtaken by the expert system.

Because the inference in the higher layers is based only on theoretical, hypothetical, and observable relations, all knowledge based on the data (e.g. homoscedasticy, Gaussian distribution of residuals, symmetry of covariance matrices) can be handled by the analysis systems at the EXECUTION level.

Commands and parameters for underlying database management and analysis systems are generated in the ACCESS layer. A knowledge based spread-sheet editor has been developed, which guarantees that only observed data may be edited and that corrections are propagated to all related observational units and to all derived attributes. Because both programm calls and all requests for updating, aggregating, and retrieving data are handled at this level (WITTKOWSKI 1988b), data structures are automatically generated according to the programs requirements.

At the SEMANTICS level appropriate data structures and methods are planned and selected, respectively. Knowledge on the lattice structure of observational units is utilized to distinguish between observational and non-observational relations, between dependent and independent replications, and between identical and replicated data (WITTKOWSKI 1988b). Knowledge on statistical methods is used to find a method (statistical model) that conforms both to the lattice structure of observational units and the hypothesis formalized in terms of influence types (WITTKOWSKI 1986). Thus it is guaranteed that the methods chosen reflect the problem under consideration and that these methods are applied to sufficient data subsets.

In the STRATEGY layer, knowledge of statistical concepts is utilized to check consistency of the model, and to help the user formalizing primary or secondary hypotheses. Only those hypotheses will be accepted that conform to the experimental design. If the desired level of significance and size for a given difference between groups are also specified, the system can help the user in determining the sample size. At this level, the user interacts with the system through the same type of structured interface that is used in the knowledge based spread-sheet editor during input of data.

Future implementations will use DOMAIN level knowledge for further assistence of the user both for building models (selecting appropriate subsets of relevant variables) and in building designs (determining the required sample size). Multiple tests and explanation of confirmatory or exploratory results can then be handled at the DIALOG level. We expect that the interaction at these higher levels will require natural language procedures (WITTKOWSKI 1988a).

DISCUSSION

Statistical expert systems differ from other expert systems (e.g. for medical diagnosis) in that different types of knowledge have to be considered. First we have knowledge of data management and statistics. Even this knowledge has more (three) levels of abstraction than domain knowledge in many other applications. Then we have knowledge of the domain, where models are build and where results of statistical analyses are to be applied (e.g. medicine or agriculture). Thus single monolithic systems for different domains would not only be difficult to construct but it would also be inefficient to maintain and expand these systems.

The proposed approach differs from other approaches (see GALE 1986; HAUX 1986; DE ANTONI 1986; PHELBS 1987 for references) in that decisions are primarily based on hypothetical, theoretical and observable relations. Empirical distribution of the data (observed relations) affects the inference only at the lowest level. It has been shown (WITTKOWSKI 1987) that this distinction between formal and actual relations is a necessary attribute of expert systems for testing statistical hypotheses. We have now demonstrated that this approach also allows for an effective bottom-up approach in the design of expert systems:

- Knowledge from higher levels of abstractions and formalized dialog procedures may be used to reduce the expenditure of time during knowledge acquisition.
- The implementation is facilitated, since the techniques of knowledge representation can be tuned to the special considerations of different types of knowledge.
- The path of actions in such a multi-layered system reflects the iterative process in statistical problem solving (Hand 1985). The relation among knowledge bases and stages in the problem solving process reduces the amount of rules that have to be simultaneously considered by the expert system.
- A well-defined control strategy leads not only to improved response time but also to transparent decisions. Thus the behavior of the system will become better understandable to the user and it can be verified, whether it conforms to appropriate statistical strategies.

The expert system approach presented leads to considerable reduction in the amount of information to be entered during analysis, the probability of selecting semantically meaningless data sets or inappropriate methods, and the probability of misinterpreting output of statistical analysis systems. Finally, statistical expertise is explicitely defined so that it can be discussed and misleading or wrong "heuristics" can be corrected. As a consequence the statistical strategy and thus the result of an analysis will become less dependent on the subjective opinion of a single expert. Thus application of AI-techniques may also lead to a better understanding of the concepts underlying the area of application (e.g. statistics).

REFERENCES

BOARDMAN TJ (ed. 1986) *Computer Science and Statistics: Proceedings of the 18th Symposium on the Interface.* Washington,DC: ASA

BLUM RL (1982) *Discovery and representation of causal relationships from a large time-oriented clinical database: The RX project.* Berlin: Springer

DEANTONI R, LAURO N, RIZZI A (eds. 1986) *COMPSTAT 86.* Heidelberg, FRG: Physica

ELLIMAN AD, WITTKOWSKI KM (1987) The impact of expert systems on statistical database management. *Statistical Software Newsletter* **13**:14-18

FAULBAUM F, UEHLINGER HM (eds. 1988) *Fortschritte der Statistik-Software 1.* Stuttgart, FRG: Fischer

GALE WA (ed. 1986) *Artificial intelligence and statistics.* Reading, Mass.: Addison-Wesley

HAND DJ (1985) More intelligent statistical software and statistical expert systems. *The American Statistician* **29**, 1-16.

HAUX R (ed. 1986) *Statistical expert systems.* Stuttgart, FRG: Fischer

OLDFORD RW, PETERS SC (1986) Object-oriented data representations for statistical data analysis.
In: DEANTONI R, LAURO N, RIZZI A (eds. 1986) 301-306

PHELPS B (ed. 1987) *Interactions in artificial intelligence and statistical methods.* Aldershot, GB: Gower

RAFANELLI M, KLENSIN J, SVENSSON P (eds. 1988) *Proceedings of the fourth international workshop on statistical and scientific database management.* (in press)

THISTED RA (1986) Tools for data analysis management.
In: BOARDMAN TJ (ed. 1986) 152-159

WITTKOWSKI KM (1985) *Ein Expertensystem zur Datenhaltung und Methodenauswahl für statistische Anwendungen.* Stuttgart, FRG: Dissertation (in German); Internat. Bookseller Hans Hartinger Nachf., Xantener Str. 14, D-1000 Berlin 15

WITTKOWSKI KM (1986) An expert system for testing statistical hypotheses. In: BOARDMAN TJ (ed. 1986) 438-443

WITTKOWSKI KM (1987) An expert System Approach for generating and testing statistical hypotheses. In: PHELPS B (ed. 1987) 45-59

WITTKOWSKI KM (1988a) Intelligente Benutzerschnittstellen für statistische Auswertungen.
In: FAULBAUM F, UEHLINGER HM (eds. 1988)

WITTKOWSKI KM (1988b) Knowledge based support for the management of statistical databases.
In: RAFANELLI M, KLENSIN J, SVENSSON P (eds. 1988) (in press)

An Expert System for the Interpretation of Results of Canonical Covariance Analysis

K. Momirović, J. Radaković and V. Dobrić, Zagreb

SUMMARY

The expected knowledge of mathematical and statistical properties of some data analysis method or technique in the sample of users of this method or technique usually decrease as the sample space increase. This is especially true for the apparently simple and intuitively easy understandable methods, like Canonical Covariance Analysis (Momirović, Dobrić and Karaman, 1983; Momirović and Dobrić, 1985; Dobrić, 1986). Unfortunately the knowledge of a typical user of CCA is seldom sufficient even for technical interpretation of obtained results. For that reason an expert system is written in GENSTAT to help the interpretation to user not understanding the real meaning of parameters and set of pattern, structure, crosstructure and latent variables intercorrelation and crosscorrelation matrices computed for the identification of latent variables content.

KEYWORDS

Robust methods / Expert system / Canonical covariance analysis

1. INTRODUCTION

Although probably at yet not formally proven for most of data analyst is quite evident that the expected knowledge of mathematical and statistical properties of some data analysis method or technique in the sample of users of this method or technique usually decrease as the sample space increase. This is especially true for the apparently simple and intuitively easy understandable methods, suported by largely available programming products, and for relatively new and for some reason appealing methods. The consequences of this can be divided in two not necessarilly distinct clusters:

(1) A lot of misunderstandings, wrong, and even stupid interpretation of obtained results

(2) A heavy pressure on data analysts, statistical consultants, or innocent autors of program products.

Compstat 1988

The opinion that some changes in the amount and level of mathematical, statistical and computer science education can significantly reduce the probability of both types of iritable behaviour of users of computational statistics is a little less than pure ilusion. The reason is very simple, and can be easily derived by the usage of elementary probability theory: all facts allows the conclusion, with probability close to 1, that, as the sample space of users increase, decrease not only the knowledge, but also the expected cognitive efficiency of them.

A possible way to reduce both, and especially the second type of consequences of computer revolution, is to produce some semi-expert, or true expert systems capable to interpret results of some data analysis or statistical procedure at least at the technical level. The aim of this paper is to describe such one of semi-expert systems, belonging to the BLABLA package in development in Computer centre of Institute of Kinesiology and University Computing Centre under project 1.11.01.02 (Inteligent Systems for Data Analysis). This system, named BLABLAQCR, gives technical interpretation of the results obtained by Canonical Covariance Analysis (Momirović, Dobrić and Karaman, 1983; Momirović and Dobrić, 1985; Dobrić, 1986).

2. SYSTEM DESIGN

BLABLAQCR consists of the following modules:

1. Knowledge base, including
   - 1.1 rules for the interpretation of results obtained by Canonical Covariance Analysis
   - 1.2 comments about the different normally possible outcome of such type of analysis
   - 1.3 comments about marginal and degenerative cases
2. Computing modul, with
   - 2.1 submodul for data preparation and declaration of all data structures

2.2 submodul for computing the basic solution

2.3 submoduls for computing identification structures in both data sets

3. Interpreter, consisting of

   3.1 the detector of abnormal cases

   3.2 the recogniser of possible normal types of results

   3.3 two educators, the first oriented to gives standard or enlarged information concerning the mathematical and statistical properties of identification structures and other results, and second, a little aggressive and no allways pleasant person, with the aim to use, according to results of Detector and/or Recogniser, the information in knowledge base, and to print or displays the technical interpretation of obtained results.

2.1 COMPUTING MODUL

The general idea of BLABLAQCR organisation and behaviour can be understood better if Computing Modul, in which Canonical Covariance Analysis algorithm is implemented, is described first.

Canonical Covariance Analysis can be formally described as follows.

Let $E=\{e_i;\ i=1,\dots,n\}$ be a sample of objects, and $V_1=\{v_j;\ j=1,\dots,m\}$, $V_2=\{v_k;\ k=1,\dots,s\}$, $m\geq s$ two sets of elliptically distributed variables. Let $Z_1=E\otimes V_1$ and $Z_2=E\otimes V_2$ be two data matrices in standard normal form, so that $R_{11}=Z_1^TZ_1$ and $R_{22}=Z_2^TZ_2$ are intercorrelation matrices, and $R_{12}=R_{21}^T=Z_1^TZ_2$ is crosscorrelation matrix.

Canonical Covariance Analysis can be defined as the solution of the problem

$$\begin{array}{l} Z_1X_{1p}=K_{1p} \\ Z_2X_{2p}=K_{2p} \end{array} \left|\ \begin{array}{l} c=K_{1p}^TK_{2p}=\max \\ X_{1p}X_{1q}=\delta_{pq} \\ X_{2p}^TX_{2q}=\delta_{pq} \\ c_p>c_{p+1} \qquad\qquad p=1,\dots,s \end{array}\right.$$

δ_{pq} denoting Kronecker δ; obviously, the problem is maxsimization of covariances, rather then correlations, of linear combinations of two sets of variables. The solution is easily obtained as singular values decomposition

$$R_{12} = \sum_{p=1}^{s} c_p x_{1p} x_{2p}^T$$

of crosscorrelation matrix between V_1 and V_2. The test statistics

$$\chi^2 = (Mc)^T (M \Psi M^T)(Mc)$$

with $c^T = (c_1, \ldots, c_s)$, M hypothesis matrix and Ψ a matrix with elements $\Psi_{pr \cdot qt} = g_{pq} g_{rt} + g_{pt} g_{rq}$,

$$G = \begin{bmatrix} W_{11} & C \\ C & W_{22} \end{bmatrix},$$

$W_{11} = X_1^T R_{11} X_1$, $W_{22} = X_2^T R_{22} X_2$, $C = X_1^T R_{12} X_2$, has assymptotically χ^2 distribution with $(s-1)$ degrees of freedom (Lužar, 1985). Let first $w < s$ canonical covariances are significant. Denote now by $X_1 = (x_{1p})$, $X_2 = (x_{2p})$, $p=1,\ldots,w$, and $C = X_1^T R_{12} X_2 = (c_p)$, the diagonal matrix of first w cocariances between K_{1p} and K_{2p}, $p=1,\ldots,w$, and by $W_{11} = X_1^T R_{11} X_1$ and $W_2 = X_2^T R_{22} X_2$ covariances among K_{1p} and K_{2p}, respectively. Identification structures for $K_1 = Z_1 X_1$ and $K_2 = Z_2 X_2$ can be defined as structure matrices $F_1 = Z_1^T K_1 D_1^{-1} = R_{11} X_1 D_1^{-1}$ and $F_2 = Z_2^T K_2 D_2^{-1} = R_{22} X_2 D_2^{-1}$, $D_1^2 = \mathrm{diag}\, W_1$ and $D_2^2 = \mathrm{diag}\, W_2$; intercorrelation matrices $H_1 = D_1^{-1} K_1^T K_1 D_1^{-1} = D_1^{-1} W_1 D_1^{-1}$ and $H_2 = D_2^{-1} K_2^T K_2 D_2^{-1} = D_2^{-1} W_2 D_2^{-1}$; pattern matrices $A_1 = F_1 H_1^{-1}$ and $A_2 = F_2 H_2^{-1}$ and crosscorrelation matrix $\rho = D_1^{-1} C D_2^{-1}$. In some instances (of course; if $w > 1$) orthonormal transformations $P_1 = A_1 T_1$ and $P_2 = A_2 T_2$ can be very usefull for the identification of latent generators of crossassociation between V_1 and V_2.

2.2 KNOWLEDGE BASE

Although conceptually very simple, because can be understood as a generalization of Principal Component Analysis (Gredelj, Momirović and Dobrić, 1986), or a method of interbattery factoring, or a robust method for Canonical Correlation Analysis (Momirović and Dobrić, 1985), Canonical Covariance Analysis is sometimes misunderstood by users with superfluous or to restricted knowledge of multivariate statistical methods.

As an example of interpretation rule, consider the optional comment* of the crosscorrelation matrix ρ: "Quasicanonical correlations can be interpreted in the following way:

(1) As correlations between linear combinations of variables from first and second set having, successively, maximal covariances

(2) As correlations between principal components of crosscorrelation matrix between first and second set of variables, defined as unstandardized linear combinations of standardized variables so that successively explain the maximum of the covariability of analyzed variables sets

(3) As robust estimates of canonical correlations between two sets of variables.

Note that crosscorrelations of quasicanonical variables with different index numbers are zero, so that crosscorrelation matrix is diagonal, but intercorrelation matrices of these variables are not diagonal in the general case."

An example of one of the possible comments of the normal output, par example $w = 3$, $0.5 \leq \rho_p \leq 0.6$, $\forall_p$, is:

> "Under applied test of significance 3 pairs of quasicanonical variables, related with covariances and therefore correlations different from zero are obtained. All obtained correlations are not only statistically significant, but also important and of approximatelly same order of magnitude, which can be considered as medium."

In one of the degenerative case ($\rho = I$) BLABLAQCR print

* Unoptional comment are short headings, in this case "Quasicanonical correlations (RHO), coefficients of alienation (ALRHO) and probability of hypothesis RHO = 0. (QRHO)."

or display the following comment:

> "Two identical variate sets are submited to analysis. I suppose that this is done deliberatelly, probably for testing purposes. If not so, I think that some mistake exists in the definition of V1 and V2 in calling program, or something is wrong with the cognitive processors of very respectable user."

Knowledge base is organized actually in four blocks, formally defined as MACRO programs. The first contains unoptinal headings, the second optional comments (generally the interpretative rules). The most important is the third MACRO, which contain subblocks with comments of the possible outcomes related to each of the main results of identification structures. The MACRO with comments of degenerative cases is organised in a similar way. Selection of text to be displayed or printed is under control of interpreter.

2.3 INTERPRETER

Interpreter is actually constructed so that contain

(1) Additional algebraic or logic expressions, transparent to user, related to sections of Computing modul

(2) A set of IF...THEN...ELSE or CASE OF statements, organized in subsets in accordance with the organization of Computing modul.

Decision rules are defined so that print of interpretation rules depends of the values of two parameters declared in calling program; first gives to interpreter information of user experience with QCR method, and second is simply a flag that can suppress BLABLA segment of QCR program. If BLABLA segment is not suppressed, outcomes of IF...THEN...ELSE or CASE OF statements select comments of degenerative or normal cases.

Detection or recognition technique is clear from the following example:

```
'CALC' MINRHO = MIN(RHOV)
'CALC' MAXRHO = MAX(RHOV)
'JUMP' L150 *((MINRHO.GE.05).AND.(MAXRHO.LE.06))
:
'LABE' L150
'LINE'  10
'PRIN' H(316)
'GOTO' L 1000
```

3. IMPLEMENTATION

BLABLAQCR is now implemented in GENSTAT 4.04, so that decision rules are realized thru complex 'JUMP' (reserved for conditional) and 'GOTO' (reserved for unconditional) statements, and not thru IF...THEN...ELSE and CASE OF statements.

Program is organized as a MACRO, using other MACRO´s; some of main MACRO´s, used by principal MACRO, use one or more "mini" MACRO´s.

REFERENCES

1. Dobrić, V. (1986):
   On a class of robust methods for multivariate data analysis. COMPSTAT 86, Proc. on Computational Statistics, Physica-Verlag, Heidelberg,pp.313 - 316.

2. Gredelj, M., Momirović, K., Dobrić, V. (1986):
   Some relations between canonical covariance analysis and principal component analysis. Proc. 9th International Symposium 'Computer at the University', 5.03 : 1 - 5.

3. Lužar, V. (1985):
   Inference in quasicanonical correlation analysis. Proc. 7th International Symposium 'Computer at the University', 518 : 1 - 6.

4. Momirović, K., Dobrić, V., Karaman, Ž. (1983):
   Canonical covariance analysis. Proc. 5th International Symposium 'Computer at the University', 463 - 473.

5. Momirović, K., Dobrić, V. (1985):
   Some relations between canonical and quasicanonical correlation analysis. Proc. 'Data Analysis and Informatics', posters. INRIA, Le Chesnay, pp. 101 - 104.

Building a Statistical Knowledge Base: A Discussion of the Approach Used in the Development of THESEUS, a Statistical Expert System

E. Bell and P. Watts, London

Keywords : Expert Systems, Statistical Analysis, ANOVA, Multiple Comparisons.

Introduction

Knowledge acquisition is one of the major problems in developing any expert system. It is recognised that different methods of knowledge acquisition should be used for eliciting different types of knowledge and that the usual approach of dialogue sessions between a domain expert and a knowledge engineer is not always appropriate.

Statistical expertise involves both technical expertise and professional or experiential expertise; professional expertise is very difficult to encompass and varies both within and between application areas. It is because of this variability in professional expertise that the concept of a local expert who can modify or extend the knowledge base became central to the design of THESEUS, a statistical expert system under development at Thames Polytechnic which provides advice and guidance in the fields of ANOVA and Multiple Comparisons.

An initial knowledge base has been built up through a program of literature reviews and statistics workshops as well as drawing on information gained during interviews with practicing statisticians. This initial knowledge base has been supplied to a number of local experts for them to modify in a manner appropriate to their own operational setting, incorporating not only their own preferred testing procedures but also such constraints as customer preferences.

Compstat 1988

In this paper different methods of knowledge acquisition and the nature of statistical expertise are discussed leading to the approach used in the development of THESEUS.

The Nature of Statistical Expertise

Statistical expertise can be divided into several different types, the boundaries may not be particularly clear but such classifications can assist in the choice of knowledge elicitation technique, Gammack and Young (1985).

FRAMEWORK : A statistician will have some form of conceptual structure in the domain which distinguishes different types of analysis. This knowledge will be used to select areas of statistics appropriate to the data being considered. For example ANOVA and regression could be two such areas.

CONCEPTS : Such concepts as hypothesis testing, population distributions, confidence intervals and degrees of freedom are necessary foundations to understanding and undertaking any analysis within the area of ANOVA.

PROCEDURAL KNOWLEDGE : This will include knowledge about the availability and requirements of specific statistical methods for analysis and assumption checking as well as knowledge about graphical representations.

HEURISTICS : These include rules of thumb used for judging the importance of effects such as violation of assumptions and how to handle them, and information about when methods such as transformations are applicable.

METHODOLOGICAL EXPERTISE : This enables the statistician to choose the most appropriate method from a range of those available. For example, selecting Dunnett's method in preference to Tukey's method when there is a control group present.

COMMUNICATION : Surrounding these different types or areas of knowledge is the expertise used in communicating effectively with the client. This involves not just establishing what the experimenter is interested in finding out, but also extracting information about the nature of the data that the statistician needs to assist in making decisions about the most appropriate analysis. This may not be regarded as knowledge but is nevertheless included here because of the influence it should have in developing the knowledge base as well as in the design of the expert system.

Each of these areas of knowledge involves both 'technical' and 'professional' knowledge. 'Technical' knowledge is hard, factual knowledge obtainable from text books and the literature. 'Professional' knowledge is judgmental, experience related and considerably more difficult to elicit and represent, covering decisions such as when to allow unequal variances to affect subsequent decisions.

Much of the existing work in statistical expert systems has been undertaken by statisticians or at least by people with a basic grounding in statistics; a great deal of technical knowledge (which comprises a part of each of the different types of knowledge) can be established by literature reviews. The professional knowledge acquisition may be facilitated by the use of more specific knowledge acquisition techniques. Interviewing methods can be helpful in understanding framework knowledge, concepts and communication expertise.

<u>Problems of Knowledge Acquisition</u>

Knowledge acquisition for expert systems has, in the past, relied heavily on informal interviews between a knowledge engineer and a domain expert. The aim is to encode the information supplied by the domain expert into some predetermined format and so develop a prototype knowledge base. The knowledge base is then refined by a cyclic process of evaluation and modification.

The knowledge engineer, who has the problem of transferring the knowledge from the domain expert to the knowledge base, also has to ensure that an appropriate and powerful enough form of knowledge representation is used. A great deal of time can be wasted trying to manipulate knowledge in order to make it fit a particular representation; this is a well-known disadvantage of expert system shells. Domain experts often find it difficult to articulate their decision making processes and face further problems of recognition and interpretation when trying to understand and evaluate the performance of the knowledge base.

Expertise in any domain will contain different types of

knowledge and thus the development of a knowledge base is the process of identifying the different types, choosing an appropriate knowledge representation scheme and then employing knowledge elicitation procedures appropriate to the situation.

Knowledge Elicitation Techniques

There are a number of methods available for aiding knowledge elicitation. INTERVIEWING METHODS are most helpful in the initial stages of knowledge acquisition for establishing the main concepts and components of the domain as well as defining the terminology used. The limitations of interviewing become more apparent when the domain expert is trying to evaluate the prototype knowledge base, trying to establish what distinguishes the performance of the expert from the performance of the system.

PROTOCOL ANALYSIS involves observing and recording the actions of the domain experts as they work through scenarios. The merit of this approach is that it gives the knowledge engineer a process to emulate. As the prototype knowledge base begins to take form, specific examples can be used to find out how the expert deals with special situations.

The basis of SCALING METHODS, which includes the repertory grid method, is to identify similarities among objects so that they can be grouped. Such methods result in values for a number of attributes used to define the objects. Cluster analysis of these attribute values enables discrimination between the objects. These methods are particularly useful where there is a number of closely related concepts, and expertise is required to discriminate between them.

CONCEPT SORTING is applicable when there is a large number of concepts within the domain. Concept sorting works by initially establishing a list of the concepts required to cover the domain and then asking the expert to sort the concepts into different groups, describing what each group has in common. This allows the concepts to be structured in an hierarchical fashion.

Protocol analysis is more appropriate for eliciting procedural knowledge and facts and heuristics, it may also be useful for understanding communication expertise. Concept sorting is really only appropriate for establishing the framework knowledge. Scaling methods could be useful for establishing concepts and understanding methodological expertise.

The Approach Used in THESEUS

As the application area chosen is small and well defined (Completely Randomised Designs and Multiple Comparisons) the knowledge acquisition does not involve the 'framework' knowledge described above but does involve all the other types. Each of the different types of knowledge involves both technical and professional expertise. Some types of knowledge such as procedural knowledge can be regarded as primarily technical in nature whereas knowledge about heuristics is mostly professional.

The knowledge acquisition for the prototype knowledge base of THESEUS was approached by using a combination of literature reviews, semi-structured interviews and workshops (protocol analysis). The aim was then to refine the prototype knowledge base with the help of practicing statisticians.

Literature reviews and small scale investigations were undertaken in order to establish a core of technical knowledge and ensure that a rational set of rules could be formulated. The review areas included hypotheses of interest to the client, choice of multiple comparison procedures, handling of outliers, use of transformations and criteria used for checking assumptions.

A series of interviews with practicing statisticians was undertaken with the purpose of gaining a general insight into the thinking that guides the statistician rather than the precise elicitation of rules. A loosely structured interview protocol was prepared to ensure that coverage of the relevant knowledge areas was complete, while allowing the contributors to describe fully, in their own ways, their approaches to data analysis. The interview schedule covered such areas as attitudes to outliers, flexibility on normality assumptions and homoscedasticity, use of transformations and the selection of test procedures. Seven such interviews were performed, the information gathered demonstrating more than anything else the enormous variability between statisticians handling similar types of study.

A series of statistical workshops was organised in which

different approaches to the analysis of data sets, supplied two weeks in advance, were presented. This enabled close examination of the rationale behind decisions about the most appropriate way to undertake the analysis.

The expert system was sent to a number of test sites where the previously interviewed statisticians were asked to evaluate the prototype knowledge base and then tó try modifying the knowledge base.

Conclusions

An academic base provides a good starting point for developing a rational prototype knowledge base containing technical expertise and some professional expertise. This prototype knowledge base can then be evaluated and modified by 'local experts'.

This approach still requires a certain level of commitment from local experts but is far less time consuming than the conventional dialogue sessions. It also takes into account the variation in approaches both within and between application areas.

The industrial trials are still going on; preliminary results are encouraging with the local experts being able to identify areas where they disagree with the prototype knowledge base and suggest alternative approaches. As the trials progress any shortcomings detected in the technical knowledge in the knowledge base will be corrected. Modification of the knowledge base, by local experts, to include alternative approaches will be monitored to assess the ease with which the knowledge representation and inference process can be understood.

References

Bell E.E., Watts P.J. & Alexander J.R. (1987), 'THESEUS : An Expert Statistical Consultant' Proc. ICOSCO-I, Turkey

Brooking A.G. (1986), 'The analysis phase in development of knowledge based systems' A.I. & Statistics , ed W. Gale, Addison-Wesley

Gammack and Young (1985), 'Psychological techniques for eliciting expert knowledge' Research and Development in Expert Systems, ed M. Bramer, Cambridge University Press

Hand, D.J. (1985), 'Choice of statistical technique', Proc. ISI Centennial Meeting, Amsterdam

PRINCE: An Expert System for Nonlinear Principal Components Analysis

I. J. Duijsens, Delft, T. J. Duijkers, Delft and G. M. van den Berg, Leiden

Objectives

Some statistical programs have so many options that users find it hard to choose and specify the analysis they want. Sometimes incorrect options are chosen and at other times possibilities are overlooked that could have been very useful. Especially programs for multivariate analysis techniques are notorious for their complexity. In recent years, statistical expert systems have been suggested as a way to solve these problems (Gale, 1986). When an expert system is developed with the explicit purpose in mind to advise on an existing statistical program, it is useful to establish a link between the system and the program. The expert system thus can give advice in statistical matters related to this program and it can generate the control language for the existing program. This implies that the conclusions drawn by the system, should not only be displayed to the user but must also be translated into actual control language. The ultimate goal of the expert system is the generation of control language that accurately reflects the demands of the user.

This paper describes PRINCE (Duijsens & Duijkers, 1988), which is a statistical expert system designed for use with the program PRINCALS (Gifi, 1981). PRINCALS is not just a program for principal components analysis, which is complicated enough in itself, but on top of that it can handle different measurement levels of the variables. Variables do not have to be treated as numerical, but can also be considered to be measured on a nominal or ordinal level. The choice of measurement level presents an additional complicating option to the user. Apart from being displayed as numbers, the results of an analysis with PRINCALS can also be presented in a graphical form using a variety of plot options.

For the novice user it is virtually impossible to decide without any help which analysis options to choose and which part of the output to look at. Furthermore, the control language of the program is rather obscure and frequently leads to errors. We wanted to develop an expert system to overcome these problems.

PRINCE's structure

PRINCE has been developed to run on an IBM PC or compatible computer and has been written in the programming language C. We preferred this language over specific A.I. languages like LISP

Compstat 1988

and PROLOG because of its greater speed and its more general standardization. In a very strict sense, PRINCE consists only of an inference engine. The knowledge that PRINCE needs, is stored in separate files so that the system itself can be considered to be built as a shell. The obvious advantage of separating the inference functions and the knowledge base is that the system does not have to be recompiled when the knowledge base is changed. Since – especially on PC's – compilation of a large program like an expert system easily takes a lot of time, independence is a real asset. A second advantage is that it is possible to develop expert systems for other programs for data analysis using PRINCE as the shell and changing only the knowledge base.

PRINCE is a rule-based expert system that tries to establish facts through backward chaining. In other words, it has a list of facts that it checks one by one to see if they are true or what their status is. PRINCE gets the information it needs either by questioning the user or by applying rules. Every fact is accompanied by a parameter indicating whether PRINCE is allowed to ask the user for information about the status of this fact. E.g., PRINCE always needs to know how many cases the data matrix contains. To establish the status of the fact "number of cases" it is allowed to ask the user for this information. If the status of a fact cannot be established by asking for it, PRINCE searches a rule that has this fact as a conclusion and tries to establish the status of the facts that form the premisses of this rule. E.g., PRINCE has a rule that says that if the user wants to print object scores and there are more than 50 cases in the data matrix, the user is advised not to print the object scores. Of course, PRINCE is not allowed to ask the user whether an advice against printing the object scores should be given. It thus checks the status of the premisses of this rule: whether the user wants to print these scores and whether there are less or more than 50 cases. PRINCE is allowed to ask the user for the status of the first premiss, but not for the status of the second premiss. However, it has a rule that says that if it knows the number of cases, it can infer whether this number is smaller or larger than 50. Since PRINCE is allowed to ask for the number of cases, the status of the second premiss can be inferred.

The facts are sorted into contexts that are treated one at a time. PRINCE starts with the facts in the first context that it needs to know and works through all of these facts before it proceeds with the facts in the next context. In this way the questions that are posed, are logically grouped together. For instance, all facts that are related to the number of variables and their measurement level belong to the same context and the questions concerning these facts are consequently not interwoven with questions on other subject matters. Some contexts are optional. If users have no need for advice on the choice of the number of dimensions of the solution, all facts relating to this subject are skipped – with the exception of the number of dimensions itself, since that is a fact that PRINCE always needs to know.

Since PRINCE operates as a shell, there is a number of separate files that are needed during a session. The three basic files are a file with facts, a file with rules and a file that contains information on the context to which each fact belongs. These files can be created and edited using a rule editor. This rule editor is a separate program that is written in dBASEIII+ and compiled with CLIPPER. Using this rule editor a knowledge engineer can add and change facts and rules in the knowledge base. The program will on request check the new knowledge base and will detect instances of circular reasoning and conclusions that may not be asked and can never be proved. Only when no faults are found, the rule editor produces files in the format that PRINCE needs.

When editing a fact, a block of text can be appended to each fact. This block serves as an explanatory help text to the person that uses PRINCE. If PRINCE asks a question and the user does not understand this question, she can ask for help. The text appended to the fact PRINCE is asking about, is then shown in a separate window on the screen. In addition to this fact-specific help, PRINCE can also use a general help file that contains definitions and examples of the main concepts in principal components analysis. Both specific and general help is obtained by pressing specified function keys that are permanently available.

PRINCE also has an option to explain why a certain question is posed and how it tries to establish a fact. If the user presses the function key connected with the why-option, PRINCE shows the conclusion it is presently working on. In other words, if the question concerns a fact that serves as a premiss to a rule, PRINCE shows the conclusion of that rule. That conclusion may be a premiss of another rule and if the user ones again chooses the why-option that next rule is displayed. When the user reaches the end of the chain, PRINCE reacts to the why-option with the message that this is a fact it just needs to know. When a user has chosen the why-option, the only information shown is the conclusion of the rule. If the user wishes to see the complete rule, the how-option can be chosen in which case all premisses are displayed.

A user may want to change answers to previously given questions. PRINCE offers this opportunity. When an answer and consequently the status of a fact is changed, PRINCE corrects all conclusions that were based on that answer and repeats questions about facts that have become ambiguous under the new information.

Acquiring knowledge

PRINCE has been developed as a prototype of an expert system that would be useful in practice. In other words, the acquisition of knowledge on principal components analysis was as important as the design of a useful structure. We have therefore sought the cooperation of several experts on this subject. Our main approach consisted of interviewing these experts.

We started with three experts and asked each of them in separate interviews which criteria, rules or heuristics they used when choosing options for PRINCALS. These interviews were written down with the rules formulated as clearly as possible, and the experts were asked to refine the rules where they could. Instead of refining the rules, however, they all showed a tendency to weaken them and to make several exceptions to criteria. Confronted with inconsistencies among them, they frequently felt unable to judge approaches that were different from their own as being unambiguously right or wrong. When these inconsistencies were presented to a fourth "ultimate" expert, who had been involved in the development of PRINCALS, he also claimed to be unable to judge them as long as no specific data set or research question was formulated. Therefore we decided to change our method of knowledge acquisition.

We organized a combined session with two experts in which they had to perform several analyses with PRINCALS on specific data sets. They were instructed to focus their attention on a limited number of problems: the number of dimensions of the analysis, the measurement level of the variables and the selection of variables to be analyzed. When the experts had different opinions on these subjects, they tried to reach consensus. Often they could not, claiming their differences of opinion to be a matter of taste. In these cases they were always able to formulate the reasons for their choices and to indicate in which situations they would choose differently. In the course of this process even the experts gained further insight in the capabilities of PRINCALS and some weaknesses in the program were found.

In trying to formulate rules for PRINCE we found that most of the knowledge we acquired, could not be turned into strict rules. There is a relatively small number of technical and logical limits to PRINCALS. For instance, it is not possible to compute category quantifications of passive variables, and if the user wants to analyze eight ordinal variables the number of dimensions of the solution cannot be larger than seven. Limits like these can be turned into absolute rules: if the user has eight ordinal variables and wants a solution in ten dimensions, the actual number of dimensions passed in the control language to PRINCALS is seven and this adaptation is displayed to the user.

Most of PRINCE's knowledge is not that absolute. E.g., it is undesirable to print object scores when there are more than 50 cases in the data matrix. However, there are situations in which the user may want to inspect all of these scores, for instance when the analysis shows one or more eccentric objects. Thus, the conclusion we draw is not that we prevent the user from printing more than 50 object scores, but that we advise against it. Most of PRINCE's conclusions are likewise framed as advices. From this it follows that the initiative is always at the side of the user. Most conclusions that could directly have been turned into control language for PRINCALS, are first displayed as feedback to the user to get explicit approval. Only when that approval is gained, is the necessary control language generated.

Since the emphasis in PRINCE is on giving advice, the help facilities are very important. As described above the user has several possibilities to get help: specific help on the current question, description of important concepts and why- and how-options to clarify PRINCE's reasoning. In addition, the user can ask PRINCE to explain the reason a certain advice is given. When an advice is displayed and the user presses a designated function key, a separate window appears that contains information about the reasoning that led to this advice.

Conclusions

Contrary to opinions that can sometimes be heard, we found that it was possible – though no small task – to write an expert system that works on a personal computer. PRINCE can handle 600 facts and 300 rules. As personal computers are growing ever faster and larger in memory size, their capabilities can only grow better.

According to our experience with this expert system, knowledge is best gained when elicited with the help of specific examples of situations that the system is supposed to handle. Analysis situations seem to be so diverse that experts are unable to formulate exhaustively the usefulness or senselessness of different options under all possible circumstances. But they are able to justify conclusively their choice of options in specific situations. We preferred to incorporate this knowledge in the system in the form of help and advice. Especially in the help facilities, a lot of concrete examples are given that enable the users to compare their own analysis problems with these models. This should make it possible for them to judge the appropriateness of the advices for their own specific problem.

As a not quite expected side-effect of knowledge acquisition we noticed that the experts refined their opinions during the interview sessions. Our request to be as specific as possible forced them to reconsider habitual choices in which process some customs turned out to be open to question.

At the moment PRINCE contains far less facts than it can handle. This is partly caused by our restriction of the acquisition of knowledge. We preferred deep knowledge on a limited number of subjects over superficial knowledge on more topics. As a consequence PRINCE lacks help and advice on subjects like missing values and relations with other techniques. We would also have liked to expand the system toward interpretation of the results of analyses done with PRINCALS. In most applications, more than one principal components analysis is performed before it is decided which analysis gives the best representation of the results. Ideally, PRINCE should be able to capture these results and judge the patterns they show. Such an expansion requires not only the addition of more rules to PRINCE, but demands an adjustment of the structure itself.

An expert system like PRINCE can play an important role in educating the users. As they work with the system they can get help on request and thus tap the knowledge of experts. When they get more experienced they will need less help, but it will always be there when they need it. And even when no help whatsoever is requested, PRINCE still serves as a friendly user interface to PRINCALS.

References

Duijsens I.J. & Duijkers T.J. (1988), *PRINCE: een expertsysteem voor de aansturing van PRINCALS*. Leiden: Department of Human & Computer Studies, University of Leiden (SWI Research Report 88-02)

Gale W.A. (ed.) (1986), *Artificial Intelligence and Statistics*. Reading, Massachusetts: Addison-Wesley.

Gifi A. (1981), *Nonlinear Multivariate Analysis*. Leiden: Department of Data Theory, University of Leiden.

Expert Systems for Non-Linear Modelling: Progress and Prospects

G. J. S. Ross, Harpenden

Summary

Expert systems can help scientists to use non-linear models effectively. The development of the model-fitting program MLP aims to provide the user with helpful advice at all stages of model choice, model fitting and interpretation of results.

Keywords Expert systems, non-linear models, non-linear regression, curve fitting, maximum likelihood estimation.

1. Introduction

Preliminary work on expert systems for non-linear modelling was described in Berzuini et al, (1986). A collaborative project to investigate the use of expert system shells as an intelligent front end to the model-fitting program MLP (Ross, 1987) resulted in the development of a systematic sequence of questions and responses which could be converted into instructions to run the program. This exercise was very useful in defining the tasks to be performed, but it was not possible, at a distance and without extra assistance, to develop a workable system that could interact fully with MLP. To exploit the full capabilities of an expert system there would have to be some way of using the results of an analysis so that the appropriate action could be taken. The radical modifications required of MLP would mean that there was little advantage in using the shell to control operations because much of the logic would have to be duplicated.

The alternative approach has been to develop 'expertise' at several levels. The user may be a complete beginner who knows neither how to use the program or even what models to fit, or someone who knows what to fit but does not know how to use the program most effectively. The aim is to provide facilities for getting real problems solved as well as teaching users more about the art of modelling. However not all problems can be anticipated by an expert system: modelling is context-dependent, and the computer cannot take decisions which require external knowledge: it can make suggestions, but the user must make the choice.

Compstat 1988

The main features of MLP which might be classified as 'expert' are as follows:

1) A MENU directive in the user language which switches from normal interactive (or batch) command language to a step-by-step question and answer system for data input, model selection, fitting and interpretation.

2) Background HELP with information on program conventions.

3) Interactive EXAMPLES that run real jobs, illustrating both the formats of standard types of input file and the scope of some more complicated analyses.

4) Diagnostic reports after model fitting, commenting on features that could be investigated or modified.

5) Explanations of general terms, and of particular features of the current analysis such as the meaning of the parameters fitted.

6) Interactive teaching aids in which tasks are set, such as the minimisation of an unknown function, or the improvement of a design by adding data or modifying the independent variable.

2. How the Menu helps the user.

MLP fits either standard models or user-defined models. The standard models have different rules, according to context, but are easy to fit if the right settings are made. User-defined models are easy to write, but not so easy to fit because the choice of parameterisation, error distribution, initial values and other operating characteristics can be very important. The menu gives an opportunity for guiding the user in a way which is not possible in normal interactive or batch operation. However users may regard the compulsory use of menus as excessively tedious if the data are extensive or the problem is standard, and so it is advantageous to be able to transfer from menu-mode to normal mode when necessary.

The directive MENU describes the manner of menu operation, then gives a list of types of job, including:

C Curve Fitting	D Distribution Fitting
A Assay Analysis	M User Models
T Teaching Aids	Q Quit Menu

to which the reply 'C' calls up the Curve Fitting menu.

The Curve Fitting menu contains sections as follows:

1. Data reading
2. Data editing
3. Data transformation
4. Curve selection
5. Error distribution
6. Basic output options
7. Extra output options
8. Fitting the curve
9. Explanations of output.

The first pass through the menu invokes one of the first three sections, in response to the question:

N No action, data already present
D New data to be read
E Edit data
T Transform data

after which it proceeds through sections 4, 5 and 6 in detail, and section 7 as many times as required. Before section 8 it asks

Ready to Fit?
Y Yes N No Q Quit Menu

If the response is 'N', or after fitting, the <u>Continuation Menu</u> is given, from which one section only need be re-entered before returning to section 8:

R Read new data	E Edit data	T Transform data
M Model change	D Distribution	O Output change
A Extra analyses	Q Quit menu	X Explain output

In this way the fitting and modification of the job afterwards is made as simple as possible, avoiding unnecessary repetition of information. The 'Return' key is used wherever possible to mean either the retention of a previous setting or the use of the standard default setting.

The Curve-Selection menu lists the standard curves fitted by MLP, but includes an extra facility to help the user who is not sure which to use. The menu gives:

Choice of Model

-1 No idea, please help

0 As before

1 Exponential 2 Two Exponentials 3 Exponential* Linear

etc.

19 Double Gaussian

31-35 Polynomials of degree 1 - 5

41-43 Splines (two-phase polynomials) of degree 1 - 3

and the response '-1' gives a small plot of the data, followed by the Shape Description menu:

L Linear (rising or falling steadily)

J Asymptotic (rising or falling but flattening off)

S Sigmoid (flat then rising then flat)

M Maximal (rising then falling)

B Bellshaped (flat, rising, falling, flat)

P Periodic (rising, falling, rising, falling)

C Compartmental models (suggested by theory)

V Changing phase (abrupt change of slope)

X Please suggest choice based on empirical analysis

The user would normally be expected to choose on the basis of the plotted curve. If 'X' is chosen the program attempts to classify the shape of the data by subdividing the range of x, computing the centroid of each subdivision, then analysing the signs and relative magnitudes of the slopes of the chords joining successive centroids. This method is not infallible but does at least suggest a curve which is likely to fit reasonably well.

If, say, 'J' is chosen, the program lists suitable curves from the Curve Selection menu, the first one being the default setting:

Three-parameter models

1 Exponential 9 Linear/Linear

Four-parameter models

4 Line + Exponential 10 Quadratic/Linear

15 Generalised hyperbola 41 Line spline

Although only one model is chosen as the primary model, others of the set may also be fitted as a sequence of alternative models, in response to a further question in the menu. The change in sums of squares due to fitting extra parameters are listed in a table. If the best fit is to be used for subsequent analysis that model may be promoted to the role of primary model.

The Explanation Menu that follows the fitting of a model includes text appropriate to the choice of model, such as:

> Exponential, Y = A + B*R**X
> The standard parameters are R, the rate of approach to the asymptote for unit change in X, the asymptote A and the scale factor B. When X=0, E(Y)= A + B. K = -LOG(R) is given so that the equation may also read Y = A + B*EXP(-K*X).

There are similarly structured menus for the other standard models in MLP, including discrete and continuous distribution fitting, probit analysis, dilution series estimation, quantitative bioassay, genetic linkage estimation, and multiple linear regression.

The menu for user models is designed to guide the user through the naming of variates and parameters and constants, the input and editing of data, defining the expectation part of the model, choosing the appropriate form of the log-likelihood function to be optimised, and then as for curve fitting, selecting output options and requests for extra analyses. Before fitting the model there is the opportunity to examine the log-likelihood function graphically, as well as setting initial estimates, step lengths and other quantities to control optimisation. To prevent the menu becoming excessively long some of the options are grouped so that their default settings may be accepted in a single operation, while allowing one or more of them to be changed.

The user-model menu has long or short explanations. Long explanations give descriptions of the theory of modelling and the way the data and parameters have to be organised. Short explanations assume that the conventions are understood, and simply asks for choices of values. It is expected that new users will start with long explanations and switch later to short explanations.

3. Background Help and Examples

Background Help facilities give general information about the program, the syntax and spelling required by the directives and functions, as an on-line aide-memoire for the user in the middle of a job, particularly in response to error reports. In MLP this is provided by a branched HELP directive which can be used at any point during an interactive session.

The EXAMPLE directive is also interactive, and allows short prepared data files to be displayed and run as real jobs. Some of these examples are to illustrate basic jobs such as simple curve fitting, while others show more advanced features such as analysis of parallelism and parameter transformations and the fitting of differential equation models. After an example has been run control returns to the keyboard so that further MLP instructions may be given to modify the example in any way required. The examples are useful as demonstrations and as test data for implementing the program on new machine ranges.

4. Diagnostic reports and feedback from results

All programs have error messages and diagnostic features. What is relevant in non-linear modelling is to consider what can be done to help the user to look again at certain aspects of the data-model combination which can lead to improvements to the current or future analyses.

If the fitting is unsuccessful, because of failure to converge, the user must be helped to decide whether it is because of poor starting values, poor parameterisation, or non-existence of solutions. Since non-convergence is signalled by the program, in menu mode this can initiate a remedial-action menu which suggests possible actions such as examining the plot of the likelihood function, reformulating the model in terms or more stable parameters, or holding certain parameters constant.

After successful fitting the program checks the users choice of step lengths, failure to exploit separable linear parameters, inactive parameters, and optionally the inflation of the variance of each parameter due to ill-conditioning of the dispersion matrix. Influential points and outliers are flagged, but these require inspection of the data and redesign of future data collection. If exact confidence limits of parameters are computed asymmetry can be noted, and a measure of non-linearity computed by comparing the predicted and observed log-likelihoods at the limit values.

5. Availability and future developments

While some of the facilities described above already exist in version 3.08 of MLP, the full range of menu-driven and diagnostic help facilities will be available in MLP 3.09. A considerable amount of testing and feedback from users will be desirable to ensure that what is clear to the author is also comprehensible to the user!

The possibility of using more sophisticated Expert System shells is still under consideration, but their advantages will have to be set against the convenience of a single program with an integrated expert system.

References

Berzuini, C., Ross, G.J.S., and Larizza, C. (1986). COMPSTAT 1986, 258-264

Ross, G.J.S. (1987) MLP 3.08 User Manual. NAG Ltd., Oxford.

Inside a Statistical Expert System: Statistical Methods Employed in the ESTES System

P. Hietala, Tampere

Summary: In this paper we describe the statistical methods and their organization in a statistical expert system called ESTES. The system is intended to provide guidance for an inexperienced time series analyst in the preliminary analysis of time series. The knowledge base (i.e. statistical methods) of the system has been organized according to the property being considered, the granularity of the analysis process, the goal of the analysis and the user experience. The explanation capability of the ESTES system provides justifications of the statistical methods used in the reasoning process.

Keywords: Expert systems; Time series analysis

1. INTRODUCTION

The ESTES (Expert System for TimE Series analysis) system is intended to provide guidance for an inexperienced time series analyst in the preliminary analysis of time series, i.e. in detecting and handling of seasonality, trend, outliers, level shifts and other essential properties of time series. The system has been implemented on Apple Macintosh™ personal microcomputers using Prolog and Pascal languages (the design philosophy, as well as the user interface and implementation details of the ESTES system can be found in Hietala (1986) and Hietala (1988)).

In the preliminary analysis of time series it is usually the case that an expert time series analyst detects the essential features of a time series just by examining its graphical representation and autocorrelation function, without any complicated calculations (see e.g. McLeod (1983)). Even in the case of an inexperienced user he/she may have plenty of useful knowledge concerning the environment of the problem in question. With this in mind, the statistical methods employed by the ESTES system are organized so that the system tries to exploit as much as possible of the knowledge or experience that the user has about the specific time series being considered. However, if there exists any conflict between the initial results computed by the system and the knowledge elicited from the user, then the ESTES system sets out to perform more extensive analysis and to apply more sophisticated statistical methods. Statistical methods employed in the initial analysis are mostly heuristic rules based on the behaviour of the specific time series and its autocorrelation function whereas during the extensive analysis phase the system usually applies statistical tests. Moreover, the employing of heuristic rules is often computationally cheaper than the corresponding statistical tests. With this kind of organization of statistical methods we strive for minimizing the number of unnecessary reasoning and calculation steps.

Compstat 1988

Statistical methods employed in the system (i.e. the knowledge concerning preliminary analysis of time series) are located in *the knowledge base* of the system. *The inference engine* exploits this knowledge in reasoning process and also facilitates the explanation capability of the system. Besides the knowledge base and the inference engine the ESTES system consists of

- *main module* which takes care of communication between other modules,
- *user dialog module* which interacts with the user,
- *graphics module* which displays graphical results,
- *time series generation module* which generates example time series, and
- *numerical computation modules* which calculate all numerical results for the other modules of the system.

The numerical computation modules have been implemented in Pascal, all the other modules in Prolog.

In this paper we describe the organization of statistical methods in the knowledge base of the ESTES system, together with the methods themselves. We also characterize the explanation facilities of the system (embodied in the inference engine) and their relationship with the statistical methods employed.

2. KNOWLEDGE BASE

2.1. Knowledge representation

We have selected *if-then rules* for representing knowledge concerning properties of time series and their handling. One reason for this is that rules have many desirable features (such as modularity, incrementability and modifiability, see e.g. Bratko (1986)). Rules in our system (see Figure 1) are either of form:

RuleName: ***if*** condition_A ***then*** conclusion_B, or,
RuleName: ***if*** condition_A ***then*** action_C.

```
remove_trend:
if      try_transformation is ok
        or
        try_seasonal_differencing is ok
        or
        try_differencing is ok
then    remove_trend. /* action */

transformations:
if      time_series exhibits increasing_variance
        and
        time_series exhibits single_seasonality
        and
        seasonal_sm_plot exhibits
           increasing_in_direct_proportion
        and
        try_log_transformation is ok
then    try_transformation is ok.
           /* conclusion */
...
```

```
variance_rule_1:
if      user_answer is yes
        and
        result_of_initial_analysis is positive
then    time_series exhibits
           increasing_variance. /* conclusion */

variance_rule_2:
if      ((user_answer is no
        and
        result_of_initial_analysis is positive
        and
        user is inexperienced)
        or
        user_answer is 'I don't know')
        and
        result_of_extensive_analysis is positive
then    time_series exhibits
           increasing_variance. /* conclusion */
...
```

Figure 1. An example of the rules in the knowledge base

The condition part of a rule may be combined (it can contain ***and*** and ***or*** operators); moreover, condition and action parts usually include invisible calls to numerical computation procedures that are implemented in Pascal. The above kind of rules are easily expressed in Prolog: in fact, they are legal Prolog clauses if we define appropriate operators (e.g. ':', 'if', 'then'). The rules of Figure 1 will be explained in more detail in Section 3.

2.2. Selection of statistical rules

One of the main criteria for choice of statistical methods, in addition to (1) *the property* (e.g. seasonality or trend) being considered, is (2) *the granularity* of the analysis process. By granularity we mean thoroughness of the statistical analysis. The ESTES system performs first an initial analysis of time series without the user. The system compares its results to the user's answers and if they are not consistent the system performs more complicated and more reliable tests. For example, if the initial analysis indicates that time series does not have trend but the user claims the contrary, then the system will employ nonparametric trend tests. If the results of this extended analysis confirm the results of the initial analysis the system just tells the user these test results and asks how to continue. So the final decisions are made by the user.

Also (3) *the goal* of the analysis process, i.e. whether the user tries to detect a property or he/she wants to handle the time series according to a detected property, determines which statistical methods are useful. On the other hand, (4) *the knowledge* possessed by the user (i.e. whether he/she is inexperienced or experienced) does not directly influence system's choice of statistical methods themselves but it determines how thoroughly user's answers concerning assumptions about time series are checked.

To sum up, the knowledge base of the ESTES system has been organized so that the selection of a class of statistical methods will be determined using the above outlined hierarchy of criteria (see also Figure 2). Within a chosen class of statistical methods, the final selection will be made according to the power of the methods, i.e. the most powerful method available is selected first. This principle is one of the necessary attributes of statistical expert systems suggested by Hand (1985).

2.3. Statistical methods in the knowledge base

One of the most popular techniques of model fitting for time series is the identification and estimation of parameters for autoregressive moving average models by Box and Jenkins (1970). However, successful model identification in this framework needs knowledge about essential properties of time series (such as seasonality, trend and outliers). Preliminary analysis of time series (i.e. detecting and handling of the above properties) can be employed to produce this knowledge. Because preliminary analysis constitutes the first phase in the model identification process, we can not use in our system sophisticated statistical methods based on the estimated model of the time series.

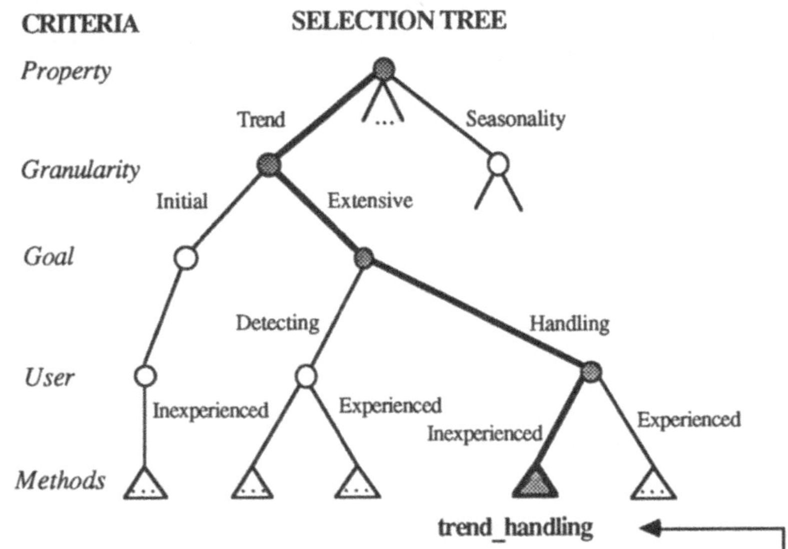

Selection of methods within the rule class "trend_handling":

R1: **if** path_to_trend_handling **then** select_most_powerful_method.
R2: **if** path_to_trend_handling **then** select_next_powerful_method.
...

Figure 2. Selecting statistical methods from the knowledge base.

Statistical methods in the ESTES system are, on one hand, *"thumb rules"* based on the behaviour of the time series and its autocorrelation function and, on the other hand, *statistical* (usually nonparametric) *tests*. "Thumb rules" are employed during the initial analysis and also in the selection of a handling method during the extensive analysis. Statistical tests are used only if there is uncertainty about the sufficiency of the initial analysis.

Figure 3 illustrates the statistical methods (together with their sources) that are currently being used by the ESTES system. They have been collected into a table with respect to the method selection criteria described above. This set of methods will be extended and refined in the near future.

3. AN EXAMPLE OF METHOD USE AND EXPLANATION: TREND HANDLING

The running example in our figures has been the handling of the property "trend". In Figure 1 we have several rules which belong to the rule class "trend_handling" (see Figure 2). These rules apply statistical methods such as transformations and differencing (see the uppermost right cell in Figure 3). Let us next assume that the user (in cooperation with the system) has detected a trend and wants to remove it. The first rule in Figure 1 states three possibilities for removing a trend. Using the first condition of the first rule the system examines if some transformation is

GRANULARITY	Initial	Extensive	
GOAL	*Detecting*	*Detecting*	*Handling*
PROPERTY			
Trend	**Interpretation of Autocorrelations:** McLeod (1983), p. 11-27; Chatfield (1985), p. 25-30	**Nonparametric Trend Tests:** Mann (1945), Cox and Stuart (1955)	**Transformations; Differencing:** McLeod (1983), p. 11-24 - 11-32; Chatfield (1985), p. 14-15, 21
Seasonality	**Interpretation of Autocorrelations:** McLeod (1983), p. 11-27; Chatfield (1985), p. 25-30	**Nonparametric Turning Point Test:** Moore and Wallis (1943)	**Seasonal Differencing:** McLeod (1983), p. 11-24 - 11-32; Chatfield (1985), p. 21
Outliers	**Inspection of Time Series:** McLeod (1983), p. 11-4; Chatfield (1985), p. 7	**Influence Matrix Method:** Chernick et. al. (1982)	**Outlier Handling Strategies:** McLeod (1983), p. 11-6
Level shifts	**Inspection of Time Series:** McLeod (1983), p. 11-4	**Nonparametric Tests for Change at a Time Point:** Pettitt (1979), Bhattacharyya and Johnson (1968)	**System Advises the User to Apply Intervention Models:** McLeod (1983), p. 9-8

Figure 3. Examples of statistical methods employed in the ESTES system

necessary. Now the first condition in the rule "transformations" ("time_series exhibits increasing_variance") leads to rules concerning variance (in the right column of Figure 1). If the user's answer about increasing variance is consistent with the results of the initial analysis (that the system has performed earlier) then no more evidence is needed. But if there is a conflict and the user is also inexperienced then more sophisticated analysis is carried out (see condition "result_of_extensive_analysis is positive" in variance_rule_2).

At any time after the ESTES system has requested for some information concerning the time series the user can ask a *'why'* question. In our example case, the user can ask: "Why does the system inquire 'Is it true: the time series exhibits increasing variance'?". To this the explanation provided by the system is short:

> "System has asked you to tell if *the time series exhibits increasing variance* in order to be able to examine if the time series requires transformations (before differencing) and remove_trend was your command".

Also, after the system has completed its reasoning process the user may ask *'how'* questions ("How has the system reached this conclusion?"). Our system's reply to this kind of questions consists of displaying a user-friendly form of its inner inference chain with explanations and justifications of those methods that are used inside the chain. For example, assume that the system advises the user to apply logarithmic transformation (see Figure 1, conclusion "try_log_transformation is ok" in the rule "transformations"). If the user now asks "How the system has reached this advise?", then the explanation can look like the following:

"*try logarithmic transformation* was the system's advise
because
remove trend was your command *and*
time series exhibits increasing variance
(was your answer and
was also result of system's initial analysis (...)) *and*
time series exhibits single seasonality with period 12
(...) *and*
seasonal 'standard deviation - mean plot' shows that standard deviation increases in direct proportion to the mean
(...)".

In addition to the above described textual explanation, the system's answer can contain displays of graphical results (e.g. standard deviation - mean plot). Moreover, if the ESTES system employs terms that are unknown to the user, he/she can ask more information about these terms - explanation via visual lexicon (Hietala (1986)).

4. CONCLUDING REMARKS

The ESTES system is an experimental research vehicle for studying the use of AI techniques in producing statistical expert systems. However, the system has not yet been tested in real-life situations and its current knowledge base is small. Therefore, in the near future our main emphasis in the development of the system will be in deepening its domain knowledge on time series.

REFERENCES:

Bhattacharyya, G.K., and Johnson, R.A. (1968). Nonparametric Tests for Shift at Unknown Time Point. *Annals of Mathematical Statistics* **39**, 1731-1743.

Box, G.E.P., and Jenkins, G.M. (1970). *Time Series Analysis Forecasting and Control*, Holden-Day, San Francisco.

Bratko, I. (1986). *Prolog Programming for Artificial Intelligence*. Addison-Wesley, Wokingham, England.

Chatfield, C. (1984). *The Analysis of Time Series: An Introduction*. Chapman and Hall Ltd, Bristol, Great Britain.

Chernick, M.R., Downing, D.J., and Pike, D.H. (1982). Detecting Outliers in Time Series Data. *Journal of the American Statistical Association* **77**, 743-747.

Cox, D.R., and Stuart, A. (1955). Some Quick Sign Tests for Trend in Location and Dispersion. *Biometrika 42*, 80-95.

Hand, D.J. (1985). Statistical Expert Systems: Necessary Attributes. *Journal of Applied Statistics* **12**, 1, 19-27.

Hietala, P. (1986). How to Assist an Inexperienced User in the Preliminary Analysis of Time Series: First Version of the ESTES Expert System. *Proceedings in Computational Statistics (COMPSTAT) 1986, 7th Symposium held at Rome 1986*, Physica Verlag, Heidelberg, 295-300.

Hietala, P. (1988). Inside a Statistical Expert System: Implementation of the ESTES System. Paper presented at the *20th Symposium on the Interface of Computing Science and Statistics*, Reston, VA, April 21-23, 1988.

Mann, H.B. (1945). Nonparametric Tests against Trend. *Econometrica* **13**, 245-259.

McLeod, G. (1983). Box-Jenkins in Practise, Vol 1. G. McLeod and Gwillym Jenkins & Partners Ltd., Lancaster.

Moore, G.H., and Wallis, W.A. (1943). Time Series Significance tests Based on Signs of Difference. *Journal of the American Statistical Association* **38**, 153-164.

Pettitt, A.N. (1979). A Nonparametric Approach to the Change-point Problem. *Applied Statistics* **28** (2), 126-135.

An Implementation of an EDA Expert System in Prolog Environment

P. Jirků, Prague

ABSTRACT

An experimental project of a system for exploratory data analysis (EDA), called GUHA-80, was described in [7]. Although discussions on the project were large and deep, the system was never implemented since the project seemed to be too complex with respect to both data structures and control of reasoning process. In the present paper we argue that a Prolog environment could be good enough for representing both object oriented data structures necessary to implement concepts defined in the project and control strategies.

Keywords: *expert systems, exploratory data analysis, reasoning strategies*

1. INTRODUCTION

Hypothesis formation in scientific research is both data processing and knowledge processing. It is a natural task to try to mechanize this activity (or at least its important parts) on the computer. Some years ago the GUHA-80 project was formulated by Hájek and Havránek in [7]. The aim of the project could be shortly expressed as follows: GUHA-80 is a project of computer-based (exploratory) data analysis on the base of empirical data. The project has been heavily inspired by D. Lenat's AM system [3], [9] which would model hypothesis formation as an open-ended cognitive activity including use of data processing procedures like "classical" GUHA programs [6], [9] or some standard packages of

Compstat 1988

of programs for data analysis. Various design choices have been formulated, a structure of knowledge (especially of concepts) elaborated, syntax of tasks fixed and lots of heuristics rules are being collected. However, the full implementation of such system seemed to be too complex, thus, two small experimental expert systems GQUANT and GSYNTR were built which help the user of GUHA programs to choose the type of association and/or to declare syntactic restrictions on hypotheses to be generated. When building an inference machine for GQUANT it turned out that it is reasonable to build a problem independent shell. It resulted in the system EQUANT.

In the present paper we are returning to the original GUHA-80 project and we argue for the fact that the Prolog environment [1] could be good enough for both representing data structures necessary to implement various kinds of knowledge (concepts, agenda of jobs, heuristic rules etc.) and implementing control strategies. We describe first a problem independent procedural production system FORBES [10] and then we discuss problems concerning representation of knowledge usually used in EDA together with implementation of various control strategies.

2. A PROBLEM INDEPENDENT PRODUCTION SYSTEM

Production systems (PS) are probably most videly used pattern directed inference systems, cf. [2], [5]. We deal with production rules that have the form

< *situation* > --> < *actions* >

where situation (conditions) is a pattern. The result of a rule application is performing an action, or a sequence of actions, which at least one time changes the actual content of the database (db). The control mechanism controls selecting of rules to be applied and it is responsible for conflict resolution. The aim is to transform a given stage of the db to reach new stages that eventually satisfy some termination condition. We shall describe now a production system FORBES (Frame-Oriented Rule-Based Expert System shell) developed by the author [10].

Let us begin with fundamental data structures used in the system, namely with rules. Rules are to be understood as a ternary predicate

```
rule(Name,Conditions,Actions).
```

Conditions as well as actions are represented as lists of expressions that follow the syntax of Prolog. Name of a rule could be used as an identifier, e.g. in the case that we want to define metarules, i.e. rules that mention other rules in their condition part. Using predefined infix notation for two operators ":" and "->", we can code production rules in a more impressive form

```
Name : [Cond1,...,CondN] -> [Action1,...,ActionM].
```

The top level interpreter of a PS is defined by a recognize-act--cycle which is typical for any PS. The definition is universal. It does not realize any restrictions onto data structures that should be used within an actual implementation. Moreover, the control mechanism, i.e. the recognize part of the recognize-act--cycle, can be easily modified.

When the rule, that should be applied, is selected the execution of actions that occur in right-hand side of the rule is fired. The corresponding *execute* procedure calls recursively the *recognize_act_cycle* procedure.

```
recognize_act_cycle :-
   Name : Conditions -> Actions,
   test(Conditions),
   execute(Actions).

test([]).
test([H|T]) :- call, test(T).
```

The procedure *recognize_act_cycle* realizes the simplest control strategy in which the first applicable rule is selected. The corresponding conditions are tested step by step. If the test succeds for all conditions in the left-hanside of the rule, the action part is sequentially executed. The process is repeated up to exhaustion, i.e. it ends when no rule is applicable or when the whole process is halted by performing a distinguished action *stop*. Note that this top level interpreter of rules works in a similar way as in OPS 5 production language [2].

A more sophisticated *recognize_act_cycle* procedure could be defined as follows

```
recognize_act_cycle                    % selection of applicable
   select,                             % rules
   select_one(Rule),                   % conflict resolution
   Rule : Conditions -> Actions,
```

```
      retractall(selected(_)),
      mk_trace(Rule),
      execute (Actions),!.
   select :-
      Rule : Conditions -> Actions,
      test(Conditions),
      add_to_selected(Rule),            % collect applicable
      fail.                             % rules
   select.
```

2. TOOLS FOR KNOWLEDGE REPRESENTATION

Since FORBES is a frame-oriented ES shell it uses special *pre-defined data structures* for expressing both conditions and actions in rules. These are namely

- concepts (that correspond to Minsky's frames or Lenat's concepts, cf. [3] and [12])
- agenda of jobs (cf. [3]).

Concepts and jobs as well as heuristic rules are stored in db as Prolog's facts (unconditional statements). Concepts are identified by their names. They have *slots*, each slot has a name. Some slots (*name, gen, specs*) are obligatory, some of them are facultative and they are listed in slot *attr* (attributes). Gen/specs slots define hierarchy of concepts in the knowledge base. (This is an analogy of AKO slot in the FRL, cf. [12].) For example the concept of hierarchical clustering could be viewed as the following frame data structure:

```
concept (name: hierarchical_clustering,
         gen:  clustering,
         specs: [single_linkage,
                 complete_linkage
                 centroid],
         attr: [type: operation,
                domain:similarity matrices]).
```

During computation additional attributes could be included.

In the system there are also built-in procedures for manipulation of concepts as e.g. *insert, delete, modify* etc. Similar primitives are defined again in OPS 5 and FRL.

Jobs in the agenda (db) are also identified by their names. Within each job its *priority* is specified under which jobs are ordered. A job can change priority of another job, thus the ordering of jobs is dynamically changed during execution. As in the case of concepts there exists also built-in actions as e.g.

```
suggest_concept(Concept),
suggest_job(Job),
execute_job(Job),
apply(Concept1,Concept2),
```

etc.

Having such built-in procedures we can express production rules in a very natural form. The running of PS is then based on both stepwise execution of jobs stored in agenda and on manipulation of concepts and their further development, including creation of new concepts, if necessary. Reasoning strategies are influenced by heuristic rules and their order in the knowledge base. It allows to implement various computational strategies in the range from *"to compute as it is possible"* to *"to compute when it is necessary"*.

Within each ES shell it is very important to have a possibility to construct an explanatory module. For this reason, it is necessary to save the information concerning the actual behaviour of PS. To reach it, we use a *mk_trace* predicate for stepwise recording of both executed actions and their results.

4. SPECIALTIES OF MECHANIZED DATA ANALYSIS

It is perfectly true, that in (E)DA we deal with such AI-structures as described in the FORBES system. However, additional data structures are necessary. These are namely input *data* and *parameters*, output data structures as *results* obtained by applying some *effective modules* (various statistical procedures) on input data and *traces* of computations. Our experience shows that it is not difficult incorporate such structures in the FORBES system. It is important that effective modules could be programmed in conventional programming languages (like Pascal or C) that are perhaps more suitable for arithmetic calculations. The problem that must be solved is a unification of input/output structures.

In spite of our first experiments concerned strongly limited tasks (in collaboration with T. Havránek we developed a knowledge base for GUHA procedures and for clustering jobs), their results support assumption that above described formalism is a good enough for developing larger computer-based system for data analysis.

Further development will concern creating larger knowledge base involving other typical tasks of EDA to learn its reasoning strategies in detail.

REFERENCES

[1] Bratko, I. : Prolog Programming for Artificial Intelligence, 1985

[2] Brownstown, L., Farrell, R., Kant, E., Martin, N.: Programming Expert Systems in OPS 5, Addison-Wesley, 1985

[3] Davis, R., Lenat, D. B.: Knowledge-Based Systems in Artificial Intelligence, McGraw-Hill, 1982

[4] Gale, A. (ed.): Artificial Intelligence and Statistics, Addison-Wesley, Reading, 1985

[5] Hájek, P., Havránek, T.: Mechanizing Hypothesis Formation, Springer-Verlag, 1978

[6] Hájek, P., Havránek, T.: GUHA-80 An Application of Artificial Intelligence to Data Analysis, Computers and Artificial Intelligence 1 (1982), 107-132

[7] Hájek, P.: The New Version of the GUHA Procedure ASSOC (Generating Hypotheses on Associations) - Mathematical Foundations, COMPSTAT 1984, 360-365

[8] Jirků, P.: FORBES - An Implementation of Frame-Oriented Rule--Based Expert System Shell in Prolog Environment, Proceedings AI'88, Prague, 1988, 93-99

[9] Lenat, D. B.: AM - An AI Approach to Discovery in Mathematics, Stanford University, STAN-CS-76-570, 1976

[10] Lenat, D. B., Brown, J. S.: Why AM and EURISKO Appear to Work, Artificial Intelligence 23 (1984), 269-294

[11] Marcus C.: Prolog Programming, Application for Database Systems, Addison-Wesley, 1986

[12] Roberts, R. C., Goldstein, I. P.: The FRL Manual, MIT AI memo 409, Sept. 1977

Automatic Acquisition of Knowledge Base from Data without Expert: ESOD (Expert System from Observational Data)

J. Ivánek and B. Stejskal, Prague

The acquisition of knowledge from experts is nowadays the most demanding activity in the construction of expert systems. Hence the effort for its automation [5,11] follows. In 1984-86 we developed the ESOD expert system shell [7,8,12], in which the knowledge base is acquired automatically from observational data.

The possibilities of the applications of the ESOD system are in **empirical research**, where the researcher makes experiments or collects data to obtain the arguments for a decision about the presence or absence of laws, or in **database processing**, where data are maybe obtained automatically even for completely different purpose than looking for knowledge and creating expert systems. Typical is e.g. observation of the function of natural and technical objects for documentation (or administrative) purposes which results in gathering extensive data files. Their processing with ESOD can conduce to a creation of an expert system that predicts or identifies various conditions of objects (illness, break-down, occurence of ore, presence of a talent, increase of performance, ...).

The ESOD system [12] consists of **dialog programs** for

- interactive data input, updating and reduction,
- detection of inconsistencies in data,
- automatic knowledge base acquisition from data,
- successfulness testing,
- consulting the acquired knowledge bases.

1. Method

The system works with **categorical data** in an extent of hundreds through thousands of objects and several through tens of variables. Each one of a finite number of values of each variable is understood as a name of a **property** (category) of objects. Properties taken out from different

Compstat 1988

variables compose possible **combinations** (conjunctions) of properties.

For an application, the user gives some **goal combination** of properties **C***; variables not included in **C*** come into account as the input variables for the prediction of **C***. The system creates a knowledge base in the form of a **selected collection of implications C ==> C***, where combinations of properties on left sides are composed from values of input variables. Each implication is provided with the **validity** (conditional probability) in data and the **weight** from [0,1].

The knowledge base is constructed in a way **analogous to the creating of an axiomatic theory**. Here the state of axioms is given to the most simple statements so that all the other known statements of the domain could be inferred from them (the requirement of completeness). At the same time the redundance is removed; statements inferrable from other axioms are going not to be the axioms (the requirement of independence).

The algorithm is based on the means of **combinational data analysis** [10,13] (related to **GUHA** method [4]) and on an application of the **corrections principle** [3]. Only those implications are inserted to the knowledge base whose validity in the given data statistically significantly **differs from the weight deduced** by the inference mechanism from the corrected weights of the implications already present in the base. Therefore the resulting base does not contain redundancies and has a minimal extent.

For the time being, we are using bayesian inference mechanism (from the PROSPECTOR expert system) and simple statistical test, but the method and the system are ready to incorporate and to test various other possible choices.

The acquired knowledge base is used for **consultations** in the following way: A client gives values of some input variables (a combination of properties) and the system activates all implications from the knowledge base the left side of which are subcombinations of the given combination. Their weights are composed by the inference mechanism and resulting weight is used for **prediction of C***.

2. Testing

In contrast to the majority of classical expert systems, the **successfulness** of expert systems created from data can be thoroughly tested, even before their application. ESOD performs this rather risky activity automatically and gives objective results without taking into account the wishful thinkings of the authors or the customers.

The first way of testing consists in **dividing the data into a learning part and a testing one**. The knowledge base is then automatically obtained from the learning part and used for consultations concerning the objects from the testing part. The results of the consultations and the real presence of the goal combination of properites at the testing objects are then compared. However, this way of successfulness testing is quite dependent on the choice of the learning and testing parts of the data, and therefore it is rightful rather for a stable ordering of objects, e.g., in space or time.

The second way of successfulness testing is **crossvalidation** (jackknifing by one-leave-out method). For every object from the data, a knowledge base is created from all the other objects in data. A consultation concerning the object picked out is performed with this knowledge base, and the result is compared to the real presence of the goal combination of properties at that object. After testing all objects in the data, the total successfulness is calculated, which is in fact the average successfulness of the consultations with as many knowledge bases as the number of objects in the data. As to our opinion it is quite a reliable indicator of the **expected successfulness** of ESOD used outside the data that are the source of knowledge.

3. Results of testing

In the table we briefly mention the results of testing ESOD on data of several domains (medicine, geology, demography). We do not describe the input data in detail; ESOD has not an idea of their meaning either. However, we show **degree of inconsistency of data**, expressed by the relative number of the objects to which there is somewhere in the data another object with the same values of the input variables but differing in the presence of the goal combination of properties.

The inconsistency of the data also has a relation to the **maximal possible successfulness** reachable in the data, which could not be overcome by any classifier which is iving the same classification of the objects for the same values of the input variables. This maximum possible successfulness can be compared to the result of the **selfprediction** performed with the obtained knowledge base to the learning data [6]. The main part of the table then contains the results of the **crossvalidatory test**.

In the table the **successfulness** is judged from two complementing points of view: so-called YES/NO successfulness means in how many percent of the cases the system was right when it predicted YES or NO, respectively. The second point of view is the percentage of "+objects" (having the goal combination) and "-objects" (not having the goal combination) well classified by the system.

The results of the tests give the user a possibility to **decide whether an expert system built by means of ESOD is acceptable or not**. (We certainly would not recommend some of those mentioned in the table.)

The results of testing ESOD on various data

Data	I.	II.	III.	IV.	V.	VI.
Number of objects	891	916	167	77	839	418
Number of variables	7	12	14	12	13	17
Freq. of goal combination	33%	57%	9%	75%	37%	25%
Inconsistency of data	46%	63%	4%	3%	27%	6%
Max.poss.successfulness at prediction to these data	86%	82%	98%	99%	91%	98%
Number of implications in acquired knowledge base	60	120	7	2	282	71
Selfprediction to the learning data						
YES predicted	30%	61%	12%	95%	32%	15%
Successfulness YES	67%	74%	50%	79%	77%	89%
NO	81%	72%	96%	100%	81%	86%
at "+objects"	61%	83%	63%	100%	66%	53%
at "-objects"	85%	60%	93%	21%	88%	98%
TOTAL successfulness	77%	73%	90%	81%	80%	86%
Crossvalidatory test						
YES predicted	30%	65%	11%	91%	33%	14%
successfulness YES	62%	69%	39%	79%	69%	69%
NO	79%	59%	94%	57%	79%	81%
at "+objects"	55%	72%	44%	95%	62%	38%
at "-objects"	83%	55%	93%	21%	84%	94%
TOTAL successfulness	**74%**	**65%**	**88%**	**77%**	**76%**	**80%**

4. Experimental comparison with EXPERT-EASE

The ESOD system was also in one experiment compared to the commercial EXPERT-EASE system (version 1.1, [2]) for knowledge (**decision trees**) acquisition from examples, which is a variant of ACLS (Analog Concept Learning System) based on the ID3 algorithm [11].

The EXPERT-EASE system was tested on observational data marked I. in the table, i.e., a little out of the area of its destination. Instead of "decision examples", the system obtained results of object observations, which nevertheless conforms with the application of ACLS to detecting rules in data banks [1].

Testing by dividing the input data (891 objects) into learning part (1-450) and testing part (451-891) was used. In the A cases the learning part was left without changes, in the B case we modified it with so-called "majority rule" mentioned in the system manual: only 177 objects representing the majority in the cases of repeating and contradictions were left, which removed the inconsistencies in the learning data. Here are the **results of the testing** (for more details, see [9]):

	EXPERT-EASE		**ESOD**
Testing cases	A	B	A
- **right** classified	184 42%	255 58%	305 69%
- **badly** classified	94 21%	150 34%	136 31%
- unclassified	163 37%	36 8%	0 0%

Let us mention that the expected successfulness of the obtained knowledge base, stated by ESOD automatically by means of the crossvalidatory test during its generation from the learning part of the data, was 72%, which is in good accordance with the reached successfulness of the system in the testing part of the data.

References:

[1] ACLS-Superfile Manual. London, Southdata Ltd. 1984, 26p.

[2] EXPERT-EASE for the IBM PC and Compatibles. User Manual. Intelligent Terminals Ltd. 1983

[3] Hájek P.: Combining functions in consulting systems and dependence of premises (a remark). In: Proc. 3.ICAIICR Smolenice. Amsterdam, North-Holland 1984, p.163-166.

[4] Hájek,P. - Havránek,T.: Mechanizing Hypothesis Formation - Mathematical Foundations for a General Theory. Berlin, Springer-Verlag 1978, 396p.

[5] Hájek P. - Ivánek J.: Artificial intelligence and data analysis. In: Proc. COMPSTAT'82 Toulouse. Vienna, Physica-Verlag 1982, p.54-60

[6] Ivánek J. - Stejskal B.: Some characteristics of categorical data with regard to a given target combination of properties. (in Czech) Prague School of Economics, Dept. of Scientific and Technical Information, Working Note 13, 1987, 7p.

[7] Ivánek J. - Stejskal B.: ESOD (Expert System from Observational Data) and its testing. (in Czech) In: Metody umělé inteligence a expertní systémy III. Praha, ČSVTS-FEL-ČVUT 1987, p.141-150

[8] Ivánek J. - Stejskal B.: ESOD (Expert System from Observational Data) and possibilities of its aplication. (in Czech) In: Aplikace expertních systémů. Dům techniky ČSVTS České Budějovice 1988

[9] Ivánek J.-Stejskal B.: An aplication of the EXPERT-EASE system on observational data and a comparison with the ESOD system. (in Czech) In: AI'88, Prague, ÚISK 1988, p.125-132

[10] Ivánek J. - Stejskal B.: Combinational data analysis. In: Proc. COMPSTAT'84 Prague, Summaries of short communications and posters. Czechoslovak Academy of Sciences

[11] Michalski R.S. - Carbonell J.G. - Mitchell T.M. (ed.): Machine Learning. An Artificial Intelligence Approach. Palo Alto, Tioga 1983, 572 p.

[12] Stejskal,B. - Ivánek,J.: ESOD (Expert System from Observational Data). User manual. (in Czech) Prague School of Economics, Dept. of Scientific and Technical Information, Working Note 16, 1987,

[13] Stejskal,B. - Ivánek,J.: Software for combinational data analysis. (in Czech) Prague School of Economics, Dept. of Scientific and Technical Information, Working Note 9, 1986, 30p.

Experiments with Probabilistic Consultation Systems

S. Nordbotten, Bergen

Abstract

This paper describes simulation experiments with probabilistic consultation systems, also called expert systems. The aim was to test the relative performance of systems with inference algorithms based on different assumptions. The approach used was simulation of task generation, knowledge acquisition, and task solving. The main results found were that a rule value approach seems to be superior to a goal driven approach, that the use of the complete Bayes' formula improves the the quality of the task solutions, that ignorance of statistical dependence among antecedents makes the estimated probability of the predictions useless as a reliability indicator, but that models taking the dependencies into account can be designed.

[Keywords:Expert system, evaluation, inference, simulation]

1. Consultation Systems

The type of consultation systems considered in this paper belong to those which try to solve classification or diagnostic problems based on uncertain knowledge. They represent knowledge by probabilistic rules and aim at expressing conclusions with estimated probabilities indicating their reliability.

We use the following terminology in this paper:

<u>Antecedent</u>: A property which describes a problem by being TRUE or FALSE and which can be observed.
<u>Cluster</u>: A set of TRUE antecedents which completely describes a problem.
<u>Conclusion</u>: A property describing a solution to a problem and which can usually not be observed, but has to be derived from the cluster found. Only one conclusion can be TRUE for each problem.
<u>Incident</u>: A TRUE cluster with a TRUE conclusion. An incident is assumed to be an expert's description of problem and its conclusion following a thorough analysis.
<u>Task</u>: A TRUE cluster to which a conclusion should be predicted.
<u>Solution</u>: A detected TRUE cluster with its predicted conclusion.

In a limited domain, let D={d[1]..d[i]..d[M]} be a set of conclusions, A={a[1]..a[k]..a[K]} be a set of antecedents, and C={c[1]..c[j]..c[N]} be the set of all subsets of A, which represent all possible clusters of TRUE antecedents. The Cartesian product set E=D*C is our sample space for which

Compstat 1988

probabilities P(d[i],c[j]) are assigned. The indexes i,j and k assigned to the elements in D, C and A will in the following represent any elment in the respective sets.

From observation, training and experience we may be able to form the following three sets of rules about the incidents in the domain for:

```
IF no antecedents have been tested in an incident                    (1.1)
THEN P(d[i]) is the probability that the conclusion d[i]=TRUE.

IF d[i]=TRUE is the expert conclusion to an incident                 (1.2)
THEN P(a[k]|d[i]) is the probability that the antecedent a[k]=TRUE.

IF d[i]=FALSE is the expert conclusion to an incident                (1.3)
THEN P(a[k]|NOT d[i]) is the probability that the antecedent a[k]=TRUE.
```

Combined with the comprehensive lists of antecedents and conclusions, these three rule sets are called the basic knowledge base.

Following some selection algorithm, antecedents are tested one by one until the detected cluster of TRUE antecedents is considered satisfactory and a conclusion is predicted with an estimated probability. Usually the most probable conclusion given the detected cluster, is selected as a prediction. The conditional probabilities for the conclusions are frequently computed by reference to Bayes' Theorem and imply repeated application of:

```
P(d[i]|a[k]):=P(a[k]|d[i])*P(d[i])/                                  (1.4)
          {P(a[k]|d[i])*P(d[i])+P(a[k]|NOT d[i])*(1-P(d[i])}
```

if a[k] is found to be TRUE, or

```
P(d[i]|NOT a[k]):={1-P(a[k]|d[i]))*P(d[i])}/                         (1.5)
          {(1-P(a[k]|d[i]))*P(d[i])+(1-P(a[k]|NOT d[i]))*(1-P(d[i]))}
```

if a[k] is found to be FALSE. The computed probabilities become new P(d[i]) after testing of each new antecedent. The denominator in these formulas are, however, approximations and require knowledge of the rules in the set (1.3) which may be difficult to learn even from extensive training.

These objections can be met by applying instead the complete formulas:

```
P(d[i]|a[k]):=P(a[k]|d[i])*P(d[i])/                                  (1.6)
          {SUM [i] P(a[k]|d[i])*P(d[i])}
```

and

```
P(d[i]|NOT a[k]):={1-P(a[k]|d[i]))*P(d[i])}/                         (1.7)
          {SUM [i] (1-P(a[k]|d[i]))*P(d[i]))}
```

Application of Bayes' formula assumes that the antecedents are statistically independent, an assumption which will probably be satisfied very seldom in real life applications. One way to meet this problem is to extend our knowledge base with the two rule sets to become the advanced knowledge base:

```
IF d[i]=TRUE is the expert conclusion to an incident                 (1.8)
THEN P(c[k]|d[i]) is the probability that the cluster c[k]=TRUE

IF d[i]=FALSE is the expert conclusion                               (1.9)
THEN P(c[j]|NOT d[i]) is the probability that the cluster c[j]=TRUE.
```

Obviously, obtaining reliable estimates for these rules will require both more intensive and extensive training. Assuming that the rules exist, only one cluster will be observed for each incident and the application of the following formula:

```
P(d[i]|c[k]):=P(c[j]|d[i])*P(d[i])/                                   (1.10)
       {P(c[j]|d[i])*P(d[i])+P(c[j]|NOT d[i])*(1-P(d[i]))}.
```

will meet the requirements of the Bayes' theorem in these respects.

2. Inference Models

The search for the solution of a task can be approached in several ways. These are presented as different inference models. Four of the models used in the project will be briefly described.

Model 1:

This model is almost identical to a model outlined by Naylor [NAYLOR 1983]. It uses the basic knowledge base (1.1)-(1.3). The prediction is computed using formulas (1.4) - (1.5). The strategy for selecting antecedents to be examined, is the rule value approach. According to this strategy, the antecedent which has maximum uncertainty reduction power is tested. The testing is stopped when the estimated probability for the most likely prediction will be the greatest even if the outcome of testing of all remaining antecedents should be in favour of other predictions. The model is rather complex and requires considerable computation.

Model 2:

Model 2 employes the same type of knowledge and prediction as Model 1. However, its selection strategy is related to the backward-chaining of ordinary rule based systems. It starts with the conclusion which is a priori the most probable, and continues to test antecedents which will support the prediction of this conclusion until another conclusion becomes more probable at which point the remaining antecedents supporting the second conclusion are tested. The search for antecedents to test continues until all antecedents related to the currently most probable conclusion are tested. This model is intuitively easier to understand and requires less computations than the previous model.

Model 4:

Model 4 is similar to Model 2 except for the computation of probabilities. Instead of making use of the rules (1.3) and the formulas (1.4) and (1.5), this model carries out the complete computation according to (1.6) and (1.7).

Model 5:

Model 5 applies the same selection strategy as Model 1. It makes use of cluster rules of the type (1.8) and (1.9), and bases the computation of the alternative prediction probabilities on the formula (1.10).

3. Design of Experiments

The main objectives for the experiments reported here were to investigate the following hypotheses:

1) The rule value approach of Model 1 is more efficient than the goal driven approach of Model 2.
2) The use of the complete Bayes' formula of Model 4 contributes to a higher quality in the results than the simpler formula of Model 2.
3) The ignorance of statistical dependency among the antecedents results in serious bias in the estimated prediction probabilities.
4) The introduction of the formula (1.10) in Model 5 will give probability estimates which can be used as reliability indicators.

To test these hypotheses, a set of simulation experiments was designed. For these experiments the following components were needed:

a. Two probabilistic knowledge bases
b. Two task files which also included the 'true' conclusions
c. Four implemented inference models
d. A statistical program

Each experiment included the processing of one task file by means of one implemented inference model and one knowledge base. Results for each incident were filed on one solution file for each experiment and included the predicted conclusion with estimated probability as well as other performance data, and with the 'true' conclusion transferred from the task file. Each solution file was be summarized by the statistical program.

4. Implementation of the Design

The knowledge base used was a modification of a demonstration base presented by Naylor [NAYLOR 1983]. Since the point was to have a knowledge base which could be considered representative for at least a certain class of domains, not to test the quality of its knowledge, the subject matter of the knowledge base was hidden. The basic base used had a set of 68 antecedents, 91 conclusions, 520 conditional probability rules for antecedents given the occurrence of a specific conclusion, and 520 conditional probability rules for antecedents given NOT the occurence of a specific conclusion.

Four different computer programs were developed for the experiments. The first program, GENERATOR, simulates real world generation of tasks within a specific domain. The simulation model used, was specified by the basic knowledge base. The program produces files of recorded random incidents including the 'true' solutions and antecedents according to the model used. GENERATOR was used to create two task files for the task solution program, and a training file for the training program. The first task file consisted of 400 recorded incidents with statistically independent antecedents, the second task file contained 376 recorded incidents with antecedents which were interrelated. The training file consisted of 2801 incidents with interrelated antecedents.

The second program, ESTIMATOR, simulates the training process of an expert when she is exposed to a training file. The program used the training file to produce the advanced knowledge base with cluster antecedent rules. The training resulted in identification of 518 different cluster antecedents, and 2*539 conditional probability rules for cluster antecedents.

The third program, PREDICTOR, included implementations of several different inference models. In each experiment, PREDICTOR, by means of the selected inference algorithm, predicted the conclusion for each incident in the task file used. When needed, it tested the task record for the occurence of antecedents, and recieved the answer YES if the ancedent was recorded in the respective incident record, and NO if not. When satisfied, PREDICTOR recorded in a solution file its prediction with probability, the numbers of antecedents tested and detected, and an index representing the computational work. The 'true' conclusion was also copied from the task file.

The solution files were processed by means of the fourth program, ANALYZER, which is a statistical program dedicated to processing of solution files from PREDICTOR. ANALYZER produced statistical summaries of each solution files.

The following experiments are reported here:

Experiment 1: Task file 1 processed by Model 1 algorithm.
Experiment 2: Task file 1 processed by Model 2 algorithm.
Experiment 4: Task file 1 processed by Model 4 algorithm.
Experiment 5: Task file 2 processed by Model 1 algorithm.
Experiment 6: Task file 2 processed by Model 5 algorithm.

5. Analysis and Conclusions

The output from the five experiments are summarized and described in the following table.

TABLE. Summary of results from selected simulations.

	Experiment:				
	1	2	4	5	6
Pct. of correctly predicted conclusions	91.8	74.0	86.3	92.6	78.2
Avg. probability for correct predictions	0.96	0.95	0.60	0.97	0.97
Avg. probability for incorret predictions	0.89	0.96	0.31	0.92	0.21
Avg. number of tested antecedents	39.9	10.4	40.8	38.9	38.9
Index of computation load	52.6	1.5	3.3	51.8	54.6

The table shows that the Model 1 rule value algorithm produced about 25% more correct predictions than the simpler Model 2. Model 1 on the other hand, was much more expensive requiring testing of about 4 times as many antecedents and 35 times the computational resource compared with Model 2.

Results from Experiment 4 indicate that the use of the complete Bayes' formulas had three effects. The number of correct predictions increased, a marked difference between the average probabilities of correct and incorrect predictions was introduced, and the costs increased compared with the results from Experiment 2.

Finally, experiments 5 and 6 were performed with Model 1 and Model 5 on Task file 2 with interrelated antecedents. The results indicate that Model 1 found even more correct predictions in this task file than in Task file 1. It also gave higher probabilities both for correct and incorrect predictions indicating that the probabilities have minimal value for guiding the user. Model 5, however, signalled strong differences between probabilities for correctly and incorrectly predicted conclusions. Experiment 6 results indicate that an inference algorithm which takes statistical dependencies among the antecedents into account, can give probabilities useful for evaluating the reliability of predictions.

The results can be summarized as:

1) The use of the rule value approach and the complete Bayes' formulas, both contribute to a higher percentage of correct prediction than the goal driven approach and the simpler Bayes' formulas,

2) The estimated probabilities are useless for indicating the reliability of the prediction if statistical dependency exists among the antecedents and is ignored in the inference algorithm,

3) If rules with conditional probabilities for clusters of antecedents are available, inference models can be designed which give prediction probability estimates useful for evaluating prediction reliability also when statistical dependencies among the antecedents are present.

6. Acknowledgement

The research reported in this presentation was partly done during a stay as a visiting faculty member with the University of Hawaii, Department of Information and Computer Sciences, which provided a professional environment. The visit was made possible by the financial support from the University of Bergen, the Royal Norwegian Council for Research and the Norwegian Research Council for Social Sciences. I am grateful to all these organizations which made it possible for me to carry out the work reported in this paper.

I also want to thank Associate Professor Joan C. Nordbotten, who as usual has been my chief critic. I am also grateful for discussions with Dr. Jiri Ivanek who made some constructive comments to my ideas on inference models at an early stage.

7. References

Naylor, C.: BUILD YOUR OWN EXPERT SYSTEM, Halsted Press, John Wiley and Sons, NY.1983.

Naylor, C.: HOW TO BUILD AN INFERENCE ENGINE, in Forsyth (ed.): Expert Systems-Principles and Case Studies, Chapman and Hall, London 1984.

Statistical Consultants and Statistical Expert Systems

I. W. Molenaar, Groningen

SUMMARY

This paper contains a critical evaluation of the value of expert systems for statistical consultation. Section 1 presents the steps that should ideally be taken when a researcher consults an expert statistician. Next statistical expert systems are considered. It is concluded that they can be fruitfully used for enhancement of the knowledge of the statistician, but that a researcher who consults only a computerized system rather than a human statistician will obtain suboptimal help. The technical knowledge of the statistician may perhaps be incorporated into a knowledge base, butnot the subtle dialogue of client and statistician, with its interplay of substantive and statistical considerations.

1. Statistical consultants

Statistical models are a tool for drawing valid and generalizable conclusions from data. The most popular models are based on assumptions like random sampling, normal distributions and homoscedasticity. Within this framework they lead to decisions with quantified risks about null hypotheses specified a priori, and to optimal parameter estimates with known sampling distributions. There are also more informal and exploratory methods of data analysis, more flexible as regards goal directedness and assumptions, in which showing interesting aspects of the data has priority compared to the assessment of formal claims of generalizability.

The difference is formulated by some as the question whether the data or the model must have priority. I believe that both model and data should yield priority to the substantive research goal, and to the skills and prior knowledge of the researcher. In a good research project they are the base for a first dialogue between the researcher and a statistical consultant. Taking available resources into account, this dialogue should lead to a draft research design in which well considered decisions are made about the choice of the units of analysis, the choice of variables to be included, of measurement procedures, of assignment to treatments, etc.

Next in this dialogue, and preceding the actual data collection, comes a tentative answer to the question what such data will look like. The combined skills of substantive expert and statistical expert should enable them to foresee certain problems, e.g. missing values, skewed distributions, violations of independence, an instruction misunderstood by the experimental subjects or the data collectors. Common sense, personal experience, literature search and pilot study results will produce useful items for the agenda of this discussion. It will permit timely changes in the research design, and should lead to the formulation of a draft model and a draft for the data analysis.

Once the data have been collected, some surprises may still appear. They may lead to decisions to change either the model or the data. There should be both substantive and statistical aspects to decisions like to

Compstat 1988

reject an outlying observation, to transform a variable, to choose a different significance test or to add a parameter to the model.

One then comes to the stages of parameter estimation, testing of hypotheses, assessment of model fit. This may lead to cycling back to the formulation and assessment of an improved model, but one should take into account the statistical risks of basing such an improvement decision on properties of the sample that do not hold for the population.

Finally comes the stage of reporting the conclusions. I believe that they are seldom interesting if they only hold for the actual data. The generalizations that one can validly make are based on both statistical and substantive grounds : the two spans of the bridge of inference, as Cornfield and Tukey say (1956, p. 912). The report should make clear the extent to which generalizations are possible. This involves more than the ritual of stars or P-values : sources of variation and sources of possible bias should be considered, and often also robustness issues.

Exploratory studies and exploratory data analyses have a modest place in this idealized description of the research process. They are important in pilot studies, and in assessing the extent to which model violations occur that might seriously invalidate the conclusions. Publication of results obtained in a purely exploratory manner is better delayed until a more formal cross validation study has shown their persistence.

This outline of the dialogue between a researcher and a statistician, worked out more fully in Molenaar (1988), gives a personal view on the optimal functioning of statistical methodology in a research setting. In many other settings a less intensive procedure suffices, e.g. because the current analysis is based on data already collected, or is part of a series of similar analyses for which the fundamental points have already been discussed. There are also cases in which generalization of the results is a minor concern; I believe that they are exceptions, however. At any rate the two major issues of this section remain upright : do not postpone the hard thinking that should precede the data collection, and take all important decisions after an intensive dialogue in which substantive knowledge and statistical expertise both play a major role.

2. Statistica expert systems

This section combines our view on statistical consultation with the results of expert system (ES) technology, and evaluates the role of intelligent computerized systems in applying statistical methods. An ES is a clear case of a "mixed initiative dialogue" : sometimes it acts like a "user guided system", where the user has the initiative and the system responds by giving information or carrying out operations, and sometimes like a "system guided program" in which the system decides what to do, with the user in the passive role of only answering when asked to do so.

A scan of some three dozen recent publications on the subject reveals that many of them have the form of an agenda for future action. I have found no reports on a statistical ES in actual use. This somewhat alarming state of affairs seems to prevail also for ES in other areas. Perhaps some of the effort now devoted to defining and building the systems should be diverted to studying the population of intended users and their needs. Again we need behavioral studies of the software user (Molenaar, 1984).

When a good ES would be available, the dialogue of section 1 may become a quadrilogue, with information exchanged between the client, the statistician, the ES and the data analysis system. Some people may argue that there will be only three parties involved, because the two electronic components should be powerful enough to wholly eliminate the statistician. That latter view is in line with a general goal in ES methodology of using expert knowledge where the human expert is not available. One also finds the view that the use of an ES could help to diminish the current abuse of statistical packages by uninitiated users, pointing again to a user with limited statistical knowledge interacting with a computer without consulting a statistician. Some interesting views on the use of ES in statistics can be found in Victor (1984) and Hahn (1985).

The possibility that the ES interfaces with a data analysis package is a special feature of its application to statistics. In other domains the ES relies purely on the user for gathering and providing information on the current problem. A first and primitive type of a statistical ES is a preprocessing program that interrogates the user and prepares the input for the analysis program. In more advanced cases, the ES not only prepares commands for the package, but interprets its output, shows it in suitable format to the user, and perhaps issues new commands based on that output (this is how REX interacts with the S package, see Gale 1986a Ch.8).

When the expert system has access to the data, it may be programmed to detect anomalies or violations of assumptions, and to offer suggestions how to deal with them. Our view that substantive and statistical issues should both play an essential role in all stages of the analysis, leads the idea that here an advisory role only should be assigned to the system. When there is e.g. a choice between a logarithmic, a square root or cube root transformation or no transformation, not only the observed distribution of the variable should be considered, but also earlier outcomes for the same variable, substantive theory about relations to other variables, interpretability of the results and the goal of the investigation.

Another interesting use of an ES coupled with a package is the audit file (Becker & Chambers, 1986). This contains a record of the operations and transformations carried out on the data, including calling and saving of files. By consulting this audit file, one may detect both unintended mistakes and purposive data tampering.

In the quadrilogue, the ES may act as an "electronic handbook" for enhancing knowledge. This requires specialization to a small subfield of statistics (like KENS for nonparametric statistics, see Hand 1987). From the size of the Encyclopedia of Statistics now published by Wiley, one can imagine the gigantic database required to cover all areas of statistics!

The electronic handbook can be consulted by the statistician or by a user with little knowledge of statistics; in the latter case the body of knowledge required is far smaller, and more standard, but there will be more serious terminological and pedagogical problems. (see below).

An electronic handbook has both advantages and disadvantages when compared to a physical one. Locating the relevant information is probably easier than the use of the table of contents or subject index of a book.

The ES could help the user, both in the search and in the verification whether the current passage is indeed the most relevant one, by asking some pertinent questions. A short answer could first be displayed, from which difficult words or cryptic sentences could be explained on request. Reading from the screen and electronic browsing, however, is less pleasant than using a book. When the first microfiche reading instrument was placed in the library of the Mathematics Department in Amsterdam, around 1955, many predicted that the long shelves of books would be replaced by a small microfiche cabinet within a few years. Books did not disappear, however, because microfiche reading was too clumsy. Will the same hold for electronic textbooks, or even for expert systems?

Another potential use of a statistical ES is the stage of finding the best research design for a substantive research question. Here one would need quite massive substantive knowledge, both about the research domain and about the available resources, and this would be a kind of knowledge for which formalization appears rather difficult. This application only looks promising when many consecutive analyses are small variations on a common theme (see e.g. Jockel, 1986).

The choice of the most appropriate statistical test for a given problem is another application. There have been attempts to formalize the choice of a statistical model and test into a tree-like structure of questions and answers. The computerized system STATCONS (De Greef et al., 1985) is intended to incorporate both the design and the data analysis stage, but it has not reached the stage of actual use (De Greef, oral comm. 1988). There have been other attempts as well to find the best statistical technique via a computerized dialogue. If the terminology used in the questions would be fully familiar to the user, this would be fine.

This brings me to what I believe to be the most fundamental obstacle in the use of statistical expert systems by non-statisticians. A help facility may be incorporated that explains the terminology. I doubt, however, whether such a verbal explanation is sufficient to let the novice user provide the correct answers to even simple questions like:

- does variable X have ordinal measurement level?
- do you have one sample of paired values or two independent samples?

When the KENS system (Hand, 1987) is in HELP mode and one enters "ordinal", about a dozen frames are mentioned. One of them is "definition of ordinal variables", and it has a clear explanation in 15 lines of text about the ordinal measurement level. For the second question, the novice fares less well. There are KENS frames in which paired and independent observations are used, but none of them explains the terminology. The keywords "repeated measures" and "correlated samples" elicit the answer "no matches", meaning that the system cannot find them in its search lists. A totally ignorant novice, and even one who knows that his/ her data are "two repeated measures of an ordinal variable" would not find the way to the frames about the matched pairs tests of Wilcoxon, Walsh and Fisher, unless familiar with the "matched pairs" terminology. I do not think that this terminology obstacle is specific to KENS.

One may argue that such users are too ignorant to do statistical analysis at all, or at least to do it without the guidance of a more knowledgeable person. I fully agree, but when we are concerned about preventing misuse of statistical packages, and claim that expert systems can help to achieve this, we should indeed be thinking of very ignorant users, and particularly of the dangerous subclass that is unaware of their own ignorance. When the consultant is a person, the degree of ignorance can be assessed, and the questions can be adapted, but when it is a computerized system it is hard to see exactly what terminology or what examples should be used in the explanation. The ES usually cannot decide whether the current data for two groups can be viewed as independent observations without knowing how they were obtained. A human consultant, even one with a modest knowledge of statistics, is more able to find this out than even the most sophisticated expert system.

Misuse of statistical packages occurs not always out of ignorance : some do it purposively. They are in a hurry, they need to tell their boss a simple story, they avoid consulting a professional statistician out of fear for what that expert will say. Will this type of mis-user feel inclined to consult an expert system, and benefit from doing so ?

3. Conclusions

The proliferation of personal computers and statistical packages has led to a strongly increased probability of the misuse of statistical methods. Moreover, the progress of statistical knowledge has led to many advanced models and estimation methods, for which the danger of misuse is far greater than for simple t-tests or sign tests. Note that misuse of statistics is rarely punished, and often goes unnoticed.

Intelligent computerized systems can certainly help to mitigate the problems, and the current research on statistical expert systems opens up some promising avenues. More energy should be devoted, however, to two major areas that seem to receive almost no attention up to now. The first is the study of what actually goes on during statistical consultation, and what kinds of statistical sins are frequently committed by those who do their analyses without expert help. The second is the systematic study of (prototypes of) expert systems being used by the groups for which they are intended, and their improvement based on this observation. The ES research has produced many challenging plans and ingenious formalizations for the notoriously difficult area of statistical consultation, but there is a tendency to go on announcing new and still better products rather than bring the present ones to a stage where they could be sold and used.
My intention when I began working on this paper was to use several statistical ES myself and report on my experiences, but of the ten authors I wrote to, only Hand was able to send me a floppy that worked.

Although this review paper has a skeptic undertone, I believe that ES use in statistics could be helpful. This would require that the ambitious goal of designing a system with which any fool could successfully handle any statistical problem, is traded in for the modest goal of designing a system that would enable a moderately experienced statistical consultant to play a more effective role in his/her quadrilogue with the user, the expert system, and the data analysis package.

References

Becker, R.A., & Chambers, J.M. (1986). Auditing of data analyses. In : Proc. of the 19th Symposium on the Interface, NorthHolland, 11-18.

COMPSTAT 1984 (T.Havranek, Z.Sidak & M.Novak, eds.). Heidelberg : Physica Verlag, notably p.227-236, 242-248.

COMPSTAT 1986 (F.de Antoni, N.Lauro & A.Rizzi, eds). Heidelberg : Physica Verlag, notably p.249-306.

Cornfield, J. & Tukey, J.W. (1956). Average values of mean squared in factorials. Annals of Math. Statistics 27, 907-949.

De Greef, P., et al. (1985) A rule based system for statistical consultation (in Dutch). In: SSS 1985 Proc. Symposium Stat. Software, Debets et al. (Eds.), Technisch Centrum, Universiteit van Amsterdam.

Gale, W.A. (ed.) (1986a). Artificial intelligence and statistics. Reading Addison Wesley.

Gale, W.A. (1986b). Student : a tool for constructing consultation systems in data analysis. Invited paper 15.1, 46 session of the International Statistical Institute.

Gale, W.A. (1988). Applications of artificial intelligence in statistics. to appear in : C.R.Rao (ed.), Handbook of Statistics, vol.9.

Gale, W.A. & Lubinsky, D.J. (1986). A comparison of representations for statistical strategies. In : Proc.of the 19th Symposium on the Interface, 88-96.

Hahn, G.J. (1985). More intelligent statistical software and statistical expert systems. The American Statistician 39, 1-16 (with discussion).

Hand, D.J. (1984). Statistical expert systems : design. The Statistician 33, 351-369.

Hand, D.J. (1985). Choice of statistical technique. IASC Selected papers 45-th ISI Session, p. 285-300.

Hand, D.J. (1985). Statistical expert systems : necessary attributes. Journal of Applied Statistics 12, 19-27.

Hand, D.J. (1987). A statistical knowledge enhancement system. J.Royal Stat. Society, A 150, 334-345.

Haux, R. (ed.) (1986). Expert systems in statistics. Stuttgart : Fischer.

Jockel,K.-H. (1986). Statistical expert systems and the statistical consultant. Considerations about the planning stage of clinical studies. In : Haux (1986; see above), 27-43.

Molenaar, I.W. (1984). Behavioral studies of the software user. Computational Statistics and Data Analysis 2, 1-12.

Molenaar, I.W. (1988) Formal statistics and informal data analysis, or why laziness should be discouraged. To appear in Statistica Neerlandica 42.

Pregibon, D. & Gale, W.A. (1984). REX : an expert system for regression analysis. In : COMPSTAT 1984, p.242-248.

Victor, N. (1984). Computational Statistics - tool or science ? Statistical Software Newsletter 10 (3), 105-125.

On Inference Process

Th. Westerhoff, Paderborn and P. Naeve, Bielefeld

1. Motivation

"An Expert system is a computing system capable of representing and reasoning about some knowledge-rich domain ... with a view to solving problems and giving advice." Jackson, p 1, (1986). Building such a system seems to be a manageable task considering the many expert system shells which are now available.

But anyone who is about to start building an expert system is well advised to take a word of warning, otherwise he migth find himself caught in a trap of undue problem reduction. Surely nobody willingly would accept the following definition of *intelligent*: "able to perform computer functions" Webster (1983). But the user of expert system shells could end up implicitly doing just this. The reasons for this are twofold. There are almost no papers describing the "inference process" of an expert in a way one could take as a design for building the inference component of a shell. On the other hand, expert system shells do not make their inference mechanism as explicit as one would need when modelling actual expert behaviour.

If one defines

Expert = Expert modeled in shell-elements PLUS remainder

it will turn out far too often that contrary to ones assumptions the remainder is neither small nor in any way controllable. So either for selecting a shell or for creating a new one a list of indispensable requirements is needed which when met by the shell allow for successful modelling of an expert's inference process.

This task was undertaken for the special case of a statistical expert. As an initial description language the terminology of the EMYCIN-paradigm was used to describe the expert's inference process.

The notation of "expert" used in this paper is embodid in several different sources of experience: the reflections and intuitions of a statistician (one of the author), observation by a

Compstat 1988

knowledge engineer (the other author), exploitation of similar attempts reported in the literature (especially the work of D.J. Hand) and last, but not least, the huge stock of statistical literatur revealing statistical expertise.

2. Requirements on an expert system shell

The result of the analysis of a "statistical" inference process can be divided into three main topics which are the headings of the following paragraphs.

2.1 Requirements of formalisms for representing knowledge and their use

In order to structure his knowledge, the statistician needs objects and rules. But what is characteristic of statistics is their dynamic use. Whereas the definition of objects and rules is always static, a statistician changes the ranges of values or even the values of a feature of an object in the course of consultation. Furthermore, he somtimes uses a rule for induction (a forward rule) and sometimes for deduction (a backward rule).

The dynamically changeable view of a statically defined (meta-) object is probably most specific to statistics. For example, a statistical expert changes his view of a statistical variable i.e. he adds or deletes attributes dependent on his current viewpoint. We suggest therefore, the following shell specifications for statistical expert systems:

1. The shell should admit objects and rules as knowledge representation formalisms.
2. It should be possible to change characteristics of objects (values, ranges of values) dynamically.
3. An object should allow for a dynamically changeable view of a static meta-object.
4. It should be possible to chain rules forwardly and/or backwardly.

2.2 Requirements for the inference engine

The different knowledge representation formalisms are often not crucial for the usability of an expert system shell; more crucial are the type and complexity of problem solving methods which can be implemented by it. The flexible use of inference processes is one of the main problems for expert system shells. Shells which were derived from existing expert systems by eliminating domain specific knowledge, inherit their problem solving mechanism from the domain-specific problem solving process. Therefore shells are often only usable for a reduced class of problems.

In the area of statistics one has to achieve a dynamic and flexible behaviour via the shell; there is no single predefined problem solving process for a statistical consultation. So we establish the following requirements:

5. The inference process should be controllable in such a way that any strategy can be modelled.
6. Deduced facts should be subject to deletion and revision, i.e. monotonic and non-monotonic reasoning and temporary inconsistencies should be possible.
7. The user, too, should be dynamically able to influence the process of inference.

2.3 Requirement for a facility for hypothetical reasoning

One of the most important requirements in representing statistical inference is the possibility of pursuing more than one goal at a time and drawing arguments by analogy. The following conclusions are drawn:

8. There should be possibilities to implement different worlds in a system.
9. Derived results of a world should be accessable in this world itself and in all its offspring worlds. But they may not influence parallel worlds.
10. It should be possible to jump from one world to a parallel one (sideward chaining). Therefore results of one world should be transferable to a parallel world.

3. Satisfaction of the requirements using two TWAICE-generations

In order to examine how existing expert system shells satisfy the given requirements, we looked into two "generations" of the Nixdorf expert system shell TWAICE, namely TWAICE release 2.5 and

TWAICE release 3.0. In the following, we only present features of TWAICE related to the above topics. Detailed descriptions of TWAICE can be found in Mescheder (1988) for TWAICE 2.5 and Schmitt (1988) for TWAICE 3.0. All statements about TWAICE not qualified by a release number are valid for both releases.

3.1 Knowledge representation formalisms and their use in TWAICE

In TWAICE, the taxonomy is used to structure the knowledge base by frame-based definitions of objects and attributes. Objects describe general types of topics (e.g. statistical variabale, statistical method). During a consultation certain individual objects are investigated which are called instances (e.g. the statistical variable "year"). Attributes describe features of objects and values describe their characteristics.

Rules formulate logical dependencies between the attributes of a domain. They are formulated for objects and can therefore be used repeatedly for different instances. There are backward and forward rules in TWAICE.

The knowledge sources of TWAICE, besides the obligatory rules and taxonomy, are optional tables, terms, procedures, and several user texts. In TWAICE 3.0 there are also optional facts, defaults, and databases.

As knowledge is mainly represented in objects and rules, the first requirement is fulfulled by TWAICE to the extent that inheritance of attributes between objects is not possible.

In TWAICE 2.5 all definitions of the taxonomy and rule base are static. Therefore this generation of TWAICE did not satisfy the dynamic requirements.

This is different in TWAICE 3.0, however. Independent of the static definition in the rule base, a rule can be chained forward and/or backward. Furthermore one can change the range of value of an attribute dynamically. So two of the three requirements of "dynamism" are fulfilled by TWAICE 3.0.

3.2 The inference strategy of TWAICE

The inference engine of TWAICE controlls the consultation and deduces results.

Its mode of work was iron-clad in TWAICE 2.5 and primarily used backward chaining. The predifined inference method uses monotonic logic. Furthermore it uses the principle of exhaustive evidence gathering. This means that in order to determine values of an attribute all relevant knowledge sources are taken into account.

A predifined inference strategy prevents one from influencing the course of problem solving dynamically. A particular strategy can only be reached by clever "programming by rules". This often requires a lot of awkward work for the knowledge engineer which prevents a "natural" modelling of the statistical expert.

TWAICE 2.5 was therefore unable to satisfy our requirements except in a primitive manner.

One of the most inportant differences between TWAICE 2.5 and TWAICE 3.0 is the opening of the inference component in TWAICE 3.0. Because of this opening a knowledge engineer can influence the inference process directly and modify all results of the dynamic knowledge base. This means that derived results or instances of objects may be deleted. Therefore non-monotonic reasoning and temporal inconstistencies are allowed in TWAICE 3.0. In order to influence the inference process directly a knowledge engineer is able to define his own special commands.

3.3 Hypothetical reasoning in TWAICE

In TWAICE 2.5 there is no concept analogous to a "world". Therefore hypothetical reasoning in TWAICE 2.5 is not possible.

In TWAICE 3.0 there exists the concept of "situations". Through situations one can partition the dynamic database into different worlds. TWAICE 3.0 manages a graph of these situations. Therefore situations in different branches of the graph can be independent while there are connections in the branches themselves. In particular, all results of a situation are inherited by its offspring unless explicitly deleted or changed.

The concept of "situations" in TWAICE 3.0 satisfies the requirements for hypothetical reasoning.

4. Summary

The representation of "statistical" inference requires dynamic behaviour by the main components of an expert system shell: the data structures for knowledge representation and the inference component. Furthermore the possibility for hypothetical reasoning and reasoning by analogy is mandatory.

Like probably every other expert system shell with predefined problem solving method, release 2.5 of TWAICE seemed not to be suitable for simulating statistical inference. The evolution of this shell to TWAICE 3.0 shows that by opening the inference component of the shell the direct representation of statistical inference becomes possible.

References:

Hand, D.J., Statistical expert system: necessary attributes, Journal of applied Statistics, vol 12, 1984

Hand, D.J., Statistical expert system: design, The Statistician, vol 33, 1984

Hand, D.J., Choice of statistical technique, ISI Bulletin of the International Statistical Institute, Proceedings of the 45th Session, vol LI/book 3, Amsterdam, 1985

Hand, D.J., Patterns in statistical strategy: in Artificial Intelligence & Statistics, Gale W.A. (ed), Reading, 1986

Jackson, P., Introduction to Expert Systems, Addison-Wesley, Wokingham, 1986

Mescheder, B., Functionality and operational features of the expert system shell TWAICE in: Savory, S. (ed): Artificial Intelligence and Expert Systems, Chichester, Ellis Horwood Ltd., 1988, 60-91

Schmitt, F., Knowledge Engineering mit einer Expertensystem-Shell, NCAG, 1987, to appear in KIFS, 1987

Webster's Ninth New Collegiate Dictionary, Merriam Webster, Springfield, 1983

Identification Keys, Diagnostic Tables and Expert Systems

R. W. Payne, Harpenden

SUMMARY

There are many parallels between the methodology of identification keys and diagnostic tables, and the new methodology of expert systems. For example, the standard identification key identifies specimens from a known set of taxa by applying tests sequentially in a hierarchical manner; this structure is identical to that used in the many expert systems that have a simple deterministic hierarchy of questions leading to a conclusion - which in a statistical context might take the form of recommending some form of analysis. Likewise the systems where the conclusion is determined by comparing an observed set of conditions against a theoretical set of rules use a similar method to that employed by the user of a diagnostic table. Even the expert systems where there is a network of nodes of menus or questions have their parallels in the on-line identification systems that have been developed for botanical and other biological work, and some of these on-line systems allow the user to modify the data base of taxonomic information to take account of new information about the taxa, so that the system is able to learn with experience. Consequently work on methods of constructing efficient keys and tables is very relevant also to expert systems.

Keywords: diagnostic tables; expert systems; identification keys.

1. INTRODUCTION

Payne & Preece (1980), in their review of identification keys and diagnostic tables, wrote

> "The fields of application for the various techniques are thus many and varied. The techniques are related to other methodologies developed independently for different purposes. Consequently the literature for identification and diagnosis is large and widely dispersed. It is not properly integrated, in that workers in one field of application have often shown themselves

Compstat 1988

> unaware of relevant theoretical developments made in other fields."

Much recent work on identification systems has been in the guise of expert systems, further dispersing the literature and, once again, practitioners seem unaware of the existing work. For example, the preface to one of the plethora of expert systems for IBM-PC contains the statement (d'Agapeyeff 1986).

> "Know-how Programming can be a surprise for both novices and professionals in computing. This despite the fact that it, or something very like it, has been practised in UK companies, like Unilever and ICI, since about 1981."

This paper briefly reviews the available methodology, paying particular attention to its applicability to expert systems.

2. IDENTIFICATION KEYS

Table 1 shows a key to the 13 yeasts most commonly isolated clinically, taken from Barnett, Payne & Yarrow (1983). Using the key, an unknown specimen would first be checked to see whether it grows with D-Glucuronate; if a positive result is obtained, the next test is that at index 10, Nitrate growth, with which a further positive result would identify *Cryptococcus albidus*.

The form in Table 1 is just one of many ways in which a key can be represented (see for example Payne, Walton & Barnett 1974), but in all of them there is the same inherent structure of a deterministic tree with a test at each node and identification being made at the endpoints. If it were stored as a computer program, with the answers solicited by menus at each node, it would probably be called an expert system!

After using the key, it would be desirable to check the identification against a more detailed description. Sometimes this is essential; for example the physiological tests used in the key in Table 1 cannot give complete identification at indices 3 and 11. In Barnett, Payne & Yarrow (1983) relevant descriptions, with photographs, are on pages 61-537. With a computerised key, the information would probably be accessed from a data base.

Many of the expert systems that have the structure of a deterministic tree seem to be designed by hand - as was done for identification keys before there were computers; there seems little attempt to optimise the tree, for example to minimise the expected number of questions required to reach a

	Negative	Positive
1 D-Glucuronate growth...	2	10
2 Raffinose growth	3	7
3 Starch growth..........	4	.........*Candida albicans*
		Candida tropicalis
4 Nitrate growth	5	.....*Rhodotorula glutinis*
5 Growth without Thiamin.	.........*Candida glabrata*	6
6 DL-Lactate growth......	.....*Candida parapsilosis*	..*Issatchenkia orientalis*
7 Nitrate growth.........	8	.....*Rhodotorula glutinis*
8 Urea hydrolysis........	9	.*Rhodotorula mucilaginosa*
9 DL-Lactate growth......	....*Pichia guilliermondii*	..*Kluyveromyces marxianus*
10 Nitrate growth.........	11	.....*Cryptococcus albidus*
11 Lactose growth.........	*Filobasidiella neoformans*	...*Cryptococcus laurentii*
		Trichosporon beigelii

Table 1 Key to the yeasts most commonly isolated clinically

conclusion (or identification). This optimisation has been a major concern of those interested in computer construction of keys (see, for example, Gower & Payne 1975, Payne 1975, Payne 1978, Payne & Preece 1980, Payne 1981, and Payne & Dixon 1984).

The basic data for constructing a key by computer is a table listing the responses that each of the taxa may give to each of the available tests. Table 2 shows the responses of the yeasts in Table 1 to the tests used in that key. This example may seem rather small to justify use of the computer. However, these are only 13 of the 469 yeasts and 8 of the 97 tests described in Barnett, Payne & Yarrow (1983). With construction by hand, there is the temptation to collect only the information that seems directly relevant to the key. For example, in Table 1 the responses to DL-Lactate growth are required only for the 4 yeasts that pass through indices 6 and 9. However, the availability of the extra data allows the identification to be checked, and facilitates the more flexible methods described in Section 4. Collection of the data also draws attention to deficiencies in current knowledge, providing a spur to research; however, methods are available to take special account of responses that are unknown if these are unavoidable (Payne 1981). The methods of construction can also cope with tests that are inappropriate to some of the taxa; for example tests that involve shapes of ascospores are not relevant to yeasts that do not reproduce in that way. When generated from a collection of logical rules, the data table can also draw attention to any unwarranted assumptions.

Responses to each test

1	2	3	4	5	6	7	8	Taxon
-	-	+	-	+	-	+	-	*Candida albicans*
-	-	-	-	v	-	-	-	*Candida glabrata*
-	-	-	-	-	-	+	-	*Candida parapsilosis*
-	-	+	-	v	-	+	-	*Candida tropicalis*
v	v	v	+	v	+	-	+	*Cryptococcus albidus*
+	+	v	+	v	-	v	+	*Cryptococcus laurentii*
-	v	+	+	v	-	-	+	*Filobasidiella neoformans*
-	-	-	-	+	-	+	-	*Issatchenkia orientalis*
v	+	-	-	+	-	v	-	*Kluyveromyces marxianus*
-	+	-	-	-	-	+	-	*Pichia guilliermondii*
-	v	-	-	v	+	v	+	*Rhodotorula glutinis*
-	+	-	-	v	-	-	+	*Rhodotorula mucilaginosa*
+	v	v	+	+	-	-	+	*Trichosporon beigelii*

Tests in each column

1 Lactose growth
2 Raffinose growth
3 Starch growth
4 D-Glucuronate growth
5 DL-Lactate growth
6 Nitrate growth
7 Growth without Thiamin
8 Urea hydrolysis

Table 2 Data used to construct the key in Table 1
(+ positive, - negative, v variable)

3. DIAGNOSTIC TABLES

The sequential testing used in an identification key is most appropriate in applications such as botanical identification where each test takes only a short time to observe. Physiological tests for the yeasts take up to 14 days, and so it is usual to do all the required tests simultaneously. Thus, if a specimen was known to be one of the common clinical yeasts, all the 8 tests in Table 2 would be done, and the key would merely be used to look up the identification after their results had been obtained. A diagnostic table provides an alternative means of looking up the identification. This is a table with rows and columns much like Table 1 except that the rows are arranged so that the rows of +'s and -'s are in lexiocographic order. Taxa with variable responses may then need to occur several times, for example the first row in a diagnostic table for the common clinical yeasts would be

- - - - - - - - *Candida glabrata*

later there will be a row

- - - - + - - - *Candida glabrata*

A diagnostic table may then be rather longer than a key to the same taxa, for example a diagnostic table to the common clinical yeasts takes 45 lines; however, it contains more information.

Whichever method is used to look up the identification in non-sequential testing, the appropriate measure of the efficiency of the process is the number (or cost) of the different tests in the key or table. The set of tests of minimum size or with minimum total cost are both irredundant, that is they are sets such that if any component test is deleted there will be some pair of taxa that can no longer be distinguished. Methods for constructing irredundant test sets are reviewed by Payne & Preece (1980, Section 6), and extended by Barnett, Payne & Yarrow (1983).

4. ON-LINE IDENTIFICATION SYSTEMS

Identification keys and diagnostic tables offer little choice over the tests used for an identification. However, the methods of test selection used to construct them have been incorporated into computer based identification systems (see Payne & Preece 1980, Section 4.5). For example, the Yeast Identification Program of Barnett, Payne & Yarrow (1985, 1987) allows the user to enter any set of test results that have been observed. It can then be asked to list all the yeasts that can give these results, then those whose results differ for one test, two tests, and so on. It can also suggest an irredundant set of tests to complete the identification, allowing the user control over which tests are included. Alternatively it can print descriptions of any specified yeast. The user can add further yeasts to the data base, so the program has the ability to learn, like a true expert system.

5. ALLOWING FOR ERRORS

Methods for including extra tests in a key to allow errors in testing to be detected and corrected have been devised by Payne & Preece (1977). Irredundant test sets can also be extended to guard against errors (see Payne & Preece 1980, p.290). These two methods are still deterministic. However, probabilistic identification schemes have also been devised (Payne & Preece 1980, Section 10), generally using Bayes theorem to obtain posterior probabilities for taxa, given the test results observed. Bayesian methods are also used by expert systems - as are methods for combining probabilities that do not conform to the axioms of probability, and which accordingly have been criticised for example by Spiegelhalter & Knill-Jones (1984).

6. CONCLUSION

The discussion above indicates the relevance of the methodology of identification keys and diagnostic tables to expert systems. The references listed below, particularly the review of Payne & Preece (1980), should allow the relevant literature to be traced.

REFERENCES

Barnett, J.A., Payne, R.W. & Yarrow, D. (1983). *Yeasts: characteristics and identification.* Cambridge: Cambridge University Press.

Barnett, J.A., Payne, R.W. & Yarrow, D. (1985). *Yeast identification program.* Cambridge: Cambridge University Press (Cambridge Micro Software).

Barnett, J.A., Payne, R.W. & Yarrow, D. (1987). *Yeast identification PC program.* Norwich: J.A. Barnett.

d'Agpeyeff, A.(1986). *A guide to know-how programming.* Slough: Expertech Ltd.

Gower, J.C. & Payne, R.W. (1975). A comparison of different criteria for selecting binary tests in diagnostic keys. *Biometrika,* 62, 665-671.

Payne, R.W., Walton, E. & Barnett, J.A. (1974). A new way of representing diagnostic keys. *Journal of General Microbiology,* 83, 414-414.

Payne, R.W. (1975). Genkey: a program for constructing diagnostic keys. In: *Biological Identification with Computers: Systematics Association Special Vol. No. 7,* (ed. R.J. Pankhurst), pp. 65-72. London: Academic Press.

Payne, R.W. & Preece, D.A. (1977). Incorporating checks against observer error into identification keys. *New Phytologist,* 79, 201-207.

Payne, R.W. (1978). *GENKEY: A program for constructing and printing identification keys and diagnostic tables.* Harpenden: Rothamsted Experimental Station.

Payne, R.W. & Preece, D.A. (1980). Identification keys and diagnostic tables: a review (with discussion). *Journal of the Royal Statistical Society, Series A,* 143, 253-292.

Payne, R.W. (1981). Selection criteria for the construction of efficient diagnostic keys. *Journal of Statistical Planning and Inference,* 5, 27-36.

Payne, R.W. & Dixon, T.J. (1984). A study of selection criteria for constructing identification keys. In: *COMPSTAT 1984: Proceedings in Computational Statistics,* (ed. T. Havranek, Z. Sidak & M. Novak), 148-153. Vienna: Physica-Verlag.

Spiegelhalter, D.J. & Knill-Jones, R.P. (1984) Statistical and knowledge-based approaches to clinical decision-support systems, withan application to gastroenterology (with discussion). *Journal of the Royal Statisical Society* 147, 35-77.

Languages and Packages

Adding New Statistical Techniques to Standard Software Systems: A Review

C. Payne, Oxford, N. Cleave and P. Brown, Liverpool

SUMMARY: The paper gives an informal review of the facilities available for adding new statistical methods to the major statistical systems. Evaluation is based on the capacity of these systems to incorporate new techniques developed for the problem of ecological inference.

KEYWORDS: Statistical systems, software review, ecological inference.

1. INTRODUCTION

Applied statisticians developing new techniques are faced with the problem of making their techniques generally available to the user community. A traditional approach has been to provide standalone programs, usually written in FORTRAN, to interested researchers; prospective users thus have to have some knowledge of programming and this mitigates against the wide adoption of the new methods. This approach is also tedious, error-prone, and costly for the developer and often results in inefficient and badly maintained and supported software. Statistical algorithms conforming to agreed standards, such as those provided by the Royal Statistical Society, NAG and ISML, are of course satisfactory for the expert computing statistician but are not very suitable for many applied researchers. Similarly development using software tools in a UNIX-like environment is best suited for experts.

A better approach is to make the new techniques available in a standard statistical system. This approach has various advantages. The system provides both the developer and the user with a shell for the basic activities of data input and management and the display of results and the availability of reliable and accurate algorithms and higher-level statistical and programming facilities encourages quicker and more efficient development of new algorithms. Furthermore, the long-term support and maintenance of the new algorithms is more assured if they are embedded in one of the standard systems. New techniques will also be more accessible to users if they are available in the standard systems they are familiar with.

The paper will review the facilities available for adding new methods to the major statistical systems. We first give a brief review of criteria for assessing the extensibility of the systems and describe the main methods available for implementing extensions. Evaluation is based on our investigations of the capacity of these systems to incorporate some new techniques proposed to solve the classical problem of making inferences about individual behaviour from aggregate data (Langbein and Lichtman, 1978).

2. CRITERIA

Candidate statistical systems for hosting extensions should satisfy the following criteria:

1. The system should support a wide range of data structures and provide basic operations on these structures. Structures should include scalars, vectors, multiway tables, matrices (of various shapes), character strings and possibly compound structures. Any structures derived in analyses (such as the var-cov matrix, residuals, parameter estimates and the design matrix) should be accessible in the system.

Compstat 1988

2. There should be a range of programming constructs such as looping, block-if, conditional execution until a logical condition is satisfied to allow iterative algorithms to be developed, subscripting and the facility to write macros and procedures.

3. The system should provide basic statistical algorithms such as non-linear optimisation, general(ised) linear models, distribution functions, clustering, random number generation for simulation and testing and standard operations such as calculation of eigenvalues and singular value decomposition and sums of squares and products.

4. There should be good control over graphical and printed output so that the developer can tailor output to fit requirements.

5. The developer must be able to program checks on user input, output suitable error messages and diagnostics and provide on-line help and documentation.

6. The developer must be provided with a good program development environment. This should include screen editor, access to the operating system from the system, debugging and trace facilities and the ability to execute external source code.

7. Facilities should be provided to allow the developer to customise applications including the use of menus and prompts. Use of the new algorithms should present an interface to the user as consistent as possible with the syntax and structure of the system.

8. The vendor should provide formal support for user-provided extensions to the system. This should normally take the form of an automatically loaded library with also the facility to load site and individual specific libraries.

9. The system should be available on the range of hardware and operating systems used by the potential users and have the speed and capacity to cope with the range of problems expected.

3. IMPLEMENTATION OF EXTENSIONS

The major statistical systems provide a variety of ways in which new techniques can be implemented by extensions to the standard system. The main methods are:

1. Programming in the command language. Here the system is used as a programming language to link together existing algorithms and facilities to perform new analyses. This is probably the quickest and most efficient way to provide extensions, presuming the system has the required facilities to perform the new analysi

2. Macros and procedures. These are subprograms written in the system language with parameters set by the user. The vendor may support an official library of approved additional procedures. Some systems allow the incorporation of new directives and functions using the standard command syntax; this is the most satisfactory approach as the user is presented with a consistent interface.

3. Provide hooks so that source code in a lower-level language can be incorporatec This is usually done by providing a dummy routine which allows structures to be passed from the system to the external program which can operate on these structur and pass new and modified structures back to the system. This approach requires relinking of the system with the external routines and a detailed knowledge of the internal workings of the system and is thus not a very satisfactory solution for general-purpose extension. It is best suited for site or individual specific requirements.

4. Execute external programs from within the system. This is usually done by writing structures to a file, suspending execution of the system while the external program operates on this file and then restarting the system by reading files output by the external program. This method is also not very suitable for general purpose extensions and is rather cruder to operate than method three.

The preferred solution is for the provision of libraries of macros or procedures to be properly supported by the system.

The application of these criteria to current systems indicates that many are deficient in one or more important respects. In particular, many have very limited data structures and provide poor access to structures derived in analyses. The main systems which satisfy these criteria are SAS, GENSTAT-5, GAUSS and S; SPSS-X, PSTAT and BMDP are deficient in major respects. The GLIM system also satisfies many of the criteria, as evidenced by the large number of new techniques implemented in GLIM macros, but it is a rather specialised system whose main deficiency is its lack of direct facilities for handling matrix structures. GAUSS is a matrix programming language with many built-in statistical functions ('APL as it should be') and although it does not strictly qualify as a statistical system it clearly provides a powerful environment for methodological developments. S is a high level language with a support system (including data management, graphics, many functions and macros) for the analysis of data. The S system is UNIX based and therefore not yet generally available to the user community. For these reasons the evaluation below will concentrate on a comparison of GENSTAT-5 and SAS; S and GAUSS are really for expert users.

4. EVALUATION: NEW METHODS FOR ECOLOGICAL INFERENCE

Aggregated data is available in a variety of data structures. One form is where individuals have been classified into categories of the same variable (e.g. vote at two time points and the classification is available for a number of areal units (e.g. constituencies). A second structure arises when individuals are separately classified on a number of variables (e.g. social class and vote) and the marginal distributions are available for a set of areal units at one time point. With both these structures the aggregate data is characterised by the absence of data for the internal cells of the cross-classification of the variables by areal unit. The aim of the methodology is to model the probability of an individual observation falling in each cell which may be conditional on chosen covariates measured at the aggregate level. The main techniques proposed include:

1. Ecological regression (Crewe and Payne, 1976, and many others)

This technique requires a set of multiple regressions, and the facility to carry out ridge regression when appropriate. Access to the regression structures (parameter estimates, fitted values) is needed, and graphical facilities are useful for the examination of residuals.

Both SAS and GENSTAT-5 provide standard linear regression procedures, allowing access to the parameter estimates, and also have matrix structures allowing one to write ridge regressions procedures. The Supplemental Library of SAS also provides a ridge regression procedure, supported by the SAS Institute, which facilitates fitting models with different values of the ridge constant and examining the ridge traces graphically. However, for an iterative method of determination of the ridge constant, a user written matrix procedure is required. Both packages provide comprehensive facilities for plotting data structures.

2. Entropy maximisation (Johnson and Hay, 1983)

This involves an iterative fit to a three-way table conditioning on all two-way margins which are supplied as data. In order to implement an iterative proportional fitting (IPF) algorithm, one needs the programming structures of looping and IF statements, testing until convergence.

SAS has a flexible matrix language which includes programming constructs, and it is fairly easy to write a program to carry out entropy maximisation. SAS does in fact contain an IPF algorithm, and implementation of entropy maximisation would be even easier if the developer had more flexible access to this algorithm - one is allowed only to specify cell frequencies, not just margins as required here. With GENSTAT-5 one also cannot take advantage of the IPF algorithm supplied, but the system allows a variety of data structures facilitatin the implementation of the technique. It can be programmed using variates, matrices, tables or a mixture of the three, and the difficulty lies only in deciding which is the most efficient approach. Entropy maximisation can be easily implemented with the use of matrix structures and algebra.

3. Ecological logit (Thomsen, 1987)

Thomsen's approach to estimating internal transition proportions estimates ecological correlations across 'homogenous' geographical units and then uses an iterative procedure to arrive at estimated proportions which agree with known marginals to within a specified tolerance. The inverse normal probability function is required to estimate the ecological correlations.

Both SAS and GENSTAT-5 provide the inverse normal function, amongst other statistical functions, and also permit the calculations necessary for the implementati of this technique. The required iteration is possible through the programming structures of looping and IF statements, and can most easily be programmed by making use of matrix structures, as for entropy maximisation.

4. Aggregated compound multinomial (Brown and Payne, 1986)

Given aggregate data available on two classification variables for each of a number of units, this model assumes that, given the row totals x(k) (where k denotes unit), the column totals Y(k) have an aggregated compound multinomial distribution. Asymptotically, this can be approximated by a multivariate normal distribution with log likelihood l=sum(l(K))

where -2l(k)=log(det(C(k)))+T([y(k)-T(P(k))*x(k))*H(k)*[y(k)-T(P(k))*x(k)]

(where T means transpose and *means matrix multiplication -(3.2) in Brown and Payne) with C(k)=cov(Y(k)) and H(k)=inv(C(k)).

Thus, fitting this model by maximum likelihood requires nonlinear optimisation, typically over many parameters. For this to be achieved in a reasonable time, an optimisation method using analytically specified first derivatives is desirable Specification of the objective function and the first derivatives are facilitated by having available matrix structures and algebra.

GENSTAT-5 allows a variety of structures, including matrices, and these can be used in expressions, functions and directives anywhere in a program. The FITNONLINEAR directive can be used to carry out the minimisation of a general function, and the objective function is easily specified as the result of a matrix expression. However, the minimisation algorithm used, a modified Newton method does not use analytic first derivatives, and so the procedure may be fairly slow. More critically, FITNONLINEAR cannot minimise a function of more than six parameters - a number which is typically exceeded for this model. It is possible to take advantage of the programming facilities within GENSTAT-5 and write one's own minimisation procedure, but typically the overheads in term of CPU time required are too high to make this feasible in practice. A better approach is to carry out the required minmisation outside GENSTAT-5 (eg by using the NAG Fortran Library). An external program can be linked in to the system, leaving data input and manipulation, and calculation of starting values, residuals, measures of fit, etc. to GENSTAT-5.

Unlike GENSTAT-5, SAS is programmed in steps, or procedures, each allowing the use of certain statements. This can sometimes be restrictive - for example matrix structures and expressions may only be used within the matrix procedure. SAS provides an NLIN procedure which fits non-linear regression by least squares, allowing the user to specify analytic first derivatives of the model with respect to each parameter, and the method of optimisation to be used. This procedure can be used for maximum likelihood estimation by iteratively re-weighted least squares, but a single univariate dependent variable must be specified, and thus, the parameters of the ACM cannot be estimated by this procedure. Again, the programming facilities of SAS allow one to write a minimisation program, but as with GENSTAT-5, the CPU time overheads are prohibitive. Operating system commands can be given from within SAS, so it is possible to run an external minimisation program, in a lower-level language, which uses SAS output and use the results again within SAS. However, the user needs to take care of the input/output file handling required (this is taken care of to some extent in GENSTAT-5 when using a Fortran program).

5. SUMMARY

Implementation of the ACM model clearly requires a sophisticated statistical system such as GENSTAT-5 or SAS; the other three models can be implemented in most of the major systems. GENSTAT-5 has generally superior facilities for extension compared to SAS, in particular it has better mechanisms for incorporating extensions; the choice of host system is however likely to be strongly influenced by which system is dominant in the applications area. GAUSS and S, clearly provide a powerful statistical development environment for the more advanced user.

6. REFERENCES

Brown, P. J. and Payne, C. D. (1986) Aggregate data, ecological regression and voting transitions, J. Amer.Stat.Soc, 81, 451-61.

Crewe, I. and Payne, C. (1976) Another game with nature: an ecological regression model of the two party vote ratio in 1970, Brit.J.Pol.Sci, 7, 43-81.

Johnston, R. J. and Hay, A, M. (1983) Voter transition probability estimates: an entropy maximising approach, Eur.J.Pol.Res, 11, 93-98.

Langbein, L. I. and Lichtman, A, J. (1978) Ecological Inference, Sage: Beverly Hills, CA.

Thomsen, S, R. (1987) Danish elections 1920-79, a logit approach to ecological analysis and inference, Poltica: Arhus.

FUNIGIRLS: A Prototype Functional Programming Language for the Analysis of Generalized Linear Models

R. Gilchrist and A. Scallan, London

Keywords : Functional programming; generalized linear models; GLIM.

1. Introduction

Statistical analysis is most readily carried out with the aid of a statistical package; there are of course numerous well-established packages, most now available both in main-frame and micro (usually IBM PC compatible) formats. In addition to routine analyses, the professional statistician will also wish to develop his/her own procedures. This necessitates either the use of a package with some form of programming facility (looping, branching, etc.), or the use of a high level language. Both approaches suffer from disadvantages. Packages tend not to be designed with the aim of extensibility in mind; thus even with the most powerful package 'languages' (e.g. SAS, GLIM), programming is something of an adventure (and an unstructured one at that). An illustration is the widespread industry of writing GLIM macros to bend the GLIM program into the analysis of models outside the standard GLM framework. Such programming can be very tricky, numerous publications result, but try understanding someone else's macros!

An alternative approach is to use a high level programming language to write routines to carry out the special analysis. This suffers from many problems. For example, all the data handling facilities of the package are lost; its statistical routines are not available; nor are its model definition syntax, its numerical algorithms, graph plotting, diagnostics, etc. Some high level languages, such as APL or the recent micro language GAUSS, can offer some facilities for the statistician , mainly in matrix manipulation. However, despite the long availability of APL, its lack of widespread use is evidence enough of the problems of this type of approach.

This paper describes a new approach to the design of a statistical language, namely a prototype statistical language for statistical modelling, based upon functional programming. This package, known as FUNIGIRLS (Functional interactive generalized iteratively re-weighted least squares) has been developed in conjunction with members of the GLIM Working Party and emulates the statistical facilities of GLIM 3.77 [2]. This prototype has been written in the functional programming language ML, essentially by writing a set of ML functions which the user can call very easily in a GLIM type fashion. However, users also have available a further set of pre-programmed functions, e.g. for matrix manipulation, and can moreover make full use of the ML language to write their own functions. Thus the package is fully extensible within the underlying 'host' language.

The paper discusses advantages of this functional approach but also notes some of the drawbacks of the implementation in ML. Some of these drawbacks may be surmountable by modifications to the underlying functional language or by allowing the calling of functions written in some other language (in the case of ML, the ML interpreter is written in C, so the calling of C routines should

Compstat 1988

be possible). However, some features which seem inherent in the philosophy of functional languages may be insurmountable in entirety.

At this point we may note that our approach is not dissimilar to that of Chambers and his co-workers [1] although, unlike them, we are using a proprietary language as the basis of our functional statistical language.

2.The need for a new statistical language

Statisticians are continually refining the techniques for data analysis and, indeed, inventing new methodologies. Academic statisticians require a language to test and implement these theoretical advances. Practitioners need to be able to incorporate these techniques easily into their analyses. Moreover, although there are some areas where routine analyses are the norm, many problems require a somewhat non-standard approach. The ability to write ones own code is often a necessity for many users, including the 'non-expert'. To be able to do so in a structured programming environment appears to have obvious benefits; moreover, most of our graduates are now used to using a structured language (e.g. Pascal). The statistical user of the future is likely to require a flexible, powerful programming environment, using a modern approach to computing.

At this point, we shall find it convenient to distinguish the 'language' of the statistical package from its 'user-interface'. The package of the future may well have different user-interfaces, specially designed for different types of users. Some may be 'intelligent', some will be 'graphical', with icons, pop-up menus, interactive dialogue with graphical displays, etc. But behind these interfaces will be the package language, in which the 'expert' user can program extensions to the pre-programmed facilities. Again, this package language may be different from the underlying 'host language,' in which the package language is written.

In this paper, we discuss the investigation of a functional language as the basis for the language of a statistical package. In terms of the classification above, our preliminary work has been using the Standard ML functional language [3] as the statistical language. Note that this ML based language (which we call FUNIGIRLS) would not be one of the user-interfaces, although it might appear as a dialogue box in an interface for the sophisticated user. Using the taxonomy above, our 'host language' is C (i.e. ML is written in C).

3. Why a functional language

Functional languages are claimed to increase programming efficiency and quality by promoting good programming techniques. They have nice mathematical properties which should appeal to those used to dealing with statistical theory; this should in turn make programming into more of an engineering discipline than a trial and error process.

We have experienced the above advantages in producing FUNIGIRLS. The time to produce the program was greatly reduced from the time required to produce the similar GLIM code. Moreover, the ML code is structured and hence fairly easily understandable.

4. Why Standard ML?

It is claimed that functional languages may increase speed when parallel computers become more widely available. However, as we shall mention below, speed is not one of the benefits of the current functional languages when carrying out numerical operations. Nevertheless, there are many potential advantages which come with the functional programming approach, not only in terms of the (structured) extensibility previously noted. The use of 'lists' of models comes naturally with a functional programming language, enabling the simple fitting of a number of differing models. The concept of 'no current state' can be a particular advantage in promoting good programming concepts and in avoiding errors caused by incorrect assumptions of the current state. Indeed, the strong typing of a language such as ML makes programming much less error prone; for the programmer of the package it also has the advantage of making this a great deal more simple and succinct, as all the checking is carried out automatically by the compiler.

Our particular reason for choosing ML was not based upon a scientific choice of the best functional language. Rather, the choice was determined by the fact that the instigator of the project, Mel Slater, had been using ML for a some time and it appeared to be a suitable vehicle for an initial prototype. It is readily admitted that it is necessary to evaluate other functional languages if a final product is to be contemplated.

Given our admittedly ad-hoc choice of ML, the immediate consequence is that FUNIGIRLS needs to be run on machines which support the ML language. Essentially, this is currently restricted to modern UNIX oriented workstations, such as the SUN or Whitechapel, or to the VAX. The current first prototype FUNIGIRLS is available with documentation and worked examples at a nominal price.

5. The FUNIGIRLS Language

As previously remarked, FUNIGIRLS is based loosely upon the concepts of the GLIM statistical package. The output from FUNIGIRLS is therefore very similar to that given by GLIM3.77. After fitting a (generalised) linear model, FUNIGIRLS outputs the 'deviance' and degrees of freedom. Other output quantities (e.g. residuals, estimates, variance-covariance matrix of the estimates, model) can be requested by a function Display, in a similar way to that of GLIM.

FUNIGIRLS has a range of pre-programmed functions to fit the standard generalised linear models, with provision for weighted fitting and 'offsets'. Initial values are not needed for fitting the standard models. Thus, it is extremely simple to fit, for example, regression, analysis of variance/covariance, log-linear models, logit models or components of variance models. In addition, FUNIGIRLS has pre-programmed functions for a wider range of standard models, such as composite links. The user can. moreover, easily construct his/her own fitting functions for non-standard models, by a simple modification of the main fitting function.

To illustrate how the program is used, the following are some simple examples of FUNIGIRLS commands:

5.1 Example 1

This example illustrates how FUNIGIRLS reads in some simple data, for a simple regression with one explanatory variate. The data shows the average blood pressure of a group (of size gr) of women, related to their age. As the readings 'bp' are averages, a weighted fit with weights gr seems appropriate.

```
val mean =Init(5,1,["INTERCEPT"],1.0);
```

This first command sets up a 5 by 1 matrix, ie a vector, of ones, with the name INTERCEPT, to be used in the design matrix to fit the constant term. (FUNIGIRLS requires the user to say explicitly if a constant intercept term is to be fitted).

```
val bp = Read(5,1,["PRESSURE"], [114.0,124.0,143.0,158.0,166.0]);

val age = Read(5,1,["WOMANS AGE"],[35.0,45.0,55.0,65.0,75.0]);

val gr  = Read(5,1,["WT"],[15.0,16.0,12.0,9.0,6.0]);
```

The above commands read in the data, with the group sizes. We now fit a model with Normal errors and the identity link; ie the usual regression model, with weights gr. FUNIGIRLS puts the results of a fit into a fit-tuple, here called res1. The function used to do the fitting is the weighted Normal fit function, W_Normal. Our first linear predictor is just the constant/intercept term.

```
val res1 = W_Normal gr L_I bp mean;

Display(res1,"de");
```

We now fit the age effect, together with the mean. Note that the addition of two terms in the linear predictor requires use of the operator ^+, as + is reserved for the addition of reals.

```
val res2 = W_Normal gr L_I bp (mean^+age);

Display(res2,"de");
```

5.2 Example 2

This example illustrates the generation of factor levels (equivalent to %Gl of Glim 3.77) and the definition of 'Factors'. The data is effectively, a 2-way contingency table, classified by ulcer-type and blood-group. A Poisson, log-link fit seems appropriate.

```
(*  Patients classified  by  blood type and disease. *)

val mean=Init(9,1,["ONE"],1.0)

and count=Read(9,1,["COUNT"],[983.0,383.0,2892.0,679.0,
                              416.0,2625.0,134.0,84.0,570.0])

and ulcer=Read(9,1,["ULCER TYPE"],Gl(3,1,9))

and blood=Read(9,1,["BLOOD TYPE"],Gl(3,3,9));
```

```
(* Now define ulcer and blood type as factors, remembering that
columns are counted starting from zero. Function Fac(x,i)
indicates that column i of the matrix x is to be treated as a
factor.*)

val ulcer=Fac(ulcer,0) and blood=Fac(blood,0);

(* Fit model with Poisson errors and log link *)

val res1=Poisson L_L count mean;Display(res1,"de");

val res2=Poisson L_L count (mean^+blood^+ulcer);Display(res2,"de");
```

5.3 Example 3

FUNIGIRLS also provides a fairly wide range of matrix handling routines. For example, we now show how the matrix handling facilities can be used to calculate parameter estimates in a regression situation. (note, of course, this is meely for illustration, as these facilities are provided automatically by FUNIGIRLS).

```
- (* First read in the data. Data is read in by   columns and we
need to supply the dimensions of the  matrix and label(s) for the
columns. *)

   val xmat=Read(12,2,["XMAT"],[2.26,1.81,1.78,1.54,2.06,1.74,2.56,
                                  2.29,1.80,2.32,2.04,1.88,66.0,52.0,
                                  72.0,67.0,69.0,71.0,88.0,68.0,59.0,
                                 73.0,68.0,58.0])
  and mean=Init(12,1,["ONE"],1.0)
  and y=Read(12,1,[Y"],[7.0,10.0,18.0,4.0,10.0,13.0,21.0,12.0,9.0,
                          65.0,20.0,31.0]);

  (* Now use the matrix facilities to  find the estimates of  the
     regression parameters. Create a full design  matrix, i.e.  a
     column of ones followed by  the  x-variables  using  the Cat
     function*)

val x=Cat(mean,xmat);

  (* Calculate the inverse of x'x  and use  this to calculate the
     parameter  estimates.  The  matrix  transpose  and inversion
     functions  are  T(.) and Inv(.) respectively.  The  operator
     %* indicates standard matrix multiplication. *)

val Ixtx=Inv( T(x) %* x ); val beta=Ixtx %* T(x) %* y; Print beta;
```

6. Discussion

The above examples give some idea of the power of the functional programming approach, as a basis for statistical language. It should be stressed that these are merely examples; the use of a functional language enables programs to be written accurately and efficiently, so extensions to the pre-programmed functions can be easily incorporated, enabling the language to be customised for any user requirement. At the same time, the use of functional programming is likely to be confined to the computer literate. It is clearly not a suitable medium for the naive user. However, for the expert statistician, well-used to mathematical manipulation, it has many attractions.

For the general user, the functional language approach would be undesirable and, indeed, unnecessary. There is no need for the general user to use, or even to be aware of the underlying user language of the package. User-interfaces can be designed which are suitable for any type of user. These may include a 'window' which allows the user to program in the underlying functional language, but this window need never be opened by most users. They could merely be presented by some form of interactive menu-driven environment, which runs the underlying language. Moreover, the use of the functional approach makes it comparatively straightforward to write such a windowing environment.

Disadvantages do exist with functional languages. One of the current practical problems is the lack of arrays in such languages. This necessitates writing arrays as lists, with consequential large overheads in processing time for numerical calculations. Thus FUNIGIRLS is somewhat slow, even for moderate size problems. Computer scientists claim that functional languages will be fast when parallel processing becomes the norm; in the meantime and even then, the need for arrays for statistical calculations seems fairly crucial. Some way forward looks possible here, as we are expecting the release soon of a functional language similar to ML but with arrays. We will be converting FUNIGIRLS to this language, hopefully with improvements in calculation speed.

A further problem with functional languages is more philosophical, namely the lack of any concept of 'state' in a functional language. This contrasts, for example, with GLIM where an underlying state is set up and maintained until altered. For example, in GLIM the default state is Normal errors, identity link. The use of the ERROR P directive changes the state to Poisson errors, log-link. Any fits after that directive are all assumed to be Poisson, log-link. In a functional language, the type of fit required must be stipulated again each time. This has considerable advantages in terms of transparency of the program (we have seen many students using GLIM who do not realise that they cannot use LINK before ERROR, as ERROR resets the state of LINK to its default value). On the other hand, it may be annoying for the sophisticated user to have no implied state and to have to use function Log_Fit each time. Our solution to this problem will be that the user-interface for the general user will have an implied state. On the other hand, the expert user who opens the FUNIGIRLS programming window will have no implied state.

References

1. Chambers, J,(1987), A Computing Environment for Statisticians, Statistical Research Reports No.40, A T and T Bell Laboratories.

2. Payne, C (ed). The GLIM System, Release 3.77, Oxford: Numerical Algorithms Group.

3. Wikstrom, A, (1987), Functional programming using Standard ML. Prentice Hall.

BLINWDR: An APL-Function Library for Interactively Solving the Problem of Robust and Bounded Influence Regression

R. Dutter, Vienna

The library BLINWDR represents a coordinated collection of APL-functions for robust and bounded influence regression analysis. It combines important features of robust analysis with the appealing matrix-oriented interactive programming language APL. The paper outlines the structure of the modules and their possible use. Some related problems like L_1-estimation, finding robust covariance matrices and computing high breakdown point estimates are also touched.

Keywords: statistical programs, interactive analysis, APL for statistics, robust regression, bounded influence regression.

1. Introduction

The computational contents of the function library is closely related to the FORTRAN program BLINWDR presented earlier (Dutter, 1982, 1987a). Such a single program sometimes turns out to be very rigid when considering special problems. Much flexibility and interactivity on the track of the analysis is required. In cases of sophisticated regression analyst, nonstandard situations frequently arise where the analyst quickly wants to see a special plot (e.g. standardized residuals over fitted values, focus on the structure of the largest residuals, reuse information from observations like influential points), or to do some intermediate calculations.

The programming language APL offers many appealing features to such a computer-oriented analyst. In recent years it has lost its reputation of being an odd language, not fitting in a computing environment (data exchange, input/output facilities, graphical support, interfaces to special processors and communication to other programs), needing special hardware and being almost entirely non-portable. Now it has become a very powerful tool for interactively analyzing data sets, rearranging data, performing different computations in easily controllable steps, presenting results in user-friendly forms, and for reiterating the analysis. As a matrix-oriented language, it is used naturally in linear algebra and linear regression analysis. The recent introduction of a new generation of APL (APL-2, see e.g. Lynwood, 1985) offers the use of generalized arrays as data structures where elements may be of different data types (numeric/character). This is very appealing in statistical data analysis where data types are mixed. (A sample vector usually is associated with a name.)

The presented library (or better, system of APL-objects), which is still under development, has been tested on at least two very different computer systems, which shows its portability, it is transparent, i.e. it is open for changes from the user, and it can easily be extended for more complex problems.

Compstat 1988

In this paper we first introduce quickly the problem of robust and bounded influence regression, sketch the general structure of the library, the definition of the user's problem through global variables, some possible computations with results stored in APL objects, and discuss the visual analysis and the interactive reuse of the results.

2. The Regression Problem

We consider the linear regression model

$$y_i = \theta_1 x_{i1} + ... + \theta_p x_{ip} + \epsilon_i = f_i(\theta) + \epsilon_i, \qquad i = 1, ...n,$$

or, in vector notation,

$$\mathbf{y} = X\theta + \epsilon$$

where y_i denotes the i-th observation of the dependent variable, x_{ij} the i-th value of the j-th independent variable and ϵ_i the i-th random error. X then stands for the $(n \times p)$ design matrix. As usual, we assume that θ is an unobservable vector of p parameters and the errors ϵ_i are independently distributed with mean zero and unknown variance σ^2, or more generally, σ^2/w_i with known factor w_i. This known multiplier we also call "user-specified weight", in the following.

The parameters θ are estimated by the method of least squares,

$$\sum_{i=1}^{n}(y_i - f_i(\theta))^2 w_i = min.$$

A generalization to robust and bounded influence regression is given by (see Hampel et al., 1986, Marazzi, 1987, and Dutter, 1987a)

$$g(\theta, \sigma) = \sum_{i=1}^{n} \rho(\sqrt{w_i}\frac{y_i - f_i(\theta)}{\sigma p_i}) v_i p_i = min.$$

where σ denotes a scaling factor which also has to be estimated from the data. The function ρ should reduce and limit the influence of large residuals (outliers in distribution), and should therefore be a function which goes less fast to infinite than the square function. The weights p_i and v_i are introduced in order to bound the influence of outliers in the factor space, and might only depend on the value in factor space ($\mathbf{x}_i$, say). If all weights are 1, i.e. $p_i \equiv v_i \equiv 1$ for all i, we obtain the Huber-type estimator, if $p_i \equiv 1$ a Mallows-type estimator and if $p_i \equiv v_i$ a Schweppe-type estimator.

We note some commonly used ρ-functions by their derivatives ψ:

(a) $\psi_{LS}(t) = t$ (least squares),

(b) $\psi_{L1}(t) = sgn(t)$ (L1-estimate),

(c) $\psi_H(t) = max\{-c, min(t, c)\}$ with $c > 0$ (Huber's ψ),

(d)

$$\psi_{HA}(t) = \begin{cases} t & |t| \leq a \\ a.sgn(t) & a < |t| \leq b \\ \frac{d-|t|}{d-b} a.sgn(t) & b < |t| \leq d \\ 0 & d < |t| \end{cases}$$

with $0 < a < b < d$ (Hampel's ψ).

The weights p_i and v_i are functions of the size of $\mathbf{x}_i$ in a metric defined by a (robust) covariance structure of the factor space (see Hampel, 1980, ánd Dutter, 1983, for details).

3. Features of the APL-function Library

This library consists of some initializing functions for setting up the problem by defining several global variables, a group of functions for computing the regression problems with all the considered options, a set of eventually generated variables containing the results, some functions for the retrieval of the results, and some utility functions. The structure might roughly be summarized as follows.

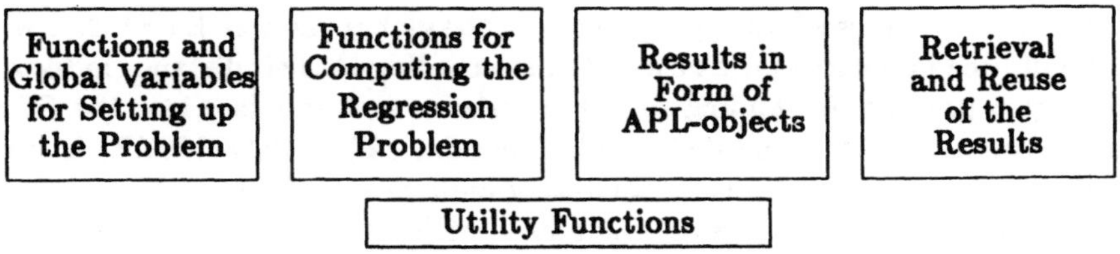

Utility functions, on one hand serve for introducing the user to the workspace (ABSTRACT and BLINWDRHOW) and for providing some help via the inline documentation (DESCRIBE). On the other hand, they may facilitate the understanding of the algorithmic calculation by invoking a row-by-row computation (STEPPING, STEPTRACE and STEPOFF). These functions may also be used in the search for bugs.

3.1 Definition of the Problem Through Global Variables

The user has to communicate to the system how his problem looks like. The first helping function is MENU where the command GLOBSET is mentioned (and many others for latter use). GLOBSET imitates a menu by asking the user for entering each necessary specification and then by setting the global variable which should be kept as long as the same data set is analyzed. Of course, the global variables can be changed individually (by hand).

A global variable for example is DATA. This vector contains a character string with the actual name of the user's data matrix which should already be in the workspace. This kind of specification allows to have more than one data set (e.g. the original and transformed one) in the workspace within the same session. We suppose that each column of the complete data matrix corresponds to the values of a variable. Parallel to such a data matrix, there should be a vector NAMES containing all the names of the variables. Additionally, there will be a vector IND which specifies the numbers of the columns of the data matrix to be used in the analysis. The first element specifies (optionally) the index of the column containing case labels. The following elements of IND specify the indices of independent variables, then that of the dependent one, and then (optionally) the index of the variable containing user-specified weights. Of course, there must be a global variable WEIGHT indicating if such weights are desired, also a variable BNULL telling if there is an intercept to be considered in the model. Further global variables indicate the type of estimation (ETYPE), the sort of loss function (LOSS) with, in general, necessary parameters.

3.2 Computations, and Results in Form of APL-Objects

The overall program is the function BLINWDR. It takes the information given through the global variables, establishes necessary arrays with the essential data, e.g. the (weighted) matrix X, the (weighted) vector Y, corresponding names NX and NY, etc., and does the basic regression computation. By calling the function GLOBAL it displays all necessary specifications of a problem. The basic regression computations are done by the functions L2REG, L1REG and ROBREG, respectively. Further, BLINWDR asks the user for the request of the (sometimes space-consuming) display of residuals and other important but long vectors, e.g. diagnostic tools (RESPRINT), of a probability plot (PROBPLOT), a plot of residuals over fitted values (YHPLOT), a plot of the observed values Y over fitted values (YYHPLOT), etc. These requests might be met by a simple return key stroke which means No, or by any other keystroke (e.g. Y) followed by return which asks for execution.

It should be noted that all regression computations are done according to the type of estimation (Huber/Mallows/Schweppe) stored in the global variable ETYPE. If weights in order to bound the influence of the observations according to their distance in factor space, are necessary, a program for finding robust covariance matrices (see also the FORTRAN program in Dutter, 1983) is needed. This function, called COVINTER, is mentioned because it is very useful as a basis for analyzing multivariate data.

Some computations in large problems are — at least for an interpreter language — time-consuming. Therefore many results are immediately stored in APL-objects, i.e. in character matrices which are very simple to display. These matrices have mnemonic names preceded by the letter 'M'. For example, the matrix MGLOBAL contains all necessary information stored in the global variables for determining the problem. Note that the display of the content is simply invoked by typing the name of the matrix MGLOBAL.

The main regression computation results are stored in MREGRESS with computed values B of θ, their estimated standard errors SB, t-values TB, the scale of the residuals, the residual mean square, the coefficient of determination, and eventually robust equivalents. The residuals RES, standardized residuals RESSTD, the estimated values YH and the diagonal elements of the hat-matrix, called PROJ, are stored in the matrix MRESPRINT (which contains at least n rows!). Remark that problems like ordering of the residuals and finding which of them are the largest, are trivial using the language APL. Many of the possible plots are stored in matrices, see MPROBPLOT, MYHPLOT, MYYHPLOT, etc.

Some important variables remain in the workspace. For example, the estimated parameters B, their standard errors SB, residuals RES, fitted values YH, the inverse of the inner product X by itself $(X^T X)^{-1}$ named XTX1, the estimated scale SIG, the coefficient of determination RR, are available, may be looked at, reused for modification of the problem setup and renamed for storing before a new model is newly computed.

3.3 Quick Displays and Reuse of the Results

As mentioned in the preceding section, important results are stored in APL-data structures. These matrices are displayed by simply typing their names. If the content (e.g. MRESPRINT) is too large for fitting on a screen, the function DISPLAY preceding the name of the matrix as parameter, may be employed.

The essential output may also be put on paper (or on system file) by using the functions PRINT, PRINTDATA and PRINTPLOTS for printing the regression results, the data and all computed plots, respectively. These functions of course, depend very much on the computing environment but they should represent the information in condensed form in usual ASCII-code for printers not being able to display the APL characters. These outputs could also be inserted in computer files with text reports on the analysis of the problem.

After having examined the results of the regression computation the user might like to change the model specification. APL is well suited for reacting to questions like "what is the value of a certain quantile of the residuals" or "a variable should better be transformed" or "put aside all observations having standardized residuals absolutely greater than a certain value" or "note all observations recognized to be of high influence". The data matrix can easily be transformed by adjusting rows or transforming columns. After this, a new data matrix would be formed, and only some global variables have to be changed before doing a new analysis.

We like also to remark that some auxiliary functions are available for cleaning (compressing) the workspace so that, for later information, only the essential part of the data and variables is kept. This opens the possibility of keeping many different stages of the analysis in the computer. The option of keeping track of all actions during the APL-session has also been considered but it did not turn out to be so important in this environment.

3.4 Bindings to Computer-intensive Programs

Large program systems in an interpreter language like APL are useful if one has a very powerful computer at his disposition, or there is the possibility of fast auxiliary processors which may be activated from the APL environment. Such auxiliary programs might be modules written in FORTRAN which may be accessible by the APL-processor, for example on a personal computer.

We have found that the simple but basic problem of singular value decomposition is better solved by a FORTRAN module because, at least once, some iterative process is involved and good FORTRAN programs are available. To write the interface is an easy task. A similar problem arises in L1-regression. Good FORTRAN programs based on linear programming algorithms are at our disposition.

With this technique of binding FORTRAN programs in the APL environment it is also possible to use modern but very computer-intensive methods with high breakdown point as e.g., the LMS-estimator implemented in PROGRESS (Rousseeuw and Leroy, 1987), or for multivariate analysis, covariance estimation based on the minimal volume ellipsoid (see the FORTRAN program PROCOVIEV, Rousseeuw and van Zomeren, 1987). These programs can be called from the environment in BLINWDR and the results used in the further analysis.

4. Final Remarks and Suggested Usage

The package represents a trial to establish a unified approach to the applied analysis of such a complex problem as robust regression. Because it is an open system and it is written according to internal guidelines so that the functions with their inline documentation should be self-explaining, everybody can modify the source code to fit his own target. So, BLINWDR could be useful when searching for ideas for new and better methods in applied regression analysis. APL is an ideal language for programming algorithms, so the presented function library could also be used as experimental package for developing other, more commercial programs.

Moreover, a preliminary version of the presented package, which by no means is complete, has been used for the analysis of a large data set of geochemical variables (Dutter, 1987b) where other — commercially available programs — were not satisfactory.

The portability of APL programs causes a certain problem because there is no generally accepted standard. We underscore our experience with the function library that runs with at least two different APL-systems — with essentially a simple change of one global variable — namely with IBM-APL/PC Version 2 and Dyalog APL, which shows a certain portability. The change to other APL-interpreters via a conversion table of the APL-characters does not seem to be too difficult.

The programming language is often criticized because of its need of special terminals for the display of the unusual symbols. For the everyday-use of the system for analysis, this is not necessary. The special symbols for assignment or operators can easily be simulated (see Karnel, 1988).

The function library is available on request from the author.

5. Acknowledgements

The principal motivation for writing APL-functions for interactive regression analysis came from difficulties when analyzing geochemical data with J. Wolfbauer and his co-workers (especially Hans Kürzl and Fritz Wurzer), but the idea of developing the library up to its present stage grew out from many discussions with Peter Huber and E.H. Huber-Buser (1988). Many stimulations came from Alfio Marazzi, Peter Rousseeuw and from Werner Stahel. I am also indebted to Gerald Karnel due to our every-second-day-discussions.

7. References

Dutter R. (1982): BLINWDR: Robust and Bounded Influence Regression. *COMPSTAT 1982.* Physica Verlag, Vienna.

Dutter R. (1983): COVINTER: A Computer Program for Computing Robust Covariances and for Plotting Tolerance Ellipses. *Res. Rep. No. 10,* Inst. Statist., Techn. Univ. Graz, Austria.

Dutter R. (1987a): BLINWDR: A FORTRAN Program for Robust and Bounded Influence Regression. In: *Statistical Data Analysis Based on the L_1-Norm and Related Methods,* Y. Dodge (ed.). North-Holland.

Dutter R. (1987b): Regressionsanalyse der geochemischen Daten, Gebiet 3, Kothgraben. *Forschungsber. TS-1987-4,* Inst. f. Statist. and Probab., T.U. Vienna.

Hampel F.R. (1980): Optimally Bounding the Gross-Error-Sensitivity and the Influence of Position in Factor Space. *Proc. Statist. Comput. Sect.,* Amer. Statist. Assoc., 59-64.

Hampel F.R., E.M. Ronchetti, P.J. Rousseeuw and W.A. Stahel (1986): *Robust Statistics. The Approach Based on Influence Functions.* J. Wiley, New York.

Huber P.J. and E.H. Huber-Buser (1988): ISP: Why a Command Language? In: *Fortschritte der Statistik-Software 1.* Faulbaum et al. (eds.). G. Fischer Verlag, Stuttgart.

Karnel G. (1988): GPSP: A General Purpose Statistical Package. *COMPSTAT 1988,* Short Communications. Physica-Verlag, Heidelberg.

Lynwood Dyalog APL (1985): *User Guide.* Dyadic Systems Limited, 30, Camp Road, Farnborogh, Hampshire, England.

Marazzi A. (1987): Solving Bounded Influence Regression Problems with ROBSYS. In: *Statistical Data Analysis Based on the L_1-Norm and Related Methods,* Y. Dodge (ed.). North-Holland.

Rousseeuw P.J. and A.M. Leroy (1987): Robust Regression and Outlier Detection. J. Wiley & Sons, New York.

Rousseeuw P.J. and B.C. van Zomeren (1987): Identification of Multivariate Outliers and Leverage Points by Means of Robust Covariance Matrices. *Rep. 87-15,* Fac. of Math. and Inform., T.U. Delft.

Computationally Intensive Methods

Exact Non-Parametric Significance Tests

C. R. Mehta, N. R. Patel and P. Senchaudhuri, Boston

1. INTRODUCTION

A very rich class of non-parametric two-sample tests are of the form

$$d(\mathbf{x}) = \sum_{i=1}^{k} a_i(m_{i-1}, x_i)$$

where the x_i's are the entries in row 1 of the 2xk contingency table $\mathbf{x}$:

x_1	x_2	...	x_k	m
x_1'	x_2'	...	x_k'	n
t_1	t_2	...	t_k	N,

$m_i = x_1+x_2+\ldots+x_i$, and $a_i(m_{i-1}, x_i)$ is a real valued function. An important special case arises when $a_i(m_{i-1}, x_i) = a_i x_i$ which, by suitable choice of scores, a_i, yields the class of linear rank tests for the two sample problem.

Inference is based on permuting the entries in the cells of the 2xk table in all possible ways subject to fixing some of the table marginals. Let Γ be the reference set of all such permutations. Let H be the null hypothesis of interest and for any $\mathbf{x}\varepsilon\Gamma$, let

$$h(\mathbf{x}) = C^{-1} \prod_{i=1}^{k} \lambda_i(m_{i-1}, x_i), \tag{1.1}$$

where $\lambda_i = (.,.)$ is a probability and the normalizing constant C is computed by the formula

$$C = \sum_{\mathbf{x}\varepsilon\Gamma} \prod_{i=1}^{k} \lambda_i(m_{i-1}, x_i).$$

For most applications $h(\mathbf{x})$ is the generalized hypergeometric function

$$h(\mathbf{x}) = \prod_{i=1}^{k} \binom{t_i}{x_i} / \binom{N}{m}.$$

Now suppose that $\mathbf{x}^*$ is the observed value of $\mathbf{x}$. Then we define the critical region of the reference set as

$$\Gamma^* = \{\mathbf{x}\varepsilon\Gamma : d(\mathbf{x}) \geq d(\mathbf{x}^*)\}.$$

The main problem is to compute the exact p-value,

$$p = \sum_{\mathbf{x}\varepsilon\Gamma^*} h(\mathbf{x}).$$

By suitable choice of the functions a_i one can describe a fairly large class of non-parametric tests in terms of permuting a 2xk contingency table. A partial list of such non-parametric tests is displayed below in Table 1.

Compstat 1988

Table 1. Functions $a_i(m_{i-1}, x_i)$ for Some Non-Parametric Tests

Test	Value of $a_i(m_{i-1}, x_i)$
Wilcoxon rank sum	$(t_1 + t_2 + \ldots + 1/2(t_i+1))x_i$
Trend	ix_i
Savage	$(\sum_{j=1}^{i} \frac{1}{r_i} - 1)x_i$ (r_i = # at risk just prior to i^{th} failure)
Median	$-\log \binom{t_i}{x_i} / \binom{N}{m}$ (t_i = # in population i, x_i = # above median
Fisher exact	$-\log \binom{t_i}{x_i} / \binom{N}{m}$ (t_i = # in population i, x_i = # of responses)
Van der Waerden	$(\Phi^{-1}(t_1 + t_2 + \ldots + 1/2(t_1 + 1))/(N+1))x_i$
Kolmogorov-Smirnov Two Sample	$\lvert (m_{i-1} + x_i)(\frac{1}{m} + \frac{1}{n}) - \frac{i}{n} \rvert$ (here $d(\mathbf{x}) = \max_{1 \le i \le k} (a_i(m_{i-1}, x_i))$)

2. Exact and Monte Carlo Algorithms

We have developed two approaches to estimating the p-value, the exact approach and the Monte Carlo approach. The exact algorithm is computationally more intensive than the Monte Carlo algorithm. In exchange for the extra numerical work, it yields the true p-value whereas the Monte Carlo method provides a point estimate of the true p-value along with an arbitrarily tight confidence interval for it. The exact algorithm is described in detail in Mehta, Patel and Tsiatis (1984) and Mehta, Patel and Wei (1988). The Monte Carlo algorithm is described in detail in Mehta, Patel and Senchaudhuri (1988). Here we provide a summary of the two methods. Refer to the papers cited above for the technical details.

Both algorithms rely on the network representation of the reference set Γ. In this representation each path through the network corresponds to one table $\mathbf{x} \varepsilon \Gamma$. The length of that path is $d(\mathbf{x})$ and its probability is $h(\mathbf{x})$. The problem of computing the p-value is then equivalent to identifying all paths through the network where lengths equal or exceed $d(\mathbf{x}^*)$, and summing their corresponding probabilities.

2.1 Network Representation of the Reference Set

It is convenient to represent Γ as a network of nodes and arcs. The network is constructed recursively in k+1 stages labelled 0,1,2,...,k. The nodes at stage j are ordered pairs of the form (j,m_j) where m_j is one possible partial sum of the form $x_1+x_2+\ldots+x_j$ for the tables in Γ. The set of all stage j nodes represents all possible partial sums up to and including column j, for every table in Γ. Arcs emanate from node (j,m_j), each arc being connected to exactly one successor node of the form $(j+1,m_{j+1})$ at stage j+1. Since the stage j+1 nodes must represent all possible partial sums, $x_1+x_2+\ldots+x_{j+1}$, for all the tables in Γ, we can write out, explicitly, the set of successor nodes to each node (j,m_j) as

$$\mathbf{R}(j,m_j) = \{(j+1,w):\max(m_j,m-\sum_{i=j+2}^{k} n_i) \le w \le \min(m,m_j+n_{j+1})\}, \tag{2.1}$$

where, throughout this paper, $m_0 = 0$, $m_k = m$, and a summation is null if its lower limit exceeds its upper limit. Beginning with the initial node (0,0) and applying (2.1) successively through stages 0,1,...,k-1, we automatically construct a network with a single terminal node (k,m) in which each path $(0,0)\rightarrow(1,m_1)\rightarrow(2,m_2)\rightarrow\ldots\rightarrow(k,m)$ represents one table $\mathbf{x}=(m_1,m_2-m_1,\ldots,m-m_{k-1})$ in Γ. Moreover the set of all such paths represents the entire reference set Γ. In this network the arc $(j,m_j)\rightarrow(j+1,m_{j+1})$ is assigned a length of $a_{j+1}(m_{j+1}-m_j)$ and a probability of $\lambda_{j+1}(m_j,m_{j+1}-m_j)$. The length of a path through the network is defined as the sum of lengths of all arcs constituting that path. Similarly the probability of a path through the network is defined as the product of probabilities of the arcs constituting that path. Thus, by specifying a path in the network, we automatically know the table, $\mathbf{x}$, it represents, together with the value of the test statistic, $t(\mathbf{x})$, and the unnormalized probability, $Ch(\mathbf{x})$, for that table. The critical region Γ^* consists of those paths whose lengths equal or exceed $d(\mathbf{x}^*)$ and the p-value is the sum of probabilities of the paths in the critical region, normalized by C, the sum of probabilities of all the paths in the network.

2.2 Processing the Network

The network is processed in two steps. First there is a backwards induction pass through the network during which we gather some important preliminary information for each network node. Armed with this information, we proceed to traverse the network in the forward direction either sampling paths at random (the Monte Carlo algorithm), or enumerating, implicitly, all paths where lengths equal or exceed $d(X^*)$ (the exact algorithm).

2.2.1 Backward Induction

Let $\Gamma(j,m_j)$ denote the set of all paths from any node (j,m_j) to the terminal node (k,m). A typical path $(j,m_j)\rightarrow(j+1,m_{j+1})\rightarrow\ldots\rightarrow(k,m)$ corresponds to the vector $\mathbf{x}_j=(m_{j+1}-m_j,m_{j+2}-m_{j+1},\ldots,m-m_{j+1})\equiv(x_{j+1},x_{j+2},\ldots,x_k)$. Therefore, $\Gamma(j,m_j)$ consists of all possible values of $\mathbf{x}_j$, the entries in columns $j+1,j+2,\ldots,k$ of the tables in Γ for which $x_{j+1}+x_{j+2}+\ldots+x_k=m-m_j$. Now define

$$\ell(j,m_j) = \min_{\mathbf{x}_j\varepsilon\Gamma(j,m_j)} \{ \sum_{i=j+1}^{k} a_i(m_{i-1}, x_i)\}, \tag{2.2}$$

$$u(j,m_j) = \max_{\mathbf{x}_j\varepsilon\Gamma(j,m_j)} \{ \sum_{i=j+1}^{k} a_i(m_{i-1}, x_i)\}, \tag{2.3}$$

$$q(j,m_j) = \sum_{\mathbf{x}_j\varepsilon\Gamma(j,m_j)} \prod_{i=j+1}^{k} \lambda_i(m_{i-1},x_i). \tag{2.4}$$

The quantities (2.2) to (2.4) are all rapidly obtained in a single backward induction pass through the network. For example,

$$\ell(j,m_j) = \min_{(j+1,w)\varepsilon\mathbf{R}(j,m_j)} \{a_{j+1}(m_j, w-m_j) + \ell(j+1,w)\}.$$

The other nodal quantities are obtained similarly.

2.2.2 Forward Processing of the Network

The above network representation provides a very efficient way to compute the exact p-value. To see this, suppose that in the process of enumerating paths through the network we have reached node (j, m_j) via the subpath $(0,0)\rightarrow(1, m_1)\rightarrow\ldots\rightarrow(j-1, m_{j-1})\rightarrow(j,m_j)$. Denote the rank length of this subpath by

$$b = \sum_{i=1}^{j} a_i(m_{i-1}, m_i- m_{i-1}).$$

Let $\mathbf{Q}$ denote the set of all complete paths from (0,0) to (k,m) that share the above subpath. Under certain conditions we can know immediately the contribution of Q to the p-value. If

$$b+u(j,m_j) < d(\mathbf{x}^*), \tag{2.5}$$

we know that every path in Q has a rank length that is less than $d(\mathbf{x}^*)$. Hence none of these paths contributes to the p-value and the entire set can be droped from further consideration. Next if

$$b+\ell(j,m_j) \geq d(\mathbf{x}^*), \tag{2.6}$$

every path in Q has a rank length equal to or greater than $d(\mathbf{x}^*)$ and the entire set of paths contributes to the p-value. The contribution of these paths is, of course, $q(j,m_j)/q(0,0)$. If neither (2.5) nor (2.6) holds, we extend the common subpath to a node $(j+1,\ m_{j+1})$ in accordance with (2.1) and proceed to verify (2.5) and (2.6) in the same manner as before.

Suppose alternatively that we wish to estimate the exact p-value by Monte Carlo methods rather than computing it exactly. We now sample tables at random from Γ. This is equivalent to traversing the network in the forward direction from (0,0) to (m,n) and randomly selecting one of the available choices of successor nodes at each stage. The selection can be made so that $\mathbf{x}\varepsilon\Gamma$ is selected with probability $h(\mathbf{x})$. This yields a crude Monte Carlo algorithm. Another possibility is importance sampling.

In importance sampling each table $\mathbf{x}\varepsilon\Gamma$ is sampled with probability $g(\mathbf{x})$ rather than $h(\mathbf{x})$. The jth such sampled table yields a realization z_j of the random variable

$$Z_j = \begin{cases} h(\mathbf{x}_j)/g(\mathbf{x}_j) & \text{if } \mathbf{x}_j\varepsilon\ \Gamma^*, \\ 0 & \text{otherwise.} \end{cases}$$

Now $\bar{z}$ is an unbiased estimate of p, since

$$E(Z) = \sum_{\mathbf{x}\varepsilon\Gamma^*} \{h(\mathbf{x})/g(\mathbf{x})\}g(\mathbf{x}) = p.$$

The 99% confidence interval for p is $\bar{z}\pm 2.58\ s_z/\sqrt{N}$. The problem is to select g so as to make s_z smaller than the corresponding variance under crude Monte Carlo.

The details of the actual sampling technique are provided in Mehta, Patel and Senchaudhuri (1988).

In Table 2 we show the advantage that importance sampling has compared to crude Monte Carlo sampling. Each p-value was estimated up to ± 0.0005 accuracy with 99% confidence. The table shows that efficiency gains ranging from 50 to 12,000 are possible.

Table 2. Comparison of Crude Monte Carlo and Importance Sampling for Various Data Sets

(Taken from Mehta, Patel and Senchaudhuri, 1988)

Line No.	Types of Hypothesis Test	p-value and 99% CI	Sampling Standard Deviation: Crude (s_u)	Importance (s_z)	Required Sample Size: Crude N_C	Importance N_I	Efficiency $s_u^2/s_z^2 \approx N_C/N_I$
1	WILCOXON	(0.193 to .0203)	.1393	.0014	5.15×10^5	52	9900
2	TREND	(.0203 to .0213)	.1427	.0013	5.43×10^5	45	12068
3	LOGRANK-CRD	(.0560 to .0570)	.2309	.0045	1.40×10^6	539	2609
4	LOGRANK-UD	(.0569 to .0579)	.2326	.0060	1.44×10^6	958	1503
5	LOGRANK-BCD	(.0517 to .0527)	.2224	.0048	1.32×10^6	613	2147
6	COMMON ODDS RATIO	(.0122 to .0132)	.1120	.0165	3.33×10^5	7249	46

3. Software Availability

All the tests listed in Table 1 and several others are available as supplementary modules within SAS, SYSTAT and EGRET. A stand-alone IBM-PC based software package with these exact procedures is also available. Contact Dr. Cyrus Mehta, Cytel Software Corporation, 137 Erie Street, Cambridge, MA 02139, telephone (617) 661-2011 for details.

References

1. Mehta C and Patel N. A network algorithm for Fisher's exact test in r×c contingency tables. JASA, 78:382, 1983.

2. Mehta C, Patel N and Senchaudhuri P. Importance Sampling: A new tool for exact permutational inference. JASA, 1988 (In Press).

3. Mehta C, Patel N and Tsiatis A. Exact significance testing for treatment equivalence for ordered categorical data. Biometrics, 40:819-825, 1984.

4. Mehta C, Patel N and Wei LJ. Computing exact permutational distributions and significance levels with restricted randomization designs. Biometrika, 75:2, 1988.

5. Zelen M. The analysis of several 2×2 contingency tables. Biometrika, 58:129-137, 1971.

Resampling Tests of Statistical Hypotheses

A. Young, Cambridge

1. Introduction

A direct approach to statistical tests of significance is possible using bootstrap techniques to simulate the null distribution of the test statistic of interest. The approach is outlined by Hinkley(1988) and by Young(1986). In this paper issues involved in the construction of such tests are considered in the context of testing the mean of a univariate population. The general method is summarized in Section 2, while questions relating to choice of reference distribution and test statistic are considered and illustrated empirically in Section 3. Section 4 discusses the importance of appropriate conditioning in resampling tests of significance.

The approach has much in common with the Monte Carlo test of Barnard(1963), though the methods considered here contrast with more conventional simulation tests through the interventionalist role played by the sample data in specification of the resampling mechanism.

2. Resampling Tests

Suppose a test statistic T is proposed for testing a statistical hypothesis H_0 about an underlying population F, with large values of T being evidence against H_0. Let $\underline{x} = (x_1, ..., x_n)$ denote the observed sample data and let $T(\underline{x}) = t_{obs}$. The resampling method is to be used to estimate the sampling distribution of T under the constraint imposed by H_0. Implementation of the test consists of choice of a set $\mathcal{F}$ of distributions satisfying H_0 and consideration of the sampling distribution of T under $\tilde{F}$, where $\tilde{F}$ is that member of the set $\mathcal{F}$ closest to, or most consistent with, the observed data. Formally, denoting by F_n the empirical distribution of the observed sample and supposing δ is a distance measure between distributions, the significance probability is computed by considering the distribution of T under sampling from $\tilde{F}$, where $\tilde{F}$ minimises $\delta(F_n, F)$ for $F \in \mathcal{F}$. The bootstrap test significance probability corresponding to t_{obs} is then

$$p = Pr\{T \geq t_{obs} \mid \tilde{F}\}.$$

Compstat 1988

In general p will be estimated by simulating B 'bootstrap data samples' from $\tilde{F}$ to produce B values $T_1, ..., T_B$ of the test statistic. The estimate of p is then

$$\hat{p} = B^{-1}\sharp\{T_i \geq t_{obs}\}.$$

The simulation size B should be chosen large enough so that $\hat{p}$ is an accurate estimate of the true bootstrap significance probability p when p is close to the nominal level of the test (c.f. Marriott, 1979). For a test of nominal size 0.05, for example, 500 to 1000 simulations would be advisable.

3. Tests on the Mean

Let $\underline{x}$ represent an i.i.d. sample from a univariate distribution F with mean μ and suppose it is required to test $H_0 : \mu = \mu_0$, for specified μ_0.
One possibilty for choice of the bootstrap reference distribution $\tilde{F}$ is to embed F_n in a class of distributions whose support is $x_1, ..., x_n$: see Efron(1981) and Owen(1988). The distribution $\tilde{F}$ will attach probabilities $w_1, ..., w_n$ to $x_1, ..., x_n$, where, since $\tilde{F} \in \mathcal{F}$, $\sum_{i=1}^{n} w_i x_i = \mu_0$, and the w_i are chosen, say, to

(1) minimize the Kullback-Leibler distance $\delta(\tilde{F}, F_n) = \sum_{i=1}^{n} w_i log(nw_i)$, or,
(2) maximize $\prod_{i=1}^{n} w_i$, or equivalently minimize $\delta(\tilde{F}, F_n) = -\sum_{i=1}^{n} n^{-1} log(nw_i)$, or
(3) minimize $\delta(\tilde{F}, F_n) = max_i \mid w_i - \frac{1}{n} \mid$.

In the first case the optimal weights w_i are given by $w_i = e^{\lambda x_i} / \sum_{i=1}^{n} e^{\lambda x_i}, i = 1, ..., n$, where λ is uniquely defined by $\sum_{i=1}^{n} x_i e^{\lambda x_i} / \sum_{i=1}^{n} e^{\lambda x_i} = \mu_0$: see Efron(1981). The distribution $\tilde{F}$ is then the 'exponentially tilted' version of F_n.
The optimal weights w_i in the second case are given by $w_i = n^{-1}\{1 + \lambda(x_i - \mu_0)\}^{-1}$, where λ is the unique solution of $\sum_{i=1}^{n}\{1+\lambda(x_i-\mu_0)\}^{-1}(x_i-\mu_0) = 0$: see Owen(1988). The distribution $\tilde{F}$ is then the non-parametric constrained maximum likelihood estimator of F.
The optimal weights in the third case may be obtained as the solution of a simple linear programming problem. Numerically, such a construction for $\tilde{F}$ is less attractive than the one- dimensional optimization required in the other two cases. In the simulation study below only the first two methods, referred to as 'TILT' and 'MLE' respectively, are considered.

A simpler method for construction of $\tilde{F}$ allows use of distributions of modified support. We might consider, for example, taking $\tilde{F}$ as a 'shifted' version of F_n, with the hypothesised mean μ_0. Then $\tilde{F}$ attaches equal weight n^{-1} to each of the points

$x_i - \bar{x} + \mu_0, i = 1, ..., n$,where $\bar{x} = n^{-1} \sum_{i=1}^{n} x_i$. This method is refered to as 'SHIFT' below.

By analogy with the usual parametric situation, T might be taken as the one-sample t-statistic. Performance of tests based on this statistic and bootstrap data distributions $\tilde{F}$ constructed by the above methods are considered for two underlying distributions, the standard normal distribution and the exponential distribution of variance 1 centered at mean 0, and for sample size 20. For both distributions a two-term Edgeworth expansion was found to give an adequate approximation to the distribution of T under $\tilde{F}$, and therefore the bootstrap significance probabilities could be estimated without simulation. For each F, 20000 replications were used to compare performances of tests of nominal size 0.05. Results are summarized in Tables 1 and 2.

$\mu_0 =$	0.0	-0.2	-0.4
TILT	0.053	0.219	0.536
MLE	0.053	0.221	0.538
SHIFT	0.048	0.204	0.513

Table 1 : Simulated powers, normal distribution.

$\mu_0 =$	0.0	-0.2	-0.4
TILT	0.038	0.230	0.686
MLE	0.039	0.234	0.673
SHIFT	0.032	0.206	0.664

Table 2 : Simulated powers, exponential distribution.

The results show that there is little to choose between the TILT and MLE procedures, but that the simpler SHIFT procedure is perhaps less powerful. In the exponential case, all three resampling tests have observed sizes substantially lower than the nominal size.

The success of a resampling test will depend crucially on the choice of test statistic T. A general procedure for choice of statistic may be based on the notion of empirical likelihood, described by Owen(1988). For the mean testing problem define the *empirical likelihood ratio* as

$$R_0 = sup \prod_{i=1}^{n} nw_i,$$

where the supremum is taken over sets of probability weights satisfying $\sum_{i=1}^{n} w_i x_i = \mu_0$. The required numerical procedure is outlined above. Theorem 1 of Owen(1988) shows that, on H_0, $T_{LR} = -2logR_0$ is asymptotically distributed as χ_1^2. A test of H_0 may be based on T_{LR}, which is an empirical version of the usual likelihood ratio statistic. Table 3 shows simulation based estimates of the null expectation and variance of T_{LR}, for the two distributions considered previously, and for a range of sample sizes n. Each figure is based on 20000 simulations.

n	Normal		Exponential	
	$E(T_{LR})$	$Var(T_{LR})$	$E(T_{LR})$	$Var(T_{LR})$
15	1.2449	4.3483	1.6319	15.0809
20	1.1546	3.2746	1.4250	9.9239
50	1.0352	2.1778	1.1271	2.7587
100	1.0057	1.9966	1.0536	2.2567
200	1.0041	1.9847	1.0285	2.0873

Table 3 : Null Distribution of T_{LR}.

The table indicates that the chi-squared approximation will be inadequate, without Bartlett correction, for sample sizes less than about 100. For small sample sizes, the resampling method can be used to simulate the distribution of the empirical likelihood ratio statistic under suitable $\tilde{F}$. It would be most convenient to construct $\tilde{F}$ by the maximum likelihood procedure described above.

Performance of empirical likelihood ratio tests of nominal size 0.05 was studied for the two underlying distributions, using 1000 simulations from $\tilde{F}$ to estimate the bootstrap significance probability p, for each of 500 replications. In the normal case, the estimated powers were $0.052, 0.164$ and 0.396 for $\mu_0 = 0.0, -0.2, -0.4$ respectively. In the exponential case the corresponding figures were $0.068, 0.102$ and 0.378. Though our figures here are based on a smaller number simulations than used previously, and are therefore less accurate estimates of true power, they would indicate a less powerful procedure than that based on the t- statistic.

4. Conditional Inference

The principle of conditioning on ancillary statistics is an important and well-developed component of statistical inference. It is clear that appropriate conditioning is also a necessary part of the implementation of resampling tests. The idea (Young, 1986) should be to condition on those features of the data under test which reflect the propensity for extreme values of the test statistic to occur for artificial reasons. Such features may or may not be quantifiable as ancillary statistics.

Here, as discussed by Hinkley and Schechtman(1987), it is important to distinguish between an experimental ancillary, which determines a performable sub-experiment, and a mathematical ancillary, which is a function of the random responses. In the parametric context, a statistic S would be said to be ancillary, and should be conditioned on, if its distribution does not depend on the parameter of interest, and therefore it contains no information on that parameter, but stratification of data samples by the value of S indicates the test statistic to have different distributions across strata.

In the mean testing problem of Section 3, it is possible that the data might represent a known mixture from two distributions with the same mean. Table 3 would indicate, for example, that the null distribution of the empirical likelihood ratio statistic might depend on the mixture proportion. This proportion is then an experimental ancillary. Conditioning could be achieved by construction of separate bootstrap data distributions from the two sub-samples.

Let $x_{(1)} < x_{(2)} < ... < x_{(n)}$ denote the order statistics of the observed sample. The configuration statistic $C = (x_{(2)} - x_{(1)}, ..., x_{(n)} - x_{(1)})$ has distribution not depending on μ. Let $A = n^{-1}\sum_{i=2}^{n}\{x_{(i)} - x_{(1)}\}^2$. Simulation shows that, for sample size $n = 20$, the empirical likelihood ratio statistic T_{LR} has null distribution which depends on A. For example, by simulating 20000 datasets from the standard normal distribution, it was found that:

$$E(T_{LR} \mid H_0, A \in [2.5, 3.0]) = 1.4578,$$

$$E(T_{LR} \mid H_0, A \in [5.5, 6.0]) = 0.9532.$$

The bootstrap test should therefore be conditioned on the observed value of A. Only samples from $\tilde{F}$ with A close to its observed value, say within 0.5 of its observed value, should be used in estimation of the significance probability p. If unconditional simulation from $\tilde{F}$ is to be used, this will mean throwing away a large proportion of bootstrap data samples. Empirically, this proportion is seen to be about 0.7 in the normal case.

A conditional bootstrap inference can be obtained using the notion, expressed by Kendall and Kendall(1980), of a resampling distribution with modified support and a simulation mechanism specifically designed to provide conditioning on A. The procedure here might be to assess significance of t_{obs} by simulating the distribution of T_{LR} under random displacements of F_n. Such displacements might be taken as normal: in order to simulate datasets representing H_0, but with the same configuration C and value of A as the data under test, the displacements should have mean $\mu_0 - \bar{x}$ and variance σ^2/n, where σ^2 is the variance of F.

This conditional procedure was applied to the two underlying distributions F considered previously, and for sample size 20 as before. In the normal case the conditional simulation of nominal size 0.05 gave powers 0.040, 0.160 and 0.442 against $\mu_0 = 0.0, -0.2$ and -0.4 respectively. The corresponding figures in the exponential case were 0.052,

0.202 and 0.536. Each figure here is based on 500 replications from F, 1000 simulations being performed for each replication.

5. References

Barnard, G. (1963). Contribution to discussion of Bartlett's paper. J.Roy.Statist.Soc. **B**, **25**, 294.

Efron, B. (1981). Nonparametric standard errors and confidence intervals. Can. J.Statist., **9**, 139-172.

Hinkley, D.V. (1988). Bootstrap methods. J.Roy.Statist.Soc. **A**, **151** (To appear).

Hinkley, D.V. and Schechtman, E. (1987). Conditional bootstrap methods in the mean-shift model. Biometrika, **74**, 85-93.

Kendall, D.G. and Kendall, W.S. (1980). Alignments in two-dimensional random sets of points. Adv.Appl.Probab., **12**, 380-424.

Marriott, F.H.C. (1979). Barnard's Monte Carlo test : how many simulations? Appl. Statist., **28**, 75-77.

Owen, A.B. (1988). Empirical likelihood ratio confidence intervals for a single functional. Biometrika, **75** (To appear).

Young, A. (1986). Conditioned data-based simulations : some examples from geometrical statistics. Int.Statist.Rev., **54**, 1-13.

Clustering Based on Neural Network Processing

H.-M. Adorf and F. Murtagh, Garching

Keywords: artificial neural networks: Hopfield nets — classification: k-min cluster, k-max cut — combinatorial optimization: approximation algorithms — computational complexity: *NP*-completeness

Introduction

Artificial neural networks — having been popular in the fifties and sixties — recently received a new wave of interest (Anderson 1986; Materna 1987; Fahlman & Hinton 1987). Neural network computing, also known as connectionism, is inspired from brain theory and is based on interconnecting a large number of simple processing elements called 'neurons', which cooperate in the computations. Essentially such a neuron adds up the weighted input on its input line, pushes the resulting net input through a (possibly non-linear) transfer function and communicates the output via its output lines to all other neurons to which it is connected. Mathematically the *static* part of a neural network is nothing but a 'network' in graph theory, i.e. a labelled weighted directed graph, whose vertices correspond to the neurons and whose weighted arcs represent the network connections. A network state corresponds to a set of weights for the vertices. The static description of a neural network is complemented by the specification of the *dynamics* governing its state changes.

When configured as auto- or hetero-associative memories, neural nets routinely perform supervised classification. Neural nets have also successfully been used for 'learning without a teacher' in demanding areas such as understanding of continuous speech and computer vision (see e.g. the March 1988 issue of Computer). Lippmann (1987) comprehensively discusses various neural network configurations for classification problems.

Many cluster criteria take the form of a *global optimization* problem. When the domains of the variables of the objective function are finite, then one may speak of *discrete* or *combinatorial optimization.* Global optimization problems with non-convex objective functions and most combinatorial optimization problems are known to be computationally difficult, and usually one has to be content with suboptimal approximations to the global optimum/optima sought (see e.g. Brucker 1979; Kannan & Korte 1984; Tovey 1985; Rinnooy Kan & Timmer 1987). It has been claimed that neural nets produce good suboptimal solutions for combinatorial optimization problems (Hopfield & Tank 1985, 1986; Jeffrey & Rosner 1986; Bounds 1987), but this has been criticised by Johnson (1987).

Compstat 1988

Design of a neural net for unsupervised classification

In the following we assume that we have to cluster a data set for which we are given a dissimilarity matrix. We will adopt the *k-min cluster* criterion (Sahni & Gonzalez 1976) which uses the overall *intra-cluster dissimilarity* as the objective function to be minimized. We show that a Hopfield-type neural net (Hopfield 1982, 1984) optimizing this criterion can be obtained essentially by 'identifying' the network connection matrix with the dissimilarity matrix, a construction inspired by the work of Hopfield & Tank (1985, 1986) who designed a neural network for approximately solving the well-known traveling salesman optimization problem.

The dynamics of a Hopfield neural net are governed by the update equation

$$\mathbf{y} \longleftarrow \sigma(W\,\mathbf{y} + \mathbf{a}) \quad , \tag{1}$$

where $\mathbf{y}$ is the network output vector of dimension m giving class assignments of all objects to all classes as explained below; $\mathbf{a}$ is an additional input (or offset) vector of dimension m; σ is a 'sigmoid' function whose components map each component of its argument into the interval $[0,1]$; and finally W is the network connection matrix of dimensions m^2 to be looked at below.

Associated with any Hopfield net is a Liapunov ('energy') function (Hopfield 1982; Feldman 1987), which in case of binary-valued neurons takes the form

$$\mathcal{E} := -\frac{1}{2}\mathbf{y}^T W \mathbf{y} - \mathbf{y}\,\mathbf{a} = -\frac{1}{2}\sum_{k,l} W_{kl}\, y_k\, y_l - \sum_{l} y_l\, a_l \quad . \tag{2}$$

The state of a Hopfield net with *symmetric* connection matrix provably converges to a local energy minimum, which corresponds to a solution (more precisely: an approximation) of the problem at hand.

The design of a neural network for problem solving proceeds in three steps: (i) devise an interpretation of neuron states, (ii) specify the connection matrix and (iii) devise some network dynamics appropriate to solve the given problem or approximate a solution.

A neural net as a classifier

We first address the question how a neural net can be viewed as a classifier. Let the number of objects under consideration be n and the number of classes into which these objects are to be grouped be k. Suppose that a neural net is laid down in a rectangular array of $n \times k$ neurons; thus each of the $m = n \times k$ neurons is uniquely labeled by a pair of indices $i\omega$. Now identify row i with object i to be classified and column ω with its potential class C_ω. This gives the required interpretation of the network states.

Representing the cluster data

Now we turn to the problem of devising a network connection matrix, which captures the cluster problem instance. If the objective function is at most quadratic, it can be straightforwardly interpreted as a neural net energy function from which a network connection matrix can be read off. We will consider two classes whenever it simplifies the presentation. The results obtained immediately carry over to the cluster problem with more than two classes.

When dissimilarities (d_{ij}) and a number k of desired classes are given, the k-min cluster criterion consists of minimizing the overall ***intra-cluster dissimilarity***. Let C_ω denote class ω; then the k-min criterion can be written as

$$\min_{\chi} \sum_{\omega} \mathcal{D}_{\omega\omega} \qquad \text{where} \qquad \mathcal{D}_{\omega\omega'} := \sum_{i \in C_\omega} \sum_{i' \in C_{\omega'}} d_{ii'} = \sum_{i} \sum_{i'} d_{ii'} \chi_{i\omega} \chi_{i'\omega'} \quad . \tag{3}$$

Here we have introduced the *characteristic function* $\chi_{i\omega}$ of a classification

$$\chi_{i\omega} := \begin{cases} 1, & \text{if object } i \text{ is in class } C_\omega; \\ 0 & \text{otherwise.} \end{cases} \tag{4}$$

It is implicitely understood that the decomposition represented by χ is disjoint, i.e. no object is simultaneously associated with two classes. We see that each component $\chi_{i\omega}$ effectively specifies the state $y_{i\omega}$ of a binary neuron: *iff* neuron $i\omega$ is on (off), object i is (is not) in class C_ω.

Let us further introduce the *total dissimilarity function* $\mathcal{D} := \sum_{\omega,\omega'} \mathcal{D}_{\omega\omega'}$, which is a constant for a given cluster problem. In the case of $k = 2$ classes it can be written as $\mathcal{D} = \mathcal{D}_{11} + \mathcal{D}_{12} + \mathcal{D}_{21} + \mathcal{D}_{22}$, i.e. as a sum of the overall intra-cluster dissimilarity $\mathcal{D}_{11} + \mathcal{D}_{22}$ and the overall *inter-cluster dissimilarity* $\mathcal{D}_{12} + \mathcal{D}_{21}$. Thus minima of the intra-cluster dissimilarity correspond to maxima of the inter-cluster dissimilarity.

For two classes the problem of maximizing the overall inter-cluster dissimilarity is equivalent to finding a *maximum cut* through the edges of a graph, i.e. a partition of the vertices into two classes, such that the sum of the weights of all edges running from one vertex set to the other is maximal. It is well-known (see e.g. Horowitz & Sahni 1979, p. 556) that this optimization problem is *NP-hard*, as the corresponding MAX CUT decision problem is *NP-complete* and polynomially reduces to the former. (For a discussion of computational complexity theory and in particular the notions of *NP*, *NP*-hardness, *NP*-completeness and problem-reducibility see Garey & Johnson 1979.) These results hold *a fortiori* for the case of $k > 2$ classes. Therefore, unless $P = NP$, there is scarcely any hope of finding an algorithm which *exactly* solves the k-min cluster problem in the sense that it guarantees to find a global minimum of the objective function in polynomial time. Even more irritating may seem the fact that finding an ϵ-approximation to a global optimum of the k-min cluster problem is also *NP*-hard for any $0 < \epsilon < 1$, whereas a $1/k$-approximate algorithm exists for the k-max cut problem (Sahni & Gonzalez 1976). These categorical results from computational complexity theory cannot be alleviated by deterministic parallelism, as any parallel computer can only be built from a polynomially bounded number of processors.

Now let us consider the energy function $\mathcal{E} = -\frac{1}{2} \sum_{i\omega} \sum_{i'\omega'} W_{i\omega,i'\omega'} y_{i\omega} y_{i'\omega'} - \sum_{i\omega} y_{i\omega} a_{i\omega}$. Comparing quadratic terms in this expression and eq. (3) and neglecting the factor 1/2 we read off the weights

$$W_{i\omega,i'\omega'} = -d_{ii'} \delta_{\omega\omega'} \quad . \tag{5}$$

Here we had to introduce a Kronecker delta-function, for ω and ω' are now arbitrary (and thus may be unequal). If we had chosen to take the inter-cluster dissimilarity, we would have gotten a weight matrix $W_{i\omega,i'\omega'} = d_{ii'}\bar{\delta}_{\omega\omega'}$, with $\bar{\delta}_{\omega\omega'} := 1 - \delta_{\omega\omega'}$. Clearly any linear combination of the energy functions (weight matrices) above is a valid energy function (weight matrix), too.

In an analogous manner from a given similarity matrix an objective function identical to the FORTIER/SOLOMON-criterion (Bock 1974, p. 204) can be derived.

Network control

Neural network algorithms are controlled by rules for neuron-updating. Various schemes can be envisaged, which result from combining deterministic or stochastic neuron selection with deterministic or stochastic neuron behaviour. Apart from choosing between binary valued, multiple valued or even continuous valued neurons, also the neural transfer functions have to be fixed.

We have chosen to work with continuous deterministic neurons which are updated by a deterministic parallel update rule according to the standard continuous Hopfield model (Hopfield 1984).

Implementation

In the absence of multiple processors providing true parallel neural network processing, the net was simulated on a serial multi-purpose computer.

The first component of the input vector $\mathbf{a}$ was set to a large constant b in order to arbitrarily force the first object to be in the first class. All other values of a were set to zero. This is therefore a labelling constraint only. The initial values of the output vector $\mathbf{y}$ were set to small uniformly distributed random values, in order to break potential ties. The sigmoid function was taken as a 'Boltzmann factor' $\sigma_i(x_i) = 1/(1+\exp(-x_i))$, identical for all neurons.

For the $k = 2$ and $k = 3$ class cases connection matrices

$$W^{(2)} := \begin{pmatrix} -D & D_b \\ D_b & -D \end{pmatrix} \quad \text{and} \quad W^{(3)} := \begin{pmatrix} -D & D_b & D_b \\ D_b & -D & D_b \\ D_b & D_b & -D \end{pmatrix} \tag{6}$$

were used. Here D denotes the given distance matrix and $D_b := D + \text{Diag}(-b)$, the latter being a diagonal matrix with constant $-b$ values. These b-values are introduced to enforce the constraint that the decomposition be disjoint. Therefore b has to be large compared to the distances.

An iterative solution of the update equation (1) is straightforward. To improve convergence, only a fractional update $\mathbf{y} \longleftarrow \mathbf{y} + 1/4(\tilde{\mathbf{y}} - \mathbf{y})$ was used at each iteration, where $\tilde{\mathbf{y}} := \sigma(W\,\mathbf{y} + \mathbf{a})$. The iteration was stopped when the state-vector changes fell below some small bound (ϵ-stopping criterion).

First results

Sets of two-class data were generated as follows. *Set I:* two (touching) groups of 50 uniformly distributed points in $[2,3]^2$ and $[3,4]^2$. *Set II:* two clearly separated groups of 80 and 20 (respectively) uniformly distributed points in $[2,3]^2$ and $[5,6]^2$. *Set III:* two (touching) groups of 80 and 20 uniformly distributed points in $[2,3]^2$ and $[3,4]^2$. Euclidean distances were used, normalized such that the maximum distance had value 1. b was set to 10.

Complete recovery of the known classes for sets I and II was achieved with 20 and 19 iterations respectively. In the case of set III, convergence after 28 iterations yielded correct recovery of three quarters of assignments to the larger class and all assignments to the smaller class. It is evident from these examples that well separated classes were properly handled by this approach.

In addition we tested our classifier with Fisher's well-known Iris data. This data set comprises readings on four physical attributes for three classes of Iris flower, each class consisting of 50 samples (objects). The Euclidean distance was used with standardization to zero mean and unit standard deviation. This was followed by normalization in the same manner as for the simulated data. b was set to 30. Following 40 iterations, the number of correctly classed objects for the three known classes were 48, 35 and 33. This represents a success rate of just over 77 %.

Conclusions

Stimulated by our interest in classifying the objects of the Point Source Catalog resulting from the Infra-Red Astronomical Satellite mission (Adorf & Meurs 1988; Meurs et al. 1988; Murtagh 1988) we have investigated a neural net based unsupervised classifier. We have demonstrated how the *k-min cluster* problem, based on similarities or dissimilarities, can be represented in a neural network of the Hopfield type. We have experimented with one of several possible forms of network dynamics and — given the simplicity of the approach — obtained encouraging results. Planned future work within the connectionist paradigm includes the use of a discrete stochastic neural net, currently being developed for scheduling astronomical telescopes (Johnston 1988; Johnston & Adorf 1988). It is also of interest to compare the performance of neural net based approaches with that of other approximation techniques such as simulated annealing or dynamic programming.

Acknowledgement

This work has been carried out within the Artificial Intelligence Pilot Project of the ST-ECF (Adorf & Johnston 1987).

References

Adorf, H.-M., Johnston, M.D.: 1987, "The Artificial Intelligence Pilot Project of the ST-ECF", *ST-ECF Newsl.* **8**, 4–5

Adorf, H.-M., Johnston, M.D.: 1988, "Artificial Neural Nets in Astronomy", *Proc. "Workshop Konnektionismus"*, Bonn, Feb. 1988, C. Lischka, J. Kindermann (eds.), (in press)

Adorf, H.-M., Meurs, E.J.A.: 1988, "Supervised and Unsupervised Classification — The Case of IRAS Point Sources", in: *Large-Scale Structures in the Universe — Observational and Analytical Methods*, W. Seitter, H. Duerbeck, M. Tacke (eds.), Springer Verlag, Heidelberg (in press)

Anderson, J.A.: 1986, "Networks for fun and profit", *Nature* **322**, 406–407

Bock, H.H.: 1974, *Automatische Klassifikation*, Vandenhoeck & Ruprecht, Göttingen

Bounds, D.G.: 1987, "New optimization methods from physics and biology", *Nature* **329**, **No. 17**, 215–219

Brucker, P.: 1979, "*NP*-Complete Operations Research Problems and Approximation Algorithms", *Z. Operations Research* **23**, 73–94

Fahlmann, S.E., Hinton, G.E.: 1987, "Connectionist Architectures for Artificial Intelligence", *Computer* **20, No. 1**, 100–109

Feldman, J.A.: 1987, "Energy Methods in Connectionist Modelling", in: *Pattern Recognition Theory and Applications*, P.A. Devijver, J.Kittler (eds.), Springer-Verlag, Berlin–Heidelberg–New York, pp. 223–247

Garey, M.R., Johnson, D.S.: 1979, *Computers and Intractability*, W.H. Freeman and Company, San Francisco

Hopfield, J.J.: 1982, "Neural networks and physical systems with emergent collective computational abilities", *Proc. Nat. Acad. Sci.* **79**, 2554–2558

Hopfield, J.J.: 1984, "Neurons with graded response have collective computational properties like those of two-state neurons", *Proc. Nat. Acad. Sci.* **81**, 3088–3092

Hopfield, J.J., Tank, D.W.: 1985, ""Neural" Computation of Decisions in Optimization Problems", *Biol. Cybern.* **52**, 141–152

Hopfield, J.J., Tank, D.W.: 1986, "Computing with Neural Circuits: A Model", *Science* **233**, 625–633

Horowitz, E., Sahni, S.: 1979, *Fundamentals of Computer Algorithms*, Pitman Publishing Ltd., London

Jeffrey, W., Rosner, R.: 1986, "Optimization Algorithms: Simulated Annealing and Neural Network Processing", *Astrophys. J.* **310**, 473–481

Johnson, D.: 1987, "More approaches to the travelling salesman guide", *Nature* **330**, 525–525

Johnston, M.D.: 1988, "Automated Observation Scheduling for the VLT", in: *Proc. ESO Conf. on Very Large Telescopes and their Instrumentation* (in press)

Johnston, M.D., Adorf, H.-M.: 1988, "Scheduling with Neural Networks", (in preparation)

Kannan, R., Korte, B.: 1984, "Approximative Combinatorial Algorithms", in: *Mathematical Programming*, R.W. Cottle, M.L. Kelmanson, B. Korte (eds.), pp. 195–248

Lippmann, R.P.: 1987, "An Introduction to Computing with Neural Nets", *IEEE Acoustics Speech and Signal Processing Magazine, April 1987*, 4–22

Materna, T.: 1987, "Neural networks enter high speed marketplace", *Computer Technology Review* **VII, No. 7**, June 1987

Meurs, E.J.A., Adorf, H.-M., Harmon, R.T.: 1988, "Mapping the Extragalactic Sky with IRAS PSC Data", in: *Proc. Conf. "Astronomy from Large Databases"*, F. Murtagh & A. Heck (eds.), pp. 49–54

Murtagh, F.: 1988, "Multivariate data analysis: bakckground and example", in: *Large-Scale Structures in the Universe — Observational and Analytical Methods*, W. Seitter, H. Duerbeck, M. Tacke (eds.), Springer Verlag, Heidelberg (in press)

Rinnooy Kan, A.H.G., Timmer, G.T.: 1987, "Stochastic Global Optimization Methods Part I: Clustering Methods", *Mathematical Programming* **39**, 27–56

Sahni, S., Gonzalez, T.: 1976, "P-Complete Approximation Problems", *J. Assoc. Comput. Mach.* **23**, 555–565

Decision Tree Classifier for Speech Recognition

A. M. Morin, Rennes

ABSTRACT

Both the computational complexity and the massive storage make the word recognition strategies based on a word model unfeasible for large vocabulary speech recognition applications. One of the possible solutions for continuous speech recognition is to recognize the basic phonetic units of the input speech. A component of an analytic recognition system is the acoustic-phonetic processor whose purpose is to encode the speech signal into a string of discrete subword units, such as phones, diphones...In the present approach, a segmentation is done prior to recognition. For that, statistically based methods are used to detect non-stationarities in the speech signal and therefore, the nature of the segmented units is not defined in advance.
We deal with the "segments", each of them being characterized by the energy, the first six cepstral coefficients of the signal and by some indices summarizing the phoneticians knowledge. We build a binary decision tree where the leaves are the phonemes of the french language -vowels, semi-vowels, consonants-. The configuration of the tree has been settled making use of a training set for a given male speaker. We construct it in a bottom-up stepwise way, using a likelihood criteria. We can not be sure that we get the "best" tree in terms of classification rate . Annealing methods allow us to improve the original configuration.
At each node, we fit either logical rules, either probabilistic rules using a logistic regression model. The system gives good results on test samples pronounced by the same speaker.
The segmentation is expected to be reliable and speaker independent. But the characteristic variables are not speaker independent. We know that there are some invariants in speech signal particularly in the plane defined by the first two formants, whoever the speaker is. We are investigating different ways of adapting easily the rules for any speaker. It could be a small sample of artfully choosen words.
This work is supposed to be a part of a complete analytic recognition system integrating different steps such as segmentation, vectorial quantization, units recognition, lexical decoding and linguistic analyser.

KEY-WORDS : discrimination, logistic regression, segmentation, word recognition, expert system.

1- INTRODUCTION

Automatic speech recognition systems have been developed to investigate the possibility of using voice as a communication support between a user and a computer for several tasks. Such systems often employ phonemic recognition schemes where the classification of vowels plays a key role . Three areas of speech recognition can be distinguished :

1- isolated word recognition where words are separated by distinct pauses;

2- continuous speech recognition where sentences are produced continuously in a natural manner;

3- speech understanding where the aim is not transcription but understanding in the sense that the system responds correctly to a spoken instruction or request.

Commercially available products exist for isolated word recognition with vocabularies of up to several hundred words. Products which recognize continuously spoken small vocabularies are already on the market but the goal of unrestricted continuous speech recognition is far from being realized.
In France, at the CNET, a system-expert called SERAC ,expert system for acoustic-phonetic recognition , has been developped for speech recognition (Mercier, 1984). A speech recognition system, SRS, can be described as a set of components which successively perform the acoustic, phonetic, lexical and syntactic analysis in order to reconstitute the uttered sequence from the speech signal
The acoustic-phonetic processing transforms the continous speech signal in a string of discrete units such as

Compstat 1988

phonemes, syliabes This stage requires a segmentation of the continuous speech prior to recognition. And we deal with these segments, trying to recognize them.

For that, we propose an alternative approach to classification which involves a procedure for the design of a decision tree classifier. Basically, this approach is characterized by the fact that samples are subjected to a sequence of decision rules before they are assigned to one of k prespecified classes. The decision procedure described here was motivated by a practical requirement for a general and flexible pattern recognition algorithm which could be trained efficiently with a minimum of manual intervention. Each of the binary decisions is specified by two components : a discriminant function and the set of classes involved at this point of the decision tree. Classical discriminant analysis is widely associated with multivariate normal distributions with common covariance matrix in each group, that is, in our case , in each of the two subsets defined by the split at a node. This cannot be true for each subset at each node. So we choose to fit a logistic discrimination model at each vertex. This method performs well and has several distinct advantages :

it is simple to use,
we don't need a lot of parameters,
we don't need a lot of distributional assumptions,
it works with continuous and, or, discrete variables.

In section 2, we propose a definition of decision tree classification and give mathematical details on the technique for tree design. In section 3, we design a classifier for data in speech recognition, more precisely in vowel recognition. Since vowels are harder to identify than consonants, the performance of the classifier is less spectacular. The resulting decision tree has an associated probability of correct classification of 74% on the sample test. This result is as good as what we can obtain with other methods dealing with the same data, such as vector quantization for example. And this tree classifier is found to be faster than the other techniques. If we construct a tree with structured rules such as Breiman's classification trees, it tends to be too elaborate and to have too many branches even after pruning. But some of the rules of this structured tree separate very well some classes of vowels. Therefore, it would be interesting to build a tree using both kinds of rules. In section 4 , we discuss directions for future research.

2- BINARY DECISION TREE CLASSIFIER : DESCRIPTION , TREE DESIGN AND DECISION RULES

a- description

We assume that a given sample set G included in R^p and a given set C of k classes numbered 1,2,....,k, C = { 1, 2, ..., k}, have been specified. Let G_i be the set of samples associated with the class i. $G=\cup G_i$. $G_i \cap G_j = \emptyset$, $i \neq j$. We define $x(c) = (x_1(c), x_2(c),...., x_p(c))$ the measurement vector for case c. The cases are supposed to fall into one of the k classes. We want to find a rule that assign a class membership in C to every measurement vector x in G. An element $x \in G_l \subset G$ is considered to be correctly classified by decision rule d if $d(x) = l$.

The classifier construction is based on past experience summarized by a learning set A. This consists of the measurement data on n_A cases observed in the past with their actual classification.

$A = \{ (x(1), j(1)), (x(2), j(2)), (x(3), j(3)),......,(x(n_A), j(n_A))\}$.

In the binary tree classifier, the global decision rule d will be broken up into a sequence of decision rules at each node. A classifier at a vertex v will be expressed as a pair (D_v , P_v) where D_v is a set of parameters which defines the classifier at this point and P_v is an explicitly expressed set of disjoint subsets of C. Not all elements of C are involved at each node.

Our decision tree classifier satisfies the following conditions :

- the root node, v = 1, is the first level in the tree. It is unique and P_1 is equal to C.
- if v and v' are two different vertices, then P_v is different from $P_{v'}$.
- there are exactly k terminal nodes in the tree. Let T be the set of terminal nodes, then T is equal to C.

Consequently, there are exactly k-1 decision nodes and no overlap between branches.

If a sample goes to a wrong branch at any level, it can never reach the correct terminal. Although error accumulation can be reduced by introducing overlap, this is not practical because overlap propagates from level to level and this results in a substantial increase in the memory requirement.
To estimate the accuracy of the classifier, we use a test set which is independant of the learning set. In section 3, we explain how we get both, learning set and test set.

b- tree design

The following step is the tree construction. To design the tree, we prefer agglomerative techniques which start with k separate elements, that is the k classes, grouping them successively in k-1, k-2,...., 1 taxa and computing at each stage a criterion which is the measure of the relevance of the regrouping at this point. We call groups the subsets Γ_{v1} and Γ_{v2} defined by the split at a node v. A group can be a class but more generally, it is a grouping of classes.
We propose two criteria to build the tree.
The first one is Ward's criterion (Ward, 1963). Define m_{vi} the mean vector for the group Γ_{vi}, and n_{vi} the weight, generally the size of Γ_{vi}.

The Ward's criterion, $C_W(\Gamma_{v1}, \Gamma_{v2})$ is equal to $\left(\frac{n_{v1} n_{v2}}{n_{v1}+n_{v2}}\right)^{\frac{1}{2}} d(m_{v1}, m_{v2})$.

d is an euclidean distance in R^p.
One of the simplest and most widely criterion function for clustering is the sum of squared error criterion. The pair of groups whose merger increases the sum of square error criterion as little as possible at a node, is the pair for which the criterion C_W is minimum. Thus, in selecting groups to be merged, these criteria C_W takes into account the number of samples in each group as well as the distance between groups.
The second criterion C_L is a function of the likelihood of the samples at a node. Define $p_{v1}(c)$ as the probability for the case c to go to the left branch, corresponding to the group G_{v1}, of the vertex v.
$p_{v2}(c)$ is equal to $1 - p_{v1}(c)$. We make the assumption that the probability for a branch is determined by a logistic-linear function of the p variables $x_1, x_2, ...x_p$.

$$p_{v1} = \frac{\exp(\alpha_0 + \alpha' X)}{1+\exp(\alpha_0 + \alpha X)} \quad \text{where } \alpha' = (\alpha_1, \alpha_2, ..., \alpha_p).$$

At each node, the probability p_{v1} has to be estimated as a function of the p independant variables. This is a classical problem in logistic regression. The p +1 parameters $\alpha_0, \alpha_1,....., \alpha_p$ are estimated by maximum likelihood involving non-linear iterative techniques.
The C_L criterion function is what is usually called the deviance in generalized linear models. Define $p_{vi}{}^*$ the estimate of p_{vi}.

$C_L(\Gamma_{v1}, \Gamma_{v2})$ = -2 (maximized likelihood at the node v) = $-2\left(\sum_{c\in\Gamma_{v1}} p^*_{v1}(c) + \sum_{c\in\Gamma_{v2}} p^*_{v2}(c)\right)$. And we merge the groups for which C_L is minimum.

The procedure is simple ansd is the same for both criterion functions:

1- let k'=k.
2- if k' = 1, stop. (the root node has been reached)
3- find the nearest pair of of distinct groups and merge them.
4- k' = k - 1.
5- go to 2.

At the root node, $\Gamma_{11} \cup \Gamma_{12} = C$.

One of the simplest and most widely criterion function for clustering is the sum of squared error criterion. The pair of groups whose merger increases the sum of square error criterion as little as possible at a node, is the pair for which the criterion C_W is minimum. Thus, in selecting groups to be merged, these criteria C_W takes into account the number of samples in each group as well as the distance between groups.
The second criterion C_L is a function of the likelihood of the samples at a node. Define $p_{v1}(c)$ as the probability for the case c to go to the left branch, corresponding to the group G_{v1}, of the vertex v.
$p_{v2}(c)$ is equal to $1 - p_{v1}(c)$. We make the assumption that the probability for a branch is determined by a logistic-linear function of the p variables $x_1, x_2, \ldots x_p$.

$$p_{v1} = \frac{\exp(\alpha_0 + \alpha' X)}{1+\exp(\alpha_0 + \alpha X)} \quad \text{where } \alpha' = (\alpha_1, \alpha_2, \ldots, \alpha_p).$$

At each node, the probability p_{v1} has to be estimated as a function of the p independant variables. This is a classical problem in logistic regression. The p +1 parameters $\alpha_0, \alpha_1, \ldots, \alpha_p$ are estimated by maximum likelihood involving non-linear iterative techniques.
The C_L criterion function is what is usually called the deviance in generalized linear models. Define p_{vi}^* the estimate of p_{vi}.

$C_L(\Gamma_{v1}, \Gamma_{v2}) = -2$ (maximized likelihood at the node v) = $-2(\sum_{c \in \Gamma_{v1}} p_{v1}^*(c) + \sum_{c \in \Gamma_{v2}} p_{v2}^*(c))$. And we merge the groups for which C_L is minimum.
The procedure is simple ansd is the same for both criterion functions:

1- let k'=k.
2- if k' = 1, stop. (the root node has been reached)
3- find the nearest pair of of distinct groups and merge them.
4- k' = k - 1.
5- go to 2.

At the root node, $\Gamma_{11} \cup \Gamma_{12} = C$.

c-decision rules

Once the tree topology has been found, it is necessary to define the decision rules at each vertex. The objective is to allocate at each node v, a sample with observations $X \in R^p$ to one of the two groups Γ_{v1} or Γ_{v2}. Remember that not all classes of C are involved in each node. The possibility that a sample is from neither group is not considered here. It would be the case if it was not well-assigned at a previous node.
We use logistic discrimination to assign an observation X and in that case

$$\text{Prob}(\Gamma_{.1} / X) = \frac{\exp(\beta_0 + \alpha' X)}{1+\exp(\beta_0 + \alpha X)} = d(X).$$

Note that $\beta_0 = \alpha_0 + \log \frac{\pi_1}{\pi_2}$, where π_i is the proportion of sample points in the group $\Gamma_{.i}$. The training set A is used to build the tree and to estimate the parameters α. D_v is the set of parameters α at the node v. The decision rule at each node v will be :

c-decision rules

Once the tree topology has been found, it is necessary to define the decision rules at each vertex. The objective is to allocate at each node ν, a sample with observations $X \in R^p$ to one of the two groups Γ_{v1} or Γ_{v2}. Remember that not all classes of C are involved in each node. The possibility that a sample is from neither group is not considered here. It would be the case if it was not well-assigned at a previous node.
We use logistic discrimination to assign an observation X and in that case

$$\mathrm{Prob}(\Gamma_{.1} / X) = \frac{\exp(\beta_0 + \alpha'X)}{1+\exp(\beta_0 + \alpha X)} = d(X).$$

Note that $\beta_0 = \alpha_0 + \log \frac{\pi_1}{\pi_2}$, where π_i is the proportion of sample points in the group $\Gamma_{.i}$. The training set A is used to build the tree and to estimate the parameters α. D_v is the set of parameters α at the node ν. The decision rule at each node ν will be :

- assign the observation x(c) to Γ_{v1} if $d(x(c)) \geq 0$,
- assign the observation x(c) to Γ_{v2} if $d(x(c)) < 0$.

At each node, the discriminant function $d(x) = 0$ defines a decision surface which is a hyperplane.

4- EXPERIMENTAL RESULTS

The data are "units" of the continuous speech signal where sentences are produced continuously in a natural manner. The data base is generated by one male speaker and is composed of five sets of ten phonetically balanced sentences sampled at 12800Hz. The first three sets (690 samples) constitute the training set where as the last two sets (605 samples) go to make up the test set. The units are segments of the signal obtained by a sequential method which detects ruptures (R. Andre-Obrecht 1988). Each segment is characterized by its energy , six cepstral coefficients, five phoneticians' cues and a label which indicates its phonetical class. The 11 phonetical classes correspond to 11 french vowels or semi-vowels |a|, |ã|, |o|, |õ|, |i|, |j|,|Ø|, |u|, |y|,|w|, |ɛ|.
We use multiple factorial analysis (Escofier, Pagès) in order to reduce the dimensionality of the problem - p=12 initially-. The set of phoneticians cues was highly correlated to the energy and the cepstral coefficients and we don't use it in the analysis.
We obtain the same results in terms of estimated error rate on the test set, around 25%, with C_W and C_L although the tree sructures are not exactly the same. 25% seems to be the average rate obtained with different methods such as vector quantization on the same data base. On the training set, the estimated error rate is 14% - resubstitution error rate- for both criterion functions but it is known to be optimistic. By using annealing methods to improve the C_L-tree configuration, we get the C_W-tree . If the tree is not constrained to be dichotomous, at the root node , the set of classes C splits in three subsets. The gain in terms of classification rate is not significantly improved where as the memory requirement and the assignment time increase.
The most frequent confusions are between |i| and |j|, | Ø| and |ɛ|. Depending on the context, it happens that some |a| are assigned in the | Ø| class.
In a neutral position where the vocal tract can be regarded as a uniform tube, the frequencies of its resonances are called formant frequencies. In speech recognition, the first two formants F_1, F_2 are of essential importance. The splittings defined at each node of the tree have an obvious meaning in the (F_1F_2) plane. And the confusion of some vowels in classification conforms with their distribution in this plane.
After pruning, the structured classification tree has 15 terminal nodes. To get it, we did not use CART but a very similar algorithm called recursive partition (Ciampi, Thiffaut) . Some of the leaves are well-identified, |a|, |i| but most part of them are mixtures of vowels. The interesting thing happens at the first two nodes. All the vowels labelled |i| go to the right branches which lead to terminal nodes.

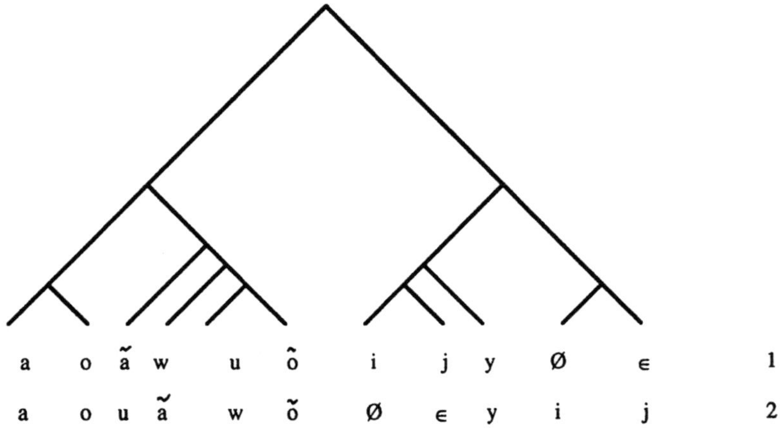

1 : Decision tree obtained using the likelihood criterion

2 : Decision tree obtained using Ward's criterion

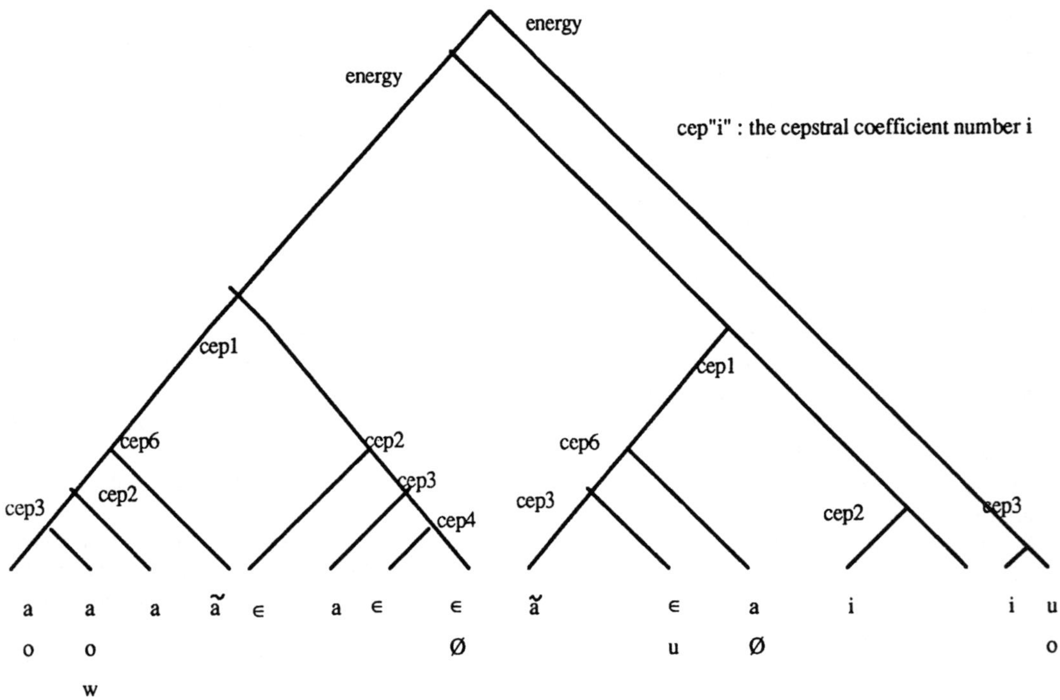

Classification tree on vowels obtained by the recursive partition algorithm

5-CONCLUSION

The problem of speech recognition is increasingly drawing the attention of scientific workers because of the potentiality for man-machine communication. The performance of a recognition system is highly dependent on the quality of segmentation. We suppose that this segmentation is reliable. We are concerned with the recognition step which is essential for the lexical decoding and the linguistic analysing.
The main advantage of the tree classifier is the gain in terms of searching time. In a balanced binary tree with k terminal nodes, it is in log (k) time units. One time unit is required for an unknown pattern to pass one level, that is one node of the tree. To improve the classification rate of the binary tree classifier, it would be worthwhile to introduce fuzzy set theory in the decision rules. We would evaluate the membership value of a sample not only for its assignment class but for the nearest neighbours of its assignment class . Doing that will be simpler than allowing group overlaps at each level.
Another problem is the robustness of the system for several speakers. Our system works only for one male speaker and must be fit for other speakers. If we suppose that the tree configuration is the same for every speaker, the idea is to adapt the α parameters at each node.

REFERENCES

André- Obrecht R. 1988 - A new statistical approach for the automatic segmentation of continuous speech signals. IEEE ASSP 36, 29-40.
Anderson J. A. 1982 - Logistic discrimination in P R KRISHNAIAH and L N KANAL ed. : Handbook of statistics, 2; pp169-192.
Breiman L., Friedman J. H., Olshen R. A., Stone C. J. 1984 - Classification and regression trees. Wadsworth Int. Co., Belmont.
Ciampi A., Thiffaut J. - 1986 - Recursive partition : a versatile method for exploratory analysis in biostatistics. Pub. McGill Unervisity, Montréal.
Escofier B., Pagès J. 1984 - L'analyse factorielle multiple. Cahier du BURO, 42, Un. Pierre et Marie Curie, Paris.
Mercier G., Gilloux M., Tarridec C., Vaissière J. 1984 - From KEAL to SERAC : a new rule-based expert system for speech recognition in NATO ADVANCES STUDIES INSTITUTE ed. : New systems and architecture for automatic speech recognition and synthesis, Bonas.
Ward J. H. 1963- Hierarchical grouping to optimize an objective function. JASA, 58, 236-244.

Algorithms

Efficient Sampling Algorithms and Balanced Samples

J. C. Deville, J. M. Grosbras and N. Roth, Paris

Need for sampling algorithms in a finite population has greatly increased in recent years. New quick methods for drawing a sample from a data file have emerged. Their principal utilizations are not, however, in the design of survey samples, but lie in methodological work involving repeated sampling, for instance bootstrap, and simulation. Then, in the case of rejective sampling, where samples are drawn until some specific property is realized, performant drawing becomes a necessary tool if the desired property has a very small probability.

The first part of this paper reviews some algorithms for sampling on a data file with equal or unequal probabilities, with main emphasis put on cost or efficiency considerations. The second part tackles another kind of efficiency, describing a method to get balanced samples using some of the algorithms shown in part one in an iterative way. The method is illustrated by a large simulation study showing that balancing at the sampling stage can nearly achieve the same efficiency as regression estimation.

I - RANDOM SAMPLING WITHOUT REPLACEMENT

I.1 - Sampling with equal probabilities

Usual algorithms for randomly selecting n records from a file containing N records may be dramatically expensive in case of large frames or large samples, especially in applications involving simulation process, for example in comparing the properties of different estimators. The two most known methods are "sequential" (SQ) and "direct access" (DA) drawing.

The SQ algorithm is based on the simple argument that if at any given time N records have not yet been processed and if n must be selected, then the next one should be choosen with probability n/N. That requires as many uniform

Compstat 1988

random deviates and tests as records in the file. In DA algorithm, n different random integers are generated between 1 and N, and corresponding records are selected. That requires an array of n pointers, and a growing number of comparisons with the size of sample (to have different integers) ; moreover, records selected are not in their initial order.

More performing methods, at first due to FAN an alii (1962), so called S- method by KNUTH (1981), consist in processing the file by steps of random range : at a given time, an integer S in choosen, following a probability distribution F(s), S records are skipped, and the next one is selected. The distribution function F(s) can be expressed in two equivalent ways, leading to two implementations :

S1 : $F(s) = P(S \leqslant s) = 1 - \binom{N-s-1}{n} \Big/ \binom{N}{n}$, $s = 0,\ldots, N-n$

S2 : $S = \lfloor \min_{1 \leqslant j \leqslant n} (N-j+1)U_j \rfloor$, where $\lfloor \cdot \rfloor$ is the floor operator, and $U_1,\ldots,U_n$

are independant uniform variables.
As in SQ and DA methods, S1 and S2 algorithms become expansive when frame (S1) or sample (S2) are very large.

S methods have been significantly improved by VITTER (1984, 1987), using variables with G and H distributions, quite close to F but easier to compute. The main features of VITTER algorithms (D) are :

D1 : Generate a random deviate U, uniformly distributed between 0 and 1, and a random deviate X following G.
D2 : If $U \leqslant h(\lfloor X \rfloor)/cg(X)$, then set $S := \lfloor X \rfloor$ and go to D4.
D3 : Otherwise if $U \leqslant f(\lfloor X \rfloor)/cg(X)$, then set $S := \lfloor X \rfloor$ and go to D4. Otherwise return to step D1.
D4 : Skip over S records, select the following one, set $N := N-S-1$ and $n := n-1$. Return to D1 if $n>0$.

$c \geqslant 1$ is a constant so that $f(\lfloor x \rfloor) \leqslant c\, g(x)$ for all x in the domain of G, and h a density so that $h(s) \leqslant f(s)$. It can be proved that the steps D1, D3 and D4 produce a random deviate S with the right law. Step D2 avoids to compute, most of the time, f $(\lfloor x \rfloor)$. VITTER gives a lot of details to optimize the implementation of the D-algorithm, and propose convenient sets for c, g and h. For instance :

$$c = \frac{n}{n-1}\,\frac{N-1}{N}\,, \quad g(s) = \frac{n-1}{N-1}\left(1-\frac{n-1}{N-1}\right)^{s}, \quad h(s) = \frac{n}{N}\left(1-\frac{n-1}{N-s}\right)^{s}$$

We have implemented the D-algorithm in the datastep language of SAS[R] System, and experimented on a IBM 4361 under VM-SP4. The sampling frame is the set of numbers between 1 and N. Times (CPU + I/O) are expressed in seconds.

Table 1 - N = 100 000

Method	n=100	n=500	n=1000
SQ	40,6	41,0	41,8
DA	1,4	8,3	24,4
S1	6,3	6,8	7,1
S2	1,3	13,8	51,9
D	1,4	2,1	2,7

Table 2 - n = 100

Method	$N=10^4$	$N=10^5$	$N=10^6$
SQ	4,4	40,6	409,0
DA	1,3	1,4	1,4
S1	1,2	6,3	56,3
S2	1,1	1,3	1,4
D	1,3	1,4	1,6

Table 3 - $N = 10^6$

Method	n=1000	n=2000	n=4000
DA	24,5	176,3	304,5
D	4,3	6,8	9,9

Clearly, the D-algorithm is to recommend for samples over 100 elements.

Another very powerful method (SG-algorithm) has been described by AHRENS and DIETER (1985). The basic idea is that if we use independant random variates G_i, following a geometric distribution with parameter p = A/N, as skip distances until the file is exhausted, the number X of records in the sample is a binomial variable with A for mean, that is equivalent to accepting each record with constant probability p. Then we can choose A superior to n, to avoid being "too short" (X<n), but not too far from n, to avoid being "too long" (X>>n) ; if too long, randomly deletions must be performed to obtain a sample with desirable size and right inclusion probabilities. Then, the main steps are as follows :

- Let A and B integers so that n<A<B. Generate random deviates G1, G2, ... from geometric distribution with mean A/N. In a first pass, G1, G2... are skip distances to select indexes in the file. Let q the number of indexes found.

- If q<n or q>B, the process is restarted. Otherwise, the number of indexes is reduced to n by randomly deleting elements and, in a second pass, the selection of records is performed.

AHRENS and DIETER give efficient values for A and B, to avoid the odds of repeating and deleting. For example : $A= n + a\sqrt{n}$ and $B= n + 4a\sqrt{n}$, where $a= \sqrt{\ln(1+n/2\pi) + \ln(1+n/2\pi)/3n}$.

The SG-method has been found to run still faster than D-algorithm, but it requires an auxiliary space to store array of B pointers.

I.2 - Sampling with unequal probabilities

HANIF and BREWER (1980) gave a review of various methods for pps random sampling. Most of them are available in practice for very small samples (n=2,3), or require severe restrictions for the inclusion probabilities π_i, or are inextricably complex.

The simplest method, and the most used, is the pps systematic algorithm. HARTLEY and RAO (1962) have provided, with tedious developments, asymptotic and approximative expressions for the pairwise inclusion probabilities π_{ij}, and so for the variance of HORVITZ-THOMPSON estimators. HAJEK (1981) questioned the accuracy of these expressions and gave a counterexample. A model for the optimality of systematic sampling is shown in IACHAN (1982).
Another simple method has been suggested by RAO, HARTLEY and COCHRAN (1962). It consists in randomly clustering the units into n groups, with sizes N_g (the N_g are as equal as it is possible), and then drawing one unit with pps from each group. The procedure works with right π_i and estimators for total and its variance are easily set up.

From our point of view, the best method has been described by SUNTER (1986). It is a sequential algorithm, as for SQ method.
At record number i let : $a_i = 1$ if record i has been chosen and 0 otherwise, $n_i = \sum_{j=1}^{i} a_i$, $p_i = \pi_i/n$, $t_i = \sum_{j=1}^{i} p_j$. Then record (i+1) would be selected with probability $\pi^*_{i+1} = (n - n_i)p_{i+1}/(1-t_i)$.

As pointed out in DEVILLE-GROSBRAS (1987), the pairwise inclusion π_{ij} derive from the following lemma : $E(n-n_i)(n-n_i-1) = n(n-1)\sum_{h=1}^{i} q_h$, where $q_h = 1 - 2p_h/(1-t_{h-1})$, $t_0 = 0$, $p_0 = 0$.

Then the π_{ij} are shown to be, for $j > i$:

$$\pi_{ij} = \pi_i \pi_j Q_i, \text{ where } Q_i = \frac{n-1}{n} \frac{1}{(1-t_{i-1})(1-t_i)} \prod_{h-1}^{i-1} q_h, \text{ and}$$

$$Q_1 = \frac{n-1}{n} \frac{1}{1-p_1}.$$

Moreover, it can be easily proved that the π_{ij} satisfy the Yates-Grundy conditions $(\pi_{ij} - \pi_i \pi_j \leqslant 0)$.

In the applications, troubles may occur if $(n - n_i)p_{i+1}/(1-t_i)$ is found to be higher than 1, especially at the end of the file where the denominator $1 - t_i = p_{i+1} + ... + p_N$ becomes small. To postpone the occurence of the problem, it is useful to sort the file by descending probabilities ; but then the problem certainly occurs when $N-i+1 = n - n_i$, if probabilities are strictly decreasing. In such a case, the best solution is in equalizing the remaining probabilities. For instance, in a file of firms, unequal probabilities mainly distinguish large firms, the small firms have nearly equal sizes, and there is no practical inconvenient to consider them with equal probabilities to complete the sample.

II - BALANCED SAMPLES

II.1 - Definition and general considerations

Let $X = (X_1, X_2,...,X_p)'$ be a vector of a numerical variables available for every unit of a population of size N. The value of X_k for unit i is $X_{k,i}$ and $X(i) = (X_{1,i}, X_{2,i},..., X_{p,i})'$. A sample s of size n drawn from the population is balanced for the set of variables (X_k) if :

$$\bar{x} = \frac{1}{n} \sum X(i) = \bar{X} = \frac{1}{N} \sum X(i)$$

Balancing a sample is an old preoccupation of survey designers. It is a way to get a "representative sample", because, in a naive sense, the sample is seen as a reduced reproduction of the population. In the framework of probabilistic sampling, balancing is not a goal and is achieved only in expectation. Balancing appears to be a condition for robust inference in the context of analytical surveys (ROYALL and HERSON-1973) and is emphasized by HAJEK (1981) to improve the precision of estimations in a classical context.

The idea of balancing a sample is closely related to regression estimation. Suppose we are in the situation where the sample means $(\bar{y}, \bar{x})$ of a variable of interest Y and of a vector of auxiliary variables X are approximately jointly normal. The regression estimator $\hat{y}_r=\bar{y}+b'(\bar{X}-\bar{x})$ works generally better than the sample mean to estimate $\bar{Y}$. If, by chance, $\bar{X} = \bar{x}$, the two estimators take the same value. An idea to keep together the simplicity and the precision of the estimator is to design samples for which the equality $\bar{X} = \bar{x}$ is achieved (at least better than for simple random sampling). If X is one dimensional there are many ways to obtain balancing (see for example YATES-1949 or THIONET-1953). We propose a new algorithm for balancing a sample when X is multidimensional.

II.2 - Description of the algorithm

The informal idea is to drop from a current sample some units lying in the direction of $\bar{x}-\bar{X}$ and to replace them by new ones coming from the opposite direction. The number of changes is computed in order to achieve balance in expectation, and the process is iterated until no further modification is needed.

We now describe the method in more precise terms. Assume, by convention that $\bar{X} = 0$. We have first to split the sampling frame into G groups (g=1 to $G \geq p$) such that the means $\bar{X}^g$ in the different groups are very different and such that the sizes N_g of every group are nearly equal. In practice, for p = 2 to 5, we can build 2^p groups according to the possible vectors of signs of $(X_1,\ldots X_p)$. A linear transformation of the variables may be performed before this operation. It is especially useful, as it is done in the following application, to make use of the principal components of X.

The algorithms consists in the following steps :

<u>Step 1</u> : Stratification with proportional allocation. In each group g, draw a simple random sample of size $n_g = n.N_g/N$. Numbers n_g are rounded off to integers using the method of step 3. Set A=n and t=1. Let $\bar{x}^g$ be the mean of the sample drawn in group g and $\bar{x}_{(t)} = \frac{1}{n} \sum_g n_g \bar{x}^g$.

Step 2 : Modification of the sample.

Search a set of integers ν_g. If $\nu_g > 0$ we shall include in the sample ν_g news units at random coming from group g. If $\nu_g < 0$, we drop at random $|\nu_g|$ units from the sample in group g. In order that the size of the sample remains equal to n and that the expected value after the modification is equal to 0, the numbers ν_g must satisfy :

$$\sum_g \nu_g = 0$$

$$\sum \nu_g \bar{X}^g = -n\, \bar{x}_{(t)}$$

For a unique determination of the ν_g, we minimize the quantity $\Sigma \nu_g^2/N_g$, the variation of allocation between groups.

Note P the diagonal matrix with elements N_g^{-1} on the diagonal, ν the vector of ν_g and Z the (p+1)xG matrix whose first p rows are the $X^{g'}$ and with $Z_{p+1,g} = 1$. We have to find ν minimizing $\nu' P \nu$ under the constraints $Z\nu = -n\tilde{x}$, where $\tilde{x} = (x'_{(t)}, 0)$. The solution is :

$$\nu = n R \tilde{x}$$

$$\text{with } R = -P^{-1} Z' (ZP^{-1} Z')^{-1}.$$

Step 3 : Rounding. Round off the number obtained in step 2 in the following way : let ν_g^- be the greatest integer such that $\nu_g^- \leqslant \nu_g$ and $r_g = \nu_g - \nu_g^-$. The numbers r_g add to an integer d<p. We draw d groups at random with probabilities equal to r_g using one of the algorithms described in part I-2. For the drawn groups $\nu_g = \nu_g^- + 1$ and for the others $\nu_g = \nu_g^-$.

Step 4 : If $\Sigma|\nu_g| \geq A$ retain the sample and end. Otherwise, set t=t+1, $A = |\nu_g|$. For each group g, if $\nu_g > 0$ draw ν_g units in the sample by simple random sampling and drop them. If $\nu_g < 0$, draw $|\nu_g|$ new units by simple random sampling in group g and add them to the sample. Compute $\bar{x}_{(t)}$ and go to Step 2.

II.3 - Some practical and theoretical aspects

It is clear that for small populations and small samples, balancing is not always possible. Moreover, if we aim to balance at best the sample, we could have to modify inclusion probabilities very drastically (think for example to the case n=1 or n=2). The good probabilistic properties of the above algorithm are in some sense asymptotical and rely on a normal approximation for quantities like $n^{-1/2}\, \bar{x}_{(t)}$.

The following properties can be shown if n is large :

* For the tth sample obtained with the algorithm :

$$E\ \bar{x}'_{(t)}\ \bar{x}_{(t)} = O\ (n^{\ 1-t/2}).$$

At each iteration the order of magnitude of x is divided by $n^{1/2}$ so that balance is achieved in the limit.

* For unit i in group g, the probability to belong to the sample is :

$\frac{n}{N}\left(1 + \frac{K_i}{n} + o(1/n)\right)$, where K_i is a finite quantity depending of $X(i) - \bar{X}^g$.

* For distinct units i and j, the probability for the couple (i,j) to belong to the sample is approximately equal to $\left(\frac{n}{N}\right)^2\left(1 + \frac{K}{n}\right)$.

Simulations show that balance is obtained with a very good approximation and that no sensible distortion appears for inclusion probabilities.

Some practical aspects are however to be taken into account. First, we have to stop the iterative procedure with some rule. Theoretically, the rounding process generates the same variance for $\bar{x}_{(t)}$ at each iteration. We have to stop when no further improvement is expected. This occurs, practically, when we would have to make more modifications at the current iteration than in the preceding.

Another aspect is the amount of computations needed by the procedure. It can be noted that matrix R in step 2 is the same for all iterations and is computed only once. Similarly, step 4 can be simplified if one uses the following method :

- in each group the sampling frame is randomly sorted before step 1. For unit of rank r, the sums of X_k up to rank r are stored as new variables.
- at each iteration, a sample requiring n_g units in group g will be composed by the first n_g units of the group. The computation of $\bar{x}_{(t)}$ is therefore straightforward.

II.4 - A simulation study

The problem of balancing a sample with respect to several variables while preserving equal inclusion probabilities has been motivated by the processing of the 1990's French census. Traditionnally, a quick estimation of the census results is published based upon $1/20^{th}$ of the dwellings. Until 1982, the sample has been drawn manually on a systematical way, occasioning report errors and unattractive work.

The alternative solution is to draw bunches of dwellings, namely districts,in order to achieve quicker processing and to avoid manual mishandling, but nonetheless increasing sampling errors. In order to offset or limit this variance increase, some information relating to the size of the districts can be used. More precisely, three different counts are available at this stage, namely : number of persons (X1), number of dwellings (X2), number of buildings (X3), which are known for each district. Despite the high correlation of these criterias, their reduction to one dimension may incur a significative loss in precision ; so classical tools such as systematic sampling on a sorted file may not be satisfactory.

The following simulations test the efficiency of the balancing method described above, in the case of two control criterias, which are the two first principal components derived from X1, X2, X3.
The different methods compared are :

- simple random sampling of districts (SRS)
- first step of balanced sampling (i.e. stratification)
- balanced sampling
- regression estimation (after initial stratification).

The variances or standard errors presented in table 4 and figure 1 are based upon 100 different samples drawn for each method.

Remarks : a) The third principal component has been neglected because of its very low contribution to the inertia of the data set. By keeping all three dimensions, we obtain results very similar to balanced sampling on two principal components.

b) The regression estimation has been implemented on the stratified samples issued from the first step of balancing in order to enhance direct comparability between the two methods.

c)The SRS samples necessarily result from another set of simulations. Hence SRS may appear artificially more efficient than the other methods, but this difference is not significative.

Thirty-six variables, including two of the control variables have been used to test the efficiency of the different strategies. The test file consists of real data

issued from the last French census and corresponds to an infra-regional level totalizing 2542 districts. All the simulations have been performed on an IBM 3090/120 E computer using SAS^R macro language and PROC MATRIX.
It appears that for nearly half of the variables considered, the standard errors, and hence confidence intervals, are reduced by more than half if compared to simple random sampling of districts (see figure 1). If compared to regression estimation, balanced sampling is slightly less efficient, especially for the variables which perform the best with regression estimation. This higher variability lies in the fact that balancing is achieved only approximately with the proposed procedure. Anyhow, from an operational point of view, the procedure allows easier tabulation, which, in the case considered here, appears as a major advantage.

Finally returning to the initial problem, balanced sampling of districts has been compared to systematic sampling of dwellings. As may be expected, the design effect due to sampling districts remains quite large for highly clustered or rare populations. Some socio-economic caracteristics, such as nationality or specific economic activity which have a very low correlation with size effects, perfom quite poorly with all the methods operating at the district level, with a loss of precision ranging from 2 to 4 in terms of variance, compared to sampling of dwellings. On the contrary, the loss of precision due to sampling districts is nearly offset for "well-behaved" variables ; thus, the variables directly connected with the number of dwellings remain slightly better estimated with systematic sampling of dwellings, whereas variables well explained by the number of persons may perform better with the balanced sampling procedure.

TABLE 4 : COMPARED EFFICIENCY OF SAMPLING METHODS
RELATIVE ESTIMATED VARIANCE (ESTIMATION ON 100 SAMPLES)

	TEST VARIABLES	BALANCED /SRS	STRATIFICATION /SRS	STRATIFICATION+ REGRESSION /SRS	BALANCED+ REGRESSION /SRS
1	ACTIFS EMPLOYES CHIMIE DE BASE	1.20	1.20	1.18	1.19
2	ACTIFS EMPLOYES PRODUCTION COMBUSTIBLES	1.14	1.16	1.14	1.14
3	ACTIFS EMPLOYES PARACHIMIE ET PHARMACIE	1.04	1.13	1.06	1.07
4	NOMBRE D ETRANGERS	0.97	1.12	1.05	0.97
5	NOMBRE D ETRANGERS ACTIFS SEXE MASCULIN	0.86	0.89	0.85	0.85
6	AGRICULTEURS EXPLOITANTS	0.78	0.94	0.86	0.81
7	CADRES	0.73	0.89	0.70	0.73
8	NOMBRE DE PERSONNES DE 75 ANS ET PLUS	0.72	0.84	0.78	0.72
9	ACTIFS EMPLOYES AGRICULTURE ET IAA	0.66	0.82	0.76	0.69
10	ARTISANS,COMMERCANTS,CHEFS D ENTREPRISE	0.65	0.71	0.63	0.67
11	NOMBRE DE LOGEMENTS RECENTS	0.56	0.74	0.61	0.57
12	NOMBRE DE FEMMES ACTIVES 30-34 ANS	0.55	0.77	0.49	0.50
13	NOMBRE DE LOGEMENTS VACANTS!SECONDAIRES	0.53	0.63	0.54	0.55
14	NOMBRE D HOMMES ACTIFS 20-24 ANS	0.48	0.45	0.44	0.44
15	NOMBRE DE FEMMES 30-34 ANS	0.47	0.79	0.41	0.43
16	NOMBRE D HOMMES 20-24 ANS	0.45	0.50	0.45	0.43
17	NOMBRE DE PERSONNES DE 55 A 74 ANS	0.39	0.50	0.35	0.34
18	ACTIFS EMPLOYES COMMERCE DE DETAIL	0.39	0.48	0.37	0.38
19	RETRAITES	0.39	0.56	0.38	0.37
20	PROFESSIONS INTERMEDIAIRES	0.28	0.57	0.27	0.26
21	NOMBRE DE PERSONNES DE 0 A 14 ANS	0.26	0.66	0.26	0.23
22	NOMBRE DE PERSONNES DE 15 A 24 ANS	0.25	0.41	0.21	0.20
23	ACTIFS EMPLOYES INDUSTRIE	0.23	0.48	0.18	0.19
24	EMPLOYES	0.23	0.43	0.18	0.18
25	OUVRIERS	0.22	0.41	0.16	0.17
26	AUTRES INACTIFS	0.22	0.59	0.18	0.18
27	ACTIFS EMPLOYES TERTIAIRE	0.19	0.43	0.13	0.14
28	NOMBRE DE PERSONNES DE 25 A 34 ANS	0.18	0.47	0.16	0.16
29	NOMBRE DE PERSONNES DE 35 A 54 ANS	0.17	0.55	0.12	0.12
30	NOMBRE DE FEMMES ACTIVES	0.16	0.43	0.10	0.11
31	NOMBRE D HOMMES ACTIFS	0.10	0.38	0.07	0.07
32	NOMBRE D HOMMES	0.09	0.43	0.06	0.06
33	NOMBRE DE LOGEMENTS	0.08	0.36	0.04	0.04
34	NOMBRE DE RESIDENCES PRINCIPALES	0.08	0.35	0.04	0.04
35	NOMBRE DE FEMMES	0.08	0.42	0.04	0.04
36	NOMBRE DE PERSONNES	0.07	0.41	0.03	0.03

FIGURE 1 : COMPARED EFFICIENCY OF SAMPLING METHODS
ESTIMATED STANDARD ERROR RELATIVE TO SRS(ESTIMATION ON 100 SAMPLES)

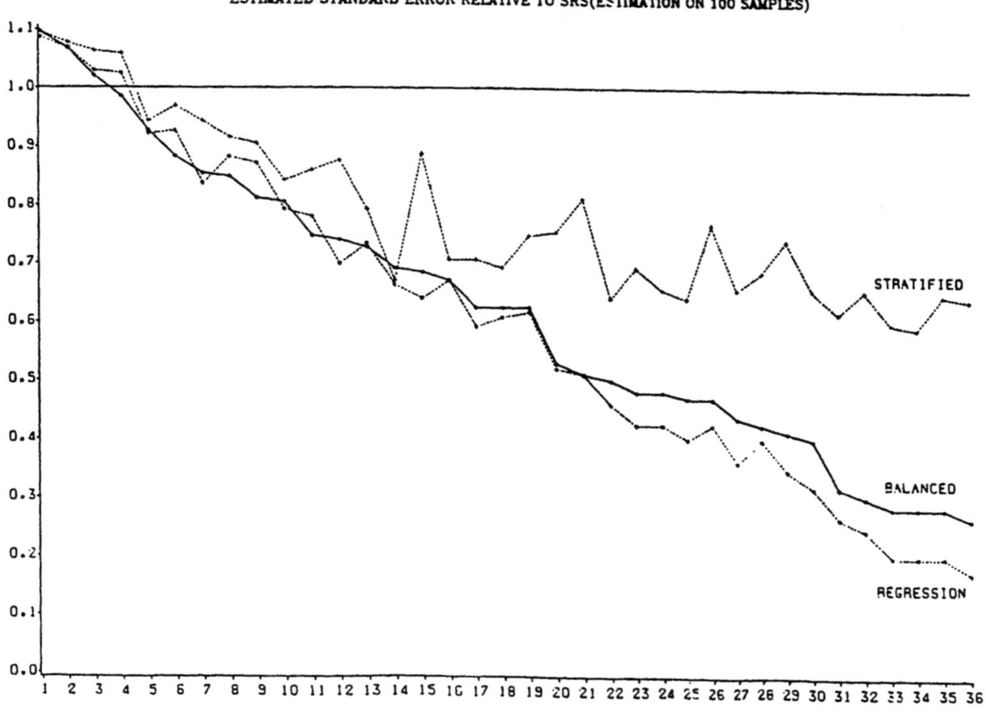

REFERENCES

AHRENS, J.H. and DIETER,U. (1985), Sequential Random Sampling.ACM. Transactions on Mathematical Software. Vol 11,N°2, 157-169.
DEVILLE,J.C, and GROSBRAS, J.M. (1987), Algorithmes de tirage. In sondages, Droesbeke, Fichet, Tassi. Edition ECONOMICA.
FAN,C.T., MULLER,M.E. and REZUCHA I. (1962), Development of sampling plans by using sequential selection techniques and digital computers.J. Am. Statist. Assoc., N°57, 387-402
HANIF, M. and BREWER, K.R.W. (1980), Sampling with unequal probabilities without replacement : A review. Int. Stat. Rev., Vol 48, 317-335.
HAJEK, J. (1981), Sampling from a finite population. DEKKER.
HARTLEY, H.O. and RAO, J.N.K. (1962), Sampling with unequal probabilities and without replacement. Ann. Math. Stat. 33, 350-374.
IACHAN, R. (1982). Systematic Sampling : A critical Review. Int. Stat. Rev., Vol 50, 293-303.
KNUTH, D.E. (1981). The Art of Computer Programming. Vol.2, Seminumerical Algorithms, 2nd ed., ADDISON-WESLEY.
RAO, J.N.K., HARTLEY, H.O. and COCHRAN, W.G. (1962). A simple procedure of unequal probability sampling without replacement. J. Roy Stat. Soc., B24, 482-491
ROYALL, R. and HERSON, J. (1973). Robust Estimation in Finite Populations. J. Am. Statist. Assoc., Vol 68, 880-893.
SUNTER, A.B. (1986), Solutions to the problem of unequal probability sampling without replacement Int. Statist. Rev. Vol 50, 33-50.
THIONET, P. (1953). La Théorie de Sondages. Etudes Théoriques N°5. INSEE Paris France.
VITTER, J.S. (1984). Faster methods for random sampling. Comm. of the ACM, Vol 7, 703-718.
VITTER, J.S. (1987). An efficient algorithm for sequential Random Sampling. ACM Transaction on Mathematical Software, Vol 13, n° 1, 58-57.
YATES, F. (1949). Sampling Methods for Censuses and Surveys. GRIFFIN.

Recursive Partition in Biostatistics: Stability of Trees and Choice of the Most Stable Classification

A. Ciampi and J. Thiffault, Montreal

1. Introduction: Stability of a model

Structures found in data by exploratory techniques are notoriously unstable. Suppose that we search for a model within a given family and that we do this on different samples from the same population, $D_0, D_1, \ldots, D_B$. When only one data set is available, one can think of D_0 as the original data set and the others as bootstrap samples from D_0. Experience shows that one can be practically sure to find different models from different samples. A striking example of this model instability is given by Gong [1], in the context of stepwise logistic regression. The problem can be expected to be even more serious for tree-structured predictors, such as the RECPAM trees [2-4] which are the main concern of this work, since the model is selected out of a family much richer than that of linear regression as usually defined.

Let T_k denote a statistical model obtained by applying a model search technique on D_k, $k=0,\ldots,B$. We will assume that a measure of agreement, $AGR(T,T';D)$ between two models T, T' on a data set D is available. It seems natural to measure the stability of T_k by taking the mean of its agreement with the other models of the sequence. We shall discuss two ways of doing so, corresponding to two conceptually distinct types of stability. It is clear that if we start from a measure of disagreement, the discussion that follows applies to the definition of instability.

One approach consists in assessing the agreement of each model T_k, $k=0,\ldots,B$, with the trees T_i, $i\neq k$, on the data set D_i. The rationale for this comparison is as follows. On each data set D_i, $i=0,\ldots,B$, the corresponding model T_i offers a description which is as close as possible to the data set even though it may "overfit" it. The agreement of T_k with T_i measured by $AGR(T_k,T_i;D_i)$ is then a measure of how well the description of D_k applies to D_i. As a result, the stability measured this way expresses the degree to which a description of a data set D_k is generalizable to the others: we shall refer to it as <u>cross-generalization stability</u> (CG-STAB). With obvious meaning of notation, we shall then write:

$$CG\text{-}STAB(T_k) = \frac{1}{B-1} \sum_{i\neq k} AGR(T_k,T_i;D_i) \qquad (1.1)$$

Another approach consists in calculating the agreement of a model T_k with

Compstat 1988

all other models T_i, $i \neq k$, on all data sets D_j with $j \neq i,k$, so that the comparison of any two models is not done on the data sets on which they have been obtained. Then the expression:

$$\text{IS-STAB}(T_k) = \frac{1}{B(B-1)} \sum_{i \neq k} \sum_{j \neq i} \text{AGR}(T_k, T_i; D_j) \tag{1.2}$$

measures how well a tree summarizes the information contained in the others and will be referred to as information-summary stability (IS-STAB).

In section 2, the RECPAM methodology is summarized and measures of agreement and disagreement for RECPAM trees are developed. The resulting measures of stability/instability are used in section 3 to select a tree out of a sequence obtained from a real data set and 10 bootstrap samples.

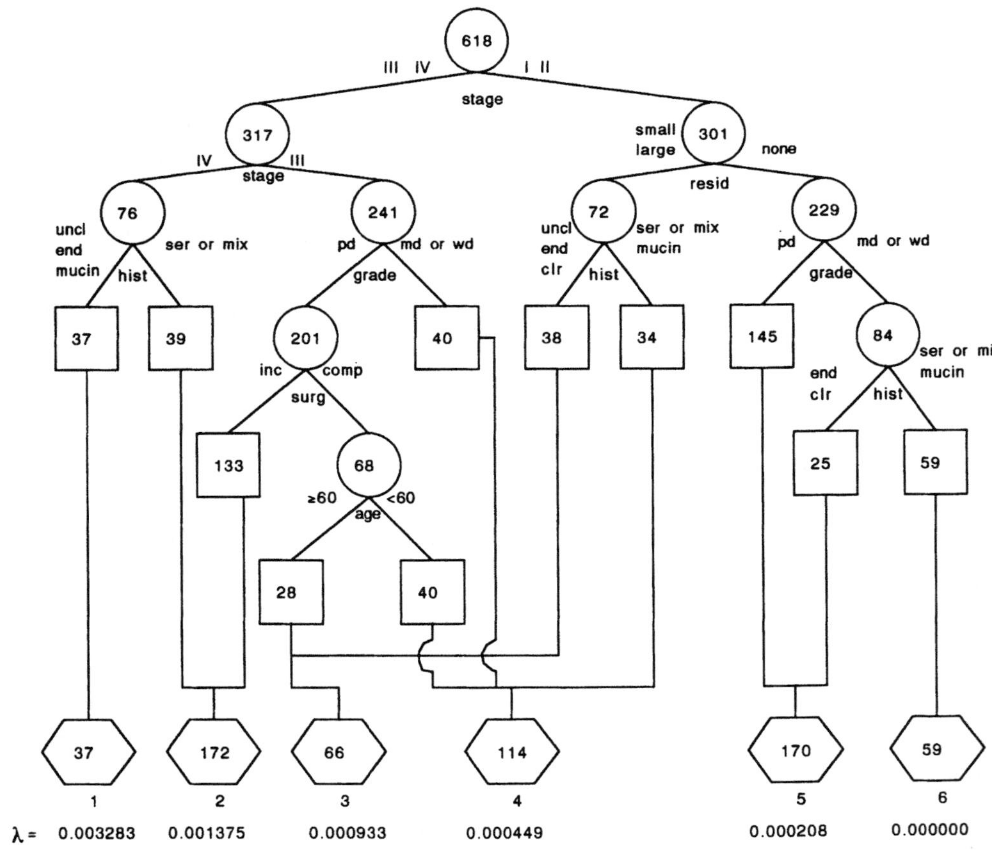

Figure 1. The most stable RECPAM tree

2. RECPAM for censored survival data

Figure 1 shows a RECPAM tree grown on bootstrap sample # 10 drawn from a data set consisting of 618 ovarian cancer patients. As we shall see in the next section, this tree can be considered the most stable. The variables of this data set include the survival time, possibly censored, and a number of prognostic factors or predictors.

At the bottom of the RECPAM diagram, hexagons represent the subpopulations of a partition of the whole population which we shall refer to as RECPAM classes. RECPAM classes are, by construction, internally homogeneous but different from each other with respect to survival. Inscribed numbers are population sizes and the λ_i's are the hazards assumed constant in time, ranked in decreasing order. At the top, a circle represents the whole population, the root node of the RECPAM tree. The path leading from the root to the RECPAM classes assigns names to them in terms of the predictor variables: for example RECPAM class 1, corresponding to the worst prognosis (largest λ) consists of all patients who have stage IV and unfavorable histology (unclassified, endometrioid or mucinous).

The upper portion of the diagram is obtained by applying the RECursive Partition (RECP) algorithm: it starts from the root node of the tree and proceeds through successive branchings of intermediate populations, also represented by circles, down to terminal nodes, represented by squares. Each branching or split is defined by a restriction on one predictor. Every split is chosen by RECP as the one that maximizes the likelihood ratio of the hypothesis that the two survival curves, assumed exponential, are different, to the hypothesis that they are identical. This portion of the diagram has been obtained by growing a large RECP tree (not shown) and then pruning it until a stopping rule is activated. The stopping rule is determined by a tuning parameter α.

The lower part of the tree, leading from the squares to the hexagons, is obtained by application of the AMalgamation algorithm to the terminal nodes of RECP. AM successively joins subpopulations with minimum likelihood ratio; termination is determined by a stopping rule governed by the same parameter α as for RECP.

The parameter α is chosen as the one which gives the lowest AIC, the Akaike Information Criterion associated to the tree as a predictor of survival. Details are discussed in [2].

Clearly, a RECPAM tree can be considered as a clustering or partition of the given population, whose sets are the RECPAM classes represented by the hexagons, but also as a predictor which summarizes the prognostic information carried by the prognostic factors contained in the data set. As such it defines an ordered variable, the rank of the λ_i's of Figure 1, which grades prognosis into classes ranging from poor (largest hazard) to good.

Let D denote a data set and T_k, T_l two RECPAM trees. We wish to compare the two trees on the data set D. Although there is a rich literature on comparison of trees, the tree structures resulting from RECPAM are far too complex for a direct applicability of the general theory. In this work, we will only compare two RECPAM trees from the point of view of their associated partitions and predictions.

To define measures of agreement and disagreement between RECPAM trees in terms of the partitions they define on D, we use Day's theory on comparison of ordered classifications [6], as developed by Faith and Belbin [6]. A measure of pair-bonding [7] of a partition is the number of pairs which are classified together in any set of a partition. We will denote by $PB(T_k;D)$ this number when calculated on the data set D for the partition defined by the tree T_k. It corresponds to Day's "complexity" [5] for a particular choice of lower valuation [6]. Two trees can be compared on D by measuring the amount of structure which is common to both classifications. This can be done by considering the intersection of the two partitions and counting the number of pairs of elements of D which are classified together in any set of this intersection, i.e. $PB(T_k \cap T_l;D)$. Such a measure is an example of Day's "organized complexity" [5]. $PB(T_k \cap T_l;D)$ is therefore a measure of the agreement between the two classifications, which we shall denote as:

$$AGR_{pb}(T_k,T_l;D) = PB(T_k \cap T_l;D) \tag{2.1}$$

A measure of their disagreement which corresponds to Day's "unorganized complexity" [5] is given by:

$$DIS_{pb}(T_k,T_l;D) = PB(T_k;D) + PB(T_l;D) - 2PB(T_k \cap T_l;D) \tag{2.2}$$

An obvious way of combining unorganized and organized complexity into another measure of disagreement is to take their difference. We then obtain the "intermediate" measure of disagreement of Faith and Belbin [6]:

$$DIS_{pbi}(T_k,T_l;D) = DIS_{pb}(T_k,T_l;D) - AGR_{pb}(T_k,T_l;D) \tag{2.3}$$

Trees as predictors can be compared in two ways. We can take the point of view that a tree gives an order of prognosis, as discussed above. Then, two trees can be compared on D by calculating the Goodman-Kruskal gamma [8] for the prognostic grading they define. Clearly this coefficient is a measure of the agreement of the predictors:

$$AGR_{pr}(T_k,T_l;D) = \gamma(T_k,T_l;D). \tag{2.4}$$

On the other hand, disagreement can be measured by :

$$DIS_{pr}(T_k,T_l;D) = \mid AIC(T_k;D) - AIC_l(T\ ;D) \mid \tag{2.5}$$

where AIC(T;D) is the AIC of model T applied to the data set D.

3. Choice of a stable tree

To illustrate the utility of the concepts developed above, we have applied them to the data set presented in section 2. Figure 2 shows a summary of the calculations. Each line corresponds to a tree.

				CG-STAB/INSTAB					IS-STAB/INSTAB					
TREE	NO. OF NODES	AIC	MEAN AIC (S.E.)	AGR_{pr}	DIS_{pr}	AGR_{pb}	DIS_{pb}	DIS_{pb1}	AGR_{pr}	DIS_{pr}	AGR_{pb}	DIS_{pb}	DIS_{pb1}	AVERAGE RANK
T_0	8	4727.78	4747.74 (38.12)	0.8524	86.077	15620.8	37039.6	21418.8	0.8483	77.769	15565.1	37227.1	21662.0	4.8
T_1	7	4594.52	4769.44 (33.00)	0.7632	38.498	12936.9	41970.8	29033.9	0.7641	26.476	12967.1	41991.0	29023.8	7.9
T_2	7	4751.63	4750.87 (38.64)	0.8140	35.647	18405.9	38477.5	20071.6	0.8054	27.740	18462.8	38352.4	19889.6	3.4
T_3	6	4769.37	4731.93 (34.66)	0.8035	23.385	17630.9	45366.3	27735.4	0.7987	33.451	17333.0	46054.4	28721.3	6.9
T_4	7	4637.99	4773.07 (37.54)	0.8270	46.482	16272.1	42052.2	25780.1	0.8241	27.757	16465.3	41590.0	25124.7	5.5
T_5	8	4605.67	4758.58 (34.36)	0.8098	28.761	13725.9	38931.5	25205.6	0.8093	27.343	13566.4	39484.1	25917.7	5.7
T_6	7	4799.00	4792.66 (39.52)	0.7240	82.169	13929.8	45474.5	31544.7	0.7199	50.991	13990.7	45372.5	31381.8	10.1
T_7	6	4757.24	4770.74 (38.00)	0.8594	56.070	19890.7	48944.5	29053.8	0.8605	28.398	19498.2	49480.9	29982.7	5.9
T_8	6	4830.98	4769.57 (37.43)	0.7667	62.275	14222.9	42869.8	28646.9	0.7731	32.705	14306.9	42752.3	28445.4	8.2
T_9	7	4523.26	4782.62 (30.68)	0.8473	44.553	15094.1	38189.2	23095.9	0.8412	28.821	15330.8	37854.7	22523.8	4.6
T_{10}	6	4906.45	4718.75 (31.05)	0.8581	20.461	19055.4	39315.6	20260.2	0.8588	30.362	19034.3	39730.5	20696.2	3.0

Figure 2. Summary of the calculations

Column (1) gives the number of RECPAM classes; column (2), the AIC of the tree on the data set on which it has been obtained ; column (3), the average AIC on all other data sets with the standard error in brackets; columns (4-8) give CG-stability/instability for the respective measurements of agreement/ disagreement; columns (8-12), the analogous IS-stability/instability; finally column (13) gives the stability rank averaged over the 10 measures of stability/instability. Although in practice one may want to base the choice of the tree on a particular measure of stability or on an informal criterion, we have chosen here RECPAM tree #10, which has best average stability rank.

Much remains to be done in order to provide good criteria of tree choice which reduce the problem of overfitting common to all automatic model-selection techniques. The measures presented in this work provide a first step towards assessing the stability of a structure discovered by exploring the data with little or no previous knowledge.

References

[1] Gong,G. (1986), "Cross-validation, the Jackknife, and the Bootstrap: Excess error estimation in forward logistic regression", Journal of the American Statistical Association, 81, 108-113.

[2] Ciampi,A., Chang,C.-H., Hogg,S.A. and McKinney,S. (1986), "Recursive Partition: A versatile method for exploratory data analysis in Biostatistics. In: Biostatistics 5, Festschrift in honor of Prof. V.M. Joshi's 70th birthday, I.B.McNeil and G.J.Umphrey, eds, Dordrechdt-Holland, 1-28.

[3] Ciampi,A. and Thiffault,J. (1988), "Recursive Partition and Amalgamation (RECPAM) for censored survival data: Criteria for tree selection", to appear in Statistical Software Newsletter, vol.14, no.2.

[4] Ciampi,A., Hogg,S.A., McKinney,S. and Thiffault,J. (1987), "RECPAM: a computer program for Recursive Partition and Amalgamation for censored survival data and other situations frequently occurring in Biostatistics. I.Methods and program features", to appear in Computer Methods and Programs in Biomedicine.

[5] Day,W.H.E. (1983), "The role of complexity in comparing classifications", Mathematical Biosciences, 66, 97-114.

[6] Faith,D.P. and Belbin,L. (1986), "Comparison of Classifications using measures intermediate between metric dissimilarity and consensus similarity", Journal of Classification, 3, 257-280.

[7] Boorman,S.A. and Arabie,P. (1972), "Structural measures and the method of sorting", In: Multidimensional Scaling: Theory and Applications in the Behavorial Sciences, vol.1, R.N. Shepard, A.K. Romney and S.B. Nerlove, eds, New York: Seminar Press, 225-249.

[8] Goodman,L.A. and Kruskal,W.H. (1954), "Measures of association for cross classifications", Journal of the American Statistical Association, 49, 723-764.

Generating Rules by Means of Regression Analysis

C. Berzuini, Pavia

ABSTRACT

This paper proposes a procedure by which the results of a regression analysis are "transformed" into a decision tree, or into decision rules. This may be useful either as a way of integrating regression analysis in a learning environment where the accumulated knowledge is represented symbolically, or simply as a way of deriving intuitive and psychologically meaningful explanations of a regression analysis. The proposed algorithm is illustrated within a clinical observational study on prognosis.

1. INTRODUCTION

A *binary regression tree* (BRT) is a simple recursive structure for expressing a sequential classification process in which a *case*, described by a set of *attributes* $\mathbf{X}=(X_1,\ldots,X_p)$, is assigned to one of a disjoint set of *classes* and a real-valued scalar *predictor function* $\phi(\mathbf{X})$ assigns to each class a range of values on the real axis. An interior node denotes a *test* on one attribute. Testing an attribute is equivalent to assessing the value of a *Boolean expression*, exg. $(X_i > cut-off)$, which may be *true* or *false*. A node is associated to a *subpopulation* of cases. To characterize it, follow the unique path from the root to that node. This path determines a value assignment to a set of Boolean variables, and hence a set of *conditions* on some or all of the X-attributes. The conjunction of these conditions logically defines the subpopulation, and constrains $\phi(\mathbf{X})$ in an interval. Consider for example the BRT in Fig. 1*c*, relative to the predictor function in Fig. 1*b*. The leftmost leaf, for example, corresponds to the condition $X_1=0$ **AND** $X_2 \leqslant 0$, which implies the constraint $-2 \leqslant \phi(\mathbf{X}) \leqslant 0$.

The leaves of a BRT correspond to classes. To classify a case we start at the root of the tree. If this is a leaf, the case is assigned to the nominated class; if it is a test, the left (respectively right) edge to a child is taken if the nominated Boolean variable is false (respectively true) for the examined case, and the process continued until a leaf is reached.

A BRT may be viewed as a generalization of the well-known notion of *decision tree*. Analogously to a decision tree, it is not difficult to "translate" a BRT into a set of *decision rules*. Decision trees and decision rules are a widely used vehicle for representing knowledge in an expert system, due to their transparency and, as far as the decision rules are concerned, modularity. Research that commenced in the late 1950s with Hunt's *Concept Learning System* [Hunt *et.al.* ,1966] has led to several reliable methods for developing decision trees from *training sets* of cases with known classes. A common approach uses *recursive splitting* (RS): a tree is grown sequentially, selecting first the test that "best" divides the sample into sets, then selecting the best test to use with each set, a.s.o. Various test selection criteria are discussed in [Payne, *et.al.* 1984]. [Breiman *et.al.*, 1984] and [Quinlan, 1986] have developed in a RS context methods handling noisy and/or incomplete data. Some use a run-time error estimate during the tree growth as a stopping criterion that prevents growing an oversized tree, others use an *a posteriori* error estimate to prune an already created decision tree. Most often, training data are assumed pre-classified. The program RECPAM for RS developed by [Ciampi *et.al.*, 1986], is able to categorize unclassified training data in such a way that output classes are well separated with respect to a (possibly censored) *response* variable.

It is natural to speculate about the relationships between recursive splitting and regression, particularly in the field of analyzing unbalanced multiply-classified data. Other authors have already discussed this topic. For example, both Ciampi *et.al.* and Breiman *et.al.* compare through case-studies the two approaches.

The decision tree representation , as long as more structured than a regression model, is more natural to understand, and expresses predictive relationships suggested by the data in a more explicit way.

However RS has its own pitfalls: unability to test hypotheses on the global sample rather than on separate portions of the data independently, marked unstability with respect to the order of the tests in the sequence, low efficiency. For example, presence of confounding variables generates a large number of "early.

This work was partially supported by C.N.R. grant no. 87.01829, MPI40% and MPI60%

Compstat 1988

splittings" after which RS "looses breath".

Since regression analysis doesn't suffer from many of those problems, why not combine the two views ? More precisely, regression can be used to to "extract" the "true" relevant relationships in the data. Then the results of regression can be structured in a tree-like form for the sake of the domain-expert s intuition. A BRT may well be used for representing the results of a regression: provided that the attributes correspond to the explanatory variables, and that the predictor function corresponds to the systematic part of the model, i.e. that which maps the attribute-space into an expected value for the response.

The starting point for this paper is a regression model selected on the basis of the training sample by means of some appropriate selection procedure (see for example [Aitkin,1978]). This paper is concerned with "mapping" the selected regression model into a BRT representation. The crucial issue is the *simplicity* of the output BRT. For a given regression model, a large number of BRTs can usually be found which provide a good "explanation" of the model, trading-off predictive efficiency with simplicity. A method and algorithms are here proposed to perform a search in the space of possible BRT representations of a given regression model.

All discussed procedures will be illustrated on a clinical observational case-study.

2. REGRESSION TERMINOLOGY

We assume that the *learning sample* consists of N pairs of observations $(Y_i;\mathbf{X}_i)$, where each $\mathbf{X}_i$ is the p-dimensional vector of attributes of the ith "example", and Y_i is a real-valued number, called **response.** The regression model represents $Y_1, \ldots, Y_N$ by random variables with, for some appropriate function g and **predictor function** ϕ:

$$\hat{Y}_i = E(Y_i) = g(\phi(\mathbf{X}_i;\beta),\alpha)' \tag{1}$$

where $\hat{Y} = E(Y_i)$ is the **expected** value of Y_i, α and β being vectors of unknown parameters. To allow parameter estimation, the regression model has to be completed by specifying the probabilistic distribution of the Y's.

The predictor function $\phi(\mathbf{X};\beta)$ is the minimal summary of the attribute vector $\mathbf{X}$ sufficient to determine Y. This function represents the starting point for the procedure described in this paper.

3. DIFFICULTIES IN INTERPRETING RESULTS OF REGRESSION

This paper is particularly concerned with the difficult problem of interpreting results of regression analysis of unbalanced multiply-classified data. These analyses are usually characterized by a complex (from a combinatorial viewpoint) *model building* phase, in which a large number of alternative formulations of models may compete. Usually there is not a clear winner in such competition. The "correct" model is somewhere in a list of "acceptable" models. An interpretation of all these models should in principle be given, any choice between them being on "external" grounds, i.e. based upon the domain-expert's intuition.

Unfortunately, a serious limitation of regression analysis is that it gives little explicit information regarding the structure of the data, and this creates interpretation difficulties.

FIGURE 1

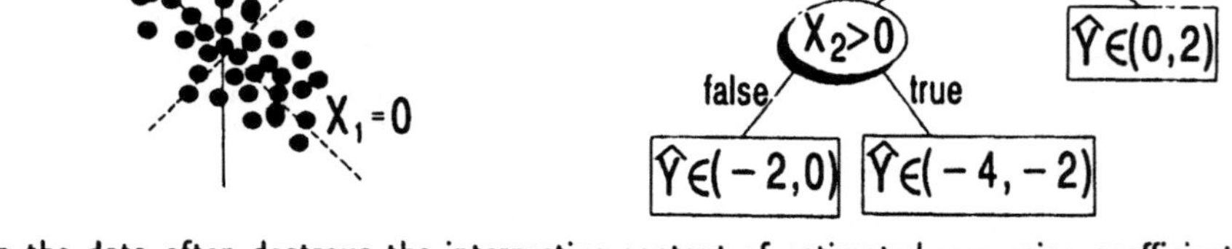

Unbalance in the data often destroys the interpretive content of estimated regression coefficients. As an example, consider the scatter diagram in Fig. 1a summarizing a fictitious data set where each sample individual is characterized by two continuous variables (y, x_2) and a binary one (x_1). These data are strongly unbalanced, since, for example, for X_2 below a certain threshold there are no data points with $X_1 = 1$. By regressing Y on (X_1, X_2), we obtain the model reported in Fig. 1b, where the coefficient of x_2 has a *negative* sign. From this one would deduce that when x_2 is very high a very low y value should be expected. But this is not what appears in Fig. 1a: marginally there is a strong correlation between y and

x_2 but of *positive* sign. The paradox is resolved by noting that the regression coefficient of x_2 *misrepresents* the actual predictive role of x_2, because the association between X_2 and Y is mostly "explained away" by X_1. The negative coefficient implies the following *conditional* statement: *for a fixed* x_1, an increasing x_2 corresponds to a decreasing expected y.

The conditional structure of predictive relationships suggested by these data is much better expressed by the BRT in Fig. 1*c* than by the predictor function. The leaves of the BRT define three classes corresponding to 3 disjoint intervals of Y. These classes admit simple logical descriptions based on X_1 and X_2. On the debit side, the BRT is less efficient than the predictor function in predicting Y.

4. TRANSFORMING THE PREDICTOR FUNCTION INTO RULES

The target of the transformation is a set of logical descriptions for a set of classes $C_1, \ldots, C_k$ associated to *disjoint* intervals $INT(C_1), \ldots, INT(C_k)$ of the expected response $\phi(\mathbf{X})$. Ideally, the description of a class C_i ($i = 1, \ldots, k$) should be fully *conjunctive*, composed of features that are *individually necessary* and *jointly sufficient* for a case to have $\phi(\mathbf{X}) \in INT(C_i)$. But this is often impossible if we want a reasonably small number of classes. As a consequence, we must accept that the description of a class C_i may have, to a certain degree:

incompleteness: it doesn't admit all cases with $\phi(\mathbf{X}) \in INT(C_i)$

disjunctions: is composed of features that may be **not** individually necessary.

The transformation takes place in four stages:

(a) growing a complete BRT: we assume that a "parsimonious" form of predictor function $\phi(\mathbf{X})$ has been selected, and, for simplicity, we assume that there are only *binary* attributes in it. From $\phi(\mathbf{X})$ a BRT such as the one in Fig. 1*c* is generated, whose leaves correspond to subpopulations within which the predictor function is constant. This kind of BRT, namely where each leave corresponds to a complete assignment of value to all attributes included in the predictor function, is called **complete.** Usually the complete BRT is too complicated to convey a simple interpretation of the data. Therefore, a "simplification" phase, described in the following, is needed.

(b) simplifying the BRT: a BRT with reduced depth and reduced number of leaves is obtained (see sec.7). The leaves define classes $C_1, \ldots, C_k$ with disjoint $INT(C_1), \ldots, INT(C_k)$, but some of their descriptions may be incomplete.

(c) amalgamating: pairs of leaves with non-significantly different expected response are fused. When they have different "parent" node, the tree becomes a *lattice*, i.e. presents multiple-connections between nodes. This introduces *disjunctions* in the class descriptions.

(d) eliminating disjunctions: if amalgamation has created multiply-connected nodes, these loops are eliminated by *re-expressing* combinations of attribute values as a single attribute, i.e. extending the description language.

5. AN APPLICATION

In order to illustrate our basic idea, we will consider a survival study on Myelofibrosis with Myeloid Metaplasia (MMM), a chronic myeloproliferative disorder. The training set comprised 138 patients with MMM consecutively seen in the Department of Internal Medicine and Medical Therapy of the University of Pavia from 1973 to 1985.

The measurement vector for each case consisted of values of the following presumed predictors of survival: age (*years*), sex, haemoglobin (*g/dl*), white blood cells count ($10^9/l$), reticulocytes count ($10^9/l$), percent of immature myeloid cells, percent of erythroblasts, spleen size (*cm*), liver size (*cm*), TEIT and bone marrow cellularity. TEIT is an estimate of the flux of iron from the plasma to erythroid tissues, and hence a measure of the expansion of erythropoiesis (red cell production). Cellularity represents the result of a histological examination of the bone marrow, and takes values on an ill-defined quantitative-scale, ranging from red cell aplasia, where hemopoietic cells are practically absent from bone marrow, to myeloid hyperplasia, where a substantial marrow area is occupied by hemopoietic cells.

Time from MMM diagnosis to death was regarded as failure time for our sample cases. For 77 cases the failure time observation was censored. Our aim was defining a prognostic classification of MMM into meaningful classes with different expected failure time.

We analyzed MMM data by means of the Cox regression model [Cox, 1972]. We considered the Cox score $\beta\mathbf{X}$ as our predictor function. By means of stepwise selection procedures the following minimal adequate predictor function was obtained:

$$\phi(A, H, C, T) = 1.9\,A + 0.85\,H + 4.8\,C + 0.05\,T + 3.9\,\neg C\;T \qquad (2)$$

where the *literals* A, H, C, T represent the following boolean variables: $A =$ (Age > 45 *yrs*), $H =$ (Hb < 13 *g/dl*), $C =$ (Cellularity $=$ aplastic), $T =$ (TEIT < 200). The value of $\phi(A, H, C, T)$ for a given case is computed by replacing literals in (2) with 1 (respectively 0) if the associated boolean variable is *true* (respectively *false*). If a literal is preceded by a $\neg$ sign (NOT), 0 and 1 must be reversed.

Note that in (2) the interaction term $(\neg CT)$ is difficult to understand

Growing the complete BRT

Fig. 2 shows the complete BRT grown from $\phi(A.H\ C.T)$. Attributes to which the domain expert attached more importance were used for top-level splittings.

Each leave of the BRT contains in the box the value of $\phi(A.H.C.T)$. and in square brackets the number of sample cases associated to it. The leaves are ordered from left to right in the figure according to increasing value of the predictor function. The BRT is unbalanced. because the leaves with an empty set of sample cases were pruned.

Simplifying and amalgamating

The BRT shown in Fig. 2 was simplified by means of the algorithm described in sec.7. and then the pair of leaves corresponding to the highest risk were amalgamated due to non-significant difference in expected survival. The result is shown in Fig. 3*a* The leaves correspond to risk classes C_1 C_4 with $INT(C_1) \cup INT(C_2) \cup \cdots = 0$. The dotted lines are not edges of the BRT: they just indicate that some classes have incomplete description. For example. the description of C_1 is: "*all cases under-45. except those with anemia and aplasia*". The description of C_2 is: "*cases above-45 with absence of anemia and marked expansion of erythropoiesis*". This latter description is incomplete since 4 cases under-45 with anemia and aplasia have $\phi(\mathbf{X}) \in INT(C_2)$. Also the description of C_3 is incomplete. Due to amalgamation. the description of C_4 is disjunctive: "*cases above-45 with anemia and one of the following: aplasia or low erythropoiesis*".

Eliminating disjunctions

A disjunction is a possible symptom of the fact that the description language is *unsufficiently expressive*. For example. the boolean variables *(PRESENCE OF APLASIA)* and *(TEIT < 200)*. forming a disjunct in the description of C_4. were recognized by the domain-expert as expressions of a common physiologic state. called *(SEVERE ERYTHROID FAILURE)*. or. shortly. *SEF* This allowed fusing nodes of the lattice in Fig. 3*a* in such a way to obtain the BRT in Fig. 3*b*.

FIGURE 2

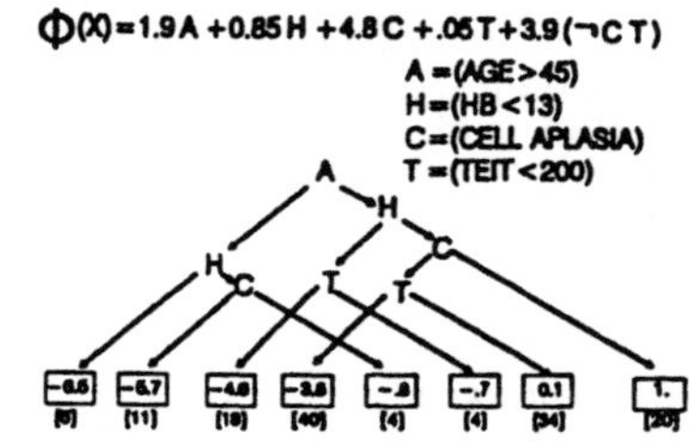

FIGURE 3*a*

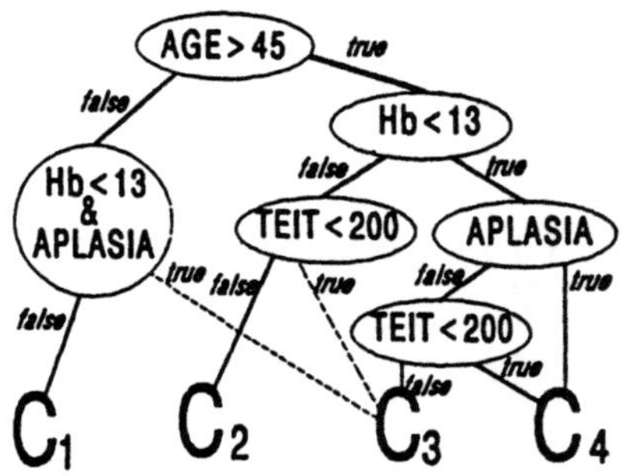

FIGURE 3*b*

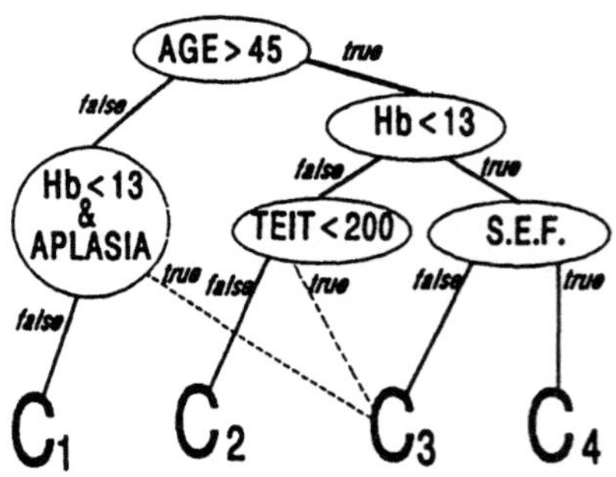

6. SIMPLIFICATION ALGORITHM

We now focus upon the second step of our proposed procedure. and formalize how the complete regression tree is simplified.

We begin with some straightforward definitions.

Let Θ be a complete BRT representing a predictor function $\phi(\mathbf{X})$.

DEFINITION 1. Two nodes of Θ are said to be **independent** when none of them is successor of the other one.

DEFINITION 2. A set of independent nodes is **complete** when each leave of Θ coincides with. or is successor of. *exactly one of them.*

DEFINITION 3. A complete set of independent nodes of Θ. linearly ordered to form a sequence $(N_1, \ldots, N_k)$. is called *I-chain.*

Many I-chains can usually be defined on a BRT. The first and last nodes of the chain are called *root* and *sink* of the chain. respectively.

Let $l_1, \ldots, l_n$ denote the leaves of Θ. If $\phi(l_i)$ denotes the value of the predictor function associated to the generic l_i. we assume for simplicity that leaves can be strictly ordered according to ϕ. and that they are indexed so that:

$$\phi(l_{i+1}) > \phi(l_i) \qquad 1 \leqslant i \leqslant n-1 \tag{3}$$

For a generic internal node N_i of Θ. let $L(N_i)$ denote the set of leaves which are descendants of N_i.

DEFINITION 4 Given two independent nodes N_i and N_j of Θ the expression :

$$L(N_j) > L(N_i) \tag{4}$$

means that for any $l_A \in L(N_j)$ and $l_B \in L(N_i)$:

$$\phi(l_A) > \phi(l_B)$$

DEFINITION 5. An l-chain is **consistent** when for any couple of nodes of the chain N_i, N_j, with $i,j \in (1,...,k)$, $j > i$, the inequality (4) is valid.

The "group-ordering condition" implicit in the definition of consistency given above guarantees that the nodes of the l-chain bear on *disjoint* intervals of the response variable, to the benefit of the characterization of associated classes.

In an inconsistent l-chain l there always exists a set $R(l)$ of sets of leaves, which are called **residual sets** with the property that if we "ignore" all leaves belonging to any $r(i) \in R(l)$, then l appears to be consistent.

To a residual set $r(l) \in R(l)$ we assign a "penalty" $PEN(r(l))$ given by the number of sample cases attached to it, weighted on the basis of a *utility* of correctly classifying them. Given an l-chain, we may look for a (not-necessarily unique) *optimal* residual set $\hat{r}(l) \in R(l)$, i.e. the $r(l)$ with minimum penalty. The problem is then that of finding

$$\hat{l} = \min_{l \in \Gamma} PEN(\hat{r}(l))$$

where Γ denotes the set of l-chains on Θ with a certain restriction on the number of chain-nodes. The set of nodes in l will correspond to the final set of classes and the logical descriptions for these classes will depend both on the structure of the tree and on the residual set.

As an example, consider Figs. 4*a*,*b*, showing two alternative l-chains on the BRT obtained from MMM data. The one in Fig. 4b is obtained by iteratively expanding the one in Fig. 4a. The two l-chains share a common residual set. In fact, if we "ignore" l_5 and l_6, we find that the sets of leaves descending from the three nodes of the first l-chain, or from the five nodes of the second l-chain, lie on disjoint intervals of the response.

The following is an algorithm for finding a sub-optimal solution.

(1) select a set N of independent nodes of Θ

(2) define an l-chain l by ranking nodes in N according to the mean value of ϕ

(3) by means of algorithm FIND-RESIDUAL:

   (a) find an optimal residual set $\hat{r}(l)$

   (b) from l and $\hat{r}(l)$ derive a set of logical class descriptions

   set $PMIN = PEN(\hat{r}(l))$

(4) expand the l-chain by selecting a node and replacing it with its immediate successors, then reapply FIND-RESIDUAL. Proceed iteratively with further expansions as long as there are expandable nodes in the chain and the minimum-penalty is not too high with respect to *PMIN*

(5) when a stable l-chain is reached, amalgamate classes which do not significantly differ in expected response.

FIG. 4

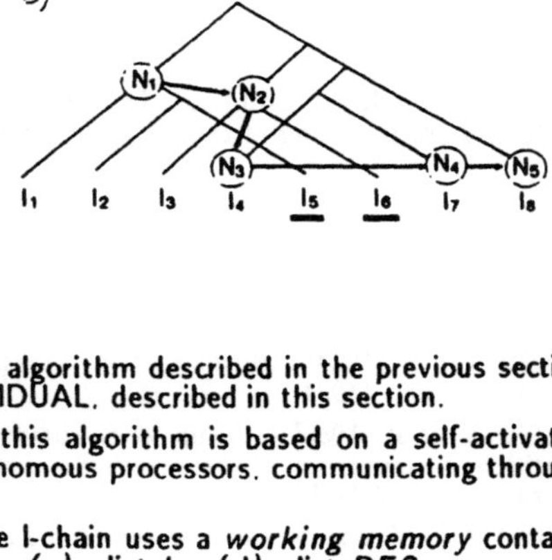

7. ALGORITHM FIND-RESIDUAL

The crucial and computationally most difficult step of the algorithm described in the previous section is step (3). This step is managed through algorithm FIND-RESIDUAL, described in this section.

In order to make it suitable for object-oriented programming, this algorithm is based on a self-activated propagation mechanism, in which the nodes are viewed as autonomous processors, communicating through "messages" sent along the links of the l-chain.

To perform its autonomous computations, each node N_i of the l-chain uses a *working memory* containing: *(a)* two *NL*-dimensional arrays α_1 and β_1, *(b)* a scalar M_i, *(c)* a list L_i *(d)* a list RES_i.

Each N_i is able to compute through a function *subs* the value of a logical variable $s_{ij} = subs(l_j, N_i)$, which is 1 (0) if l_j is (is not) a successor of N_i in Θ. Part of the process is illustrated by Fig. 5. Each node N_i of the l-chain (except the root), upon receiving from N_{i-1} a message containing α_{i-1} and β_{i-1}, computes α_i and β_i iteratively:

$$\alpha_i(j) = \max\{\alpha_{i-1}(j), \beta_{i-1}(j)\}$$

$$\beta_i(j) = \max\{\alpha_i(j-1)+s_{ij}, \beta_i(j-1)\}$$

for $j = 1,\dots,NL$, assuming $\alpha_0(j) = \beta_0(j) = \alpha_i(0) = \beta_i(0) = 0$. Then N_i pairwise compares corresponding elements in α_i and β_i. Upon finding $\alpha_i(j) \geqslant \beta_i(j)$, it sets $M_i = j$. Finally, N_i sends α_i and β_i as a message to N_{i+1}.

The above propagation process is triggered by activating the root N_1 to compute:

$$\beta_1(0)=0 \ , \ \alpha_1(j)=0 \ ,$$

$$\beta_1(j)=\max\{s_{1j}, \beta_1(j-1)\}, \quad j=1,\dots,NL$$

and to send to N_2 a message containing α_1 and β_1. After computing M_k, N_k sets $L_k = (l_{(M_k+1)}, \dots, l_{NL})$ and triggers a "backwards" propagation by sending M_k as a message to N_{k-1}. The "backwards" propagation wave ripples along the l-chain by a simple mechanism: each N_i, upon receiving M_{i+1} from N_{i+1}, computes $L_i = (l_{(M_i+1)},\dots,l_{M_{i+1}})$, looks for all leaves $l_h \in L_i$ which are not among its own successors and puts them in the list RES_i. The final residual set is obtained by joining the RES-lists.

The algorithm above generates only one residual set, which may not be the optimal one. The extension of the algorithm to generate the full set of residual sets, which may be then searched for the optimal residual set, is straightforward, but its description is too lengthy to be included. It is available from the author upon request.

Once a stable l-chain has been found, and characterized by a residual set, the generic node N_i of this l-chain will represent the class C_i. The logical description of C_i is obtained as follows. The path from the root to N_i determines a value assignment to a set of Boolean variables, and the conjunction of these assignments forms the *first part* of the logical description for C_i. Then N_i inspects leaves falling in its own RES list, and determines features that discriminate cases of C_i belonging to such leaves from cases of C_i **not** belonging to them.

FIG. 5

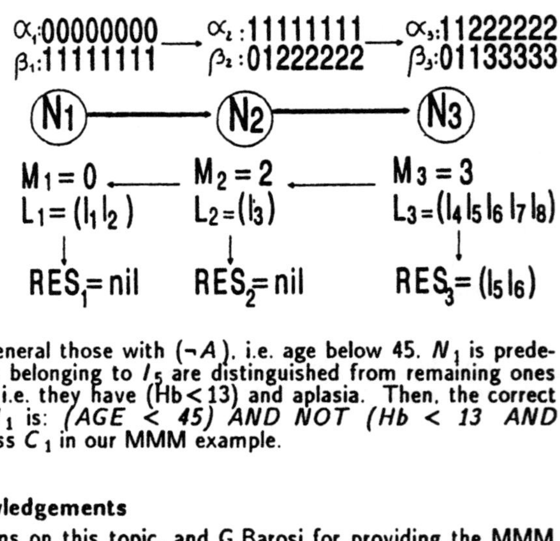

For example, cases belonging to N_1 are in general those with $(\neg A)$, i.e. age below 45. N_1 is predecessor of one residual leave: l_5. Within C_1, cases belonging to l_5 are distinguished from remaining ones because they satisfy the Boolean condition (HC), i.e. they have (Hb < 13) and aplasia. Then, the correct logical description for the class associated to N_1 is: *(AGE < 45) AND NOT (Hb < 13 AND APLASIA)*, which is exactly the description for class C_1 in our MMM example.

Acknowledgements

I wish to thank A.Ciampi for the helpful discussions on this topic, and G.Barosi for providing the MMM data and collaborating in the interpretation of results of their analysis.

REFERENCES

[Aitkin, 1978] M.Aitkin, The Analysis of Unbalanced Cross-classifications, *J.R.Statist.Soc.A*, **141**, Part 2 (1978), 195-223.

[Breiman *et.al.*, 1984] L.Breiman *et al.*, *Classification and Regression Trees*, Wadsworth International Group, Belmont, California (1984).

[A.Ciampi *et.al.*, 1986] A.Ciampi *et.al.* Stratification by stepwise regression, correspondence analysis and recursive partition: A comparison of three methods of analysis for survival data with covariates, *Computational Statistics & Data Analysis* **4** (1986), 185-204.

[Cox, 1972] D.R.Cox, Regression models and life tables, *J.Royal Stat.Soc.B*, **34** (1972), 187-208.

[Payne, *et.al.* 1984] R.W.Payne, T.J.Dixon, A study of selection criteria for constructing identification keys, *Proceedings of COMPSTAT 1984*, Physica Verlag.

[Quinlan, 1983] J.R.Quinlan, Learning efficient classification procedures and their application to chess endgames, in: *Machine Learning: an AI approach* (R.Michalski *et.al.* eds.), Palo Alto,CA,Tioga (1983).

A New Algorithm for Matched Case-Control Studies with Applications to Additive Models

T. Hastie and D. Pregibon, New Jersey

Abstract

Logistic models are commonly used to analyze matched case-control data. The standard analysis requires the computation of conditional maximum likelihood estimates. We propose a simple algorithm that uses a diagonal approximation for the (non-diagonal) weight matrix deriving from the Newton-Raphson method. The primary purpose of the new algorithm is to exploit iterative reweighted least-squares procedures for fitting general additive rather than simple linear structure.

1. Introduction.

The linear logistic regression model has become a standard tool of epidemiologists and biostatisticians for the analysis of matched case-control data. The model relates the odds of an individual with exposure $\boldsymbol{x} = (x_1, x_2, \cdots, x_m)$ developing the disease relative to an individual with baseline exposure $\boldsymbol{0}$ as

$$\log(\pi(\boldsymbol{x})) = \eta(\boldsymbol{x}) = x_1\beta_1 + x_2\beta_2 + \cdots + x_m\beta_m, \tag{1}$$

where $\pi(\boldsymbol{x})$ denotes the odds and $\eta(\boldsymbol{x})$ the linear predictor. Breslow and Day (1980) give a clear and detailed description. We establish notation which follows that of Pregibon (1984), and give a very brief outline of the methodology.

Consider a 1:R matched case-control study. Index the observations in each set from 0 to R, letting 0 denote the case. The *conditional* probability that within the kth matched set, the assignment of the $R+1$ values $\boldsymbol{x}_{rk}$ to case and controls is as observed, is given by Breslow and Day (1980) as

$$\mu_{0k} = \frac{\exp(\eta_{0k})}{\sum_{r=0}^{R}(\exp(\eta_{rk}))}. \tag{2}$$

The (conditional) likelihood of $\boldsymbol{\beta}$ given the data is simply the product, $L(\boldsymbol{\beta}) = \prod_{k=1}^{K}\mu_{0k}$, over all K matched sets.

We discuss algorithms for maximizing $L(\boldsymbol{\beta})$. Our primary goal is to develop algorithms for estimating non-parametric extensions of (1). In particular we concentrate on estimating the smooth but otherwise arbitrary functions f_j in the additive model:

$$\log(\pi(\boldsymbol{x})) = \eta(\boldsymbol{x}) = f_1(x_1) + f_2(x_2) + \cdots + f_m(x_m). \tag{3}$$

Additive models of this kind are discussed in Hastie and Tibshirani (1986) who motivate the *local scoring* algorithm for estimating the f_j as a non-parametric extension of iterative reweighted least-squares (IRLS). They concentrated, however, on situations where the weight matrix is diagonal as in the case of exponential family densities. In this case each observation has an associated weight, and the local scoring algorithm estimates the functions by *smoothing* appropriate partial residuals using a weighted scatterplot smoother.

Compstat 1988

Although the Newton-Raphson algorithm for maximizing $L(\beta)$ can be written as IRLS, we will see that the weight matrix is not diagonal. This hampers the non-parametric extension and alternative algorithms are needed.

2. Algorithms for the linear model.

2.1. The standard algorithm.

The score for β can be written as

$$S(\beta) = \sum_{k=1}^{K}\sum_{r=0}^{R}(y_{rk}x_{rk} - \mu_{rk}x_{rk}) = X^t(y - \mu), \tag{4}$$

where $y = \{y_{rk}\}$ is a vector of binary indicators of case (1) or control (0) status, and $\mu = \{\mu_{rk}\}$ is the vector of associated 'probabilities'. It is also convenient to define $U = \text{diag}(\mu_{rk})$ and μ_k, the sub-vector of probabilities (which sum to 1) for each matched set. The information matrix can be written as

$$\mathcal{I} = \sum_{k=1}^{K}\left[\sum_{r=0}^{R}\mu_{rk}x_{rk}x_{rk}^t - \left(\sum_{r=0}^{R}\mu_{rk}x_{rk}\right)\left(\sum_{r=0}^{R}\mu_{rk}x_{rk}\right)^t\right] = X^tWX, \tag{5}$$

where W is a $N \times N$ block diagonal matrix with kth block $W_k = U_k - \mu_k\mu_k^t$. Thus writing $z = X\beta^{old} + (W^{old})^-(y - \mu^{old})$, the Newton-Raphson update can be expressed as

$$\beta^{new} = \left(X^tW^{old}X\right)^{-1}X^tW^{old}z. \tag{6}$$

Rather than dealing with a GIRLS problem (generalized iterative reweighted least-squares!) of this kind, we can, with further manipulations, reduce it to the simpler IRLS (see appendix). The benefit in doing this is that standard weighted least-squares software can be used in the iterations. This becomes particularly important for the non-parametric extensions.

2.2. Delta Algorithms.

Jørgenson (1984) gave the name "delta" to modified Newton algorithms where the weight matrix W is replaced by an approximation. He showed that delta algorithms are ascent methods, so like the Newton-Raphson algorithm, convergence is guaranteed with step-size optimization.

Two useful approximations to the block matrix W are $\overline{W} = \text{diag}(\mu_{rk}(1 - \mu_{rk}))$ and $\overline{\overline{W}} = \overline{W} - \overline{W}1(1^t\overline{W}1)^{-1}1^t\overline{W}$. The first of these simply ignores all off-diagonal elements of W. The second puts values on the off-diagonal which attempt to adjust for all the matched sets simultaneously. This approximation was discovered empirically and corresponds to using the approximation $\overline{W}$ and augmenting X with a column of ones, i.e. adding the grand mean to the linear predictor. (This constant cancels in the definition of μ_{rk}.)

We demonstrate the convergence of the different algorithms on data from a study of possible physiological causes of depression (Rubin et al, 1988). The exposure variables represent concentrations of 7 hormones for 40 depressed patients and their matched controls. Figure 1 shows, on the log scale, the convergence of the deviance for the Newton-Raphson algorithm (6), and the two delta algorithms just described. Initially

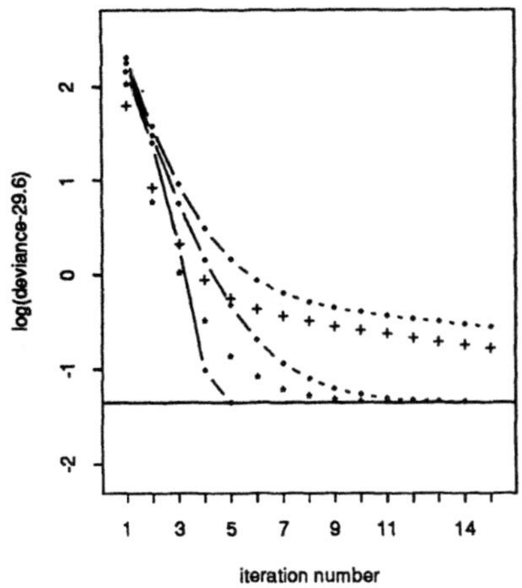

Figure 1. Convergence patterns in fitting the linear model to the depression data (7 variables). The steepest curve is the Newton-Raphson algorithm, the flattest curve is the delta algorithm, and the middle curve the delta algorithm with intercept. The +'s below the top curve show the effect of step-size optimization for that curve, as do the $*$'s below the middle curve

the delta algorithm based on $\overline{\boldsymbol{W}}$ shows a convergence pattern similar to the Newton-Raphson, but then flattens off and converges linearly. The convergence for the augmented diagonal algorithm based on $\overline{\overline{\boldsymbol{W}}}$ is considerably faster.

In order to investigate step-size optimization, we implemented an *Armijo-Goldstein* step search routine (Gill, Murray and Wright, 1980, section 4.3) which safeguards against over-stepping. Figure 1 shows the improvement due to step-size optimization to both versions of the delta algorithm. For this example the augmented version remains superior to the unaugmented one.

3. Estimation of the additive model.

In this section we derive an algorithm for estimating the components of the additive model (3). We use a penalized likelihood approach and present it in a simplified form, referring to the already large literature for details. For penalized likelihood techniques close in spirit to this work see Hastie and Tibshirani (1988) and O'Sullivan (1988).

The conditional penalized log-likelihood has two components, the likelihood component which measures fidelity of the function to the data, and the integrated squared second derivative component which measures its smoothness. As is usual for penalized problems of this kind, we can reduce the criterion for p functions to a criterion for p vectors, the jth consisting of the of the N fitted values at the data points x_{jrk}:

$$l(\boldsymbol{f}_1, \cdots, \boldsymbol{f}_m) = \sum_{k=1}^{K} \log(\mu_{0k}) - \frac{1}{2}\sum_{j=1}^{m} \lambda_j \boldsymbol{f}_j{}^t \boldsymbol{K}_j \boldsymbol{f}_j. \tag{7}$$

The smoothness penalties λ_j control the balance between goodness of fit and smoothness. The $N \times N$ penalty matrices $\boldsymbol{K}_j$ capture the second derivative operation.

A bit more algebra leads to the set of *normal equations* which need to be solved for the vectors of new fitted functions in a Newton-Raphson step:

$$\begin{pmatrix} I & S_1 & S_1 & \cdots & S_1 \\ S_2 & I & S_2 & \cdots & S_2 \\ \vdots & \vdots & \vdots & \ddots & \vdots \\ S_m & S_m & S_m & \cdots & I \end{pmatrix} \begin{pmatrix} f_1^{new} \\ f_2^{new} \\ \vdots \\ f_m^{new} \end{pmatrix} = \begin{pmatrix} S_1 z \\ S_2 z \\ \vdots \\ S_m z \end{pmatrix}, \tag{8}$$

where the S_j are $N \times N$ cubic spline smoother matrices of the form $S_{Wj} = (W + \lambda_j K_j)^{-1} W$ for smoothing against variable x_j. W is the same as in (5), and z is the adjusted dependent variable $z = \eta^{old} + (W^{old})^{-}(y - \mu^{old})$ and differs from z in (6) only in that η is now the additive predictor.

The Gauss-Seidel iterative procedure makes exceptional good use of this structure and has intuitive appeal. Also termed the *backfitting algorithm*, it proceeds as follows:

$$\begin{aligned} \textbf{Initialize:}\quad & f_j = f_j^0, j = 1, 2, \dots m, \\ \textbf{Cycle:}\quad & j = 1, 2, \dots m, 1, 2, \dots \\ & f_j = S_j(z - \sum_{k \neq j} f_k) \\ \textbf{Until:}\quad & \text{the individual functions do not change.} \\ \textbf{Finalize:}\quad & \text{set } f_j^{new} = f_j,\ j = 1, 2, \dots m. \end{aligned} \tag{9}$$

Buja et al (1988) prove that the backfitting algorithm always converges for spline smoothers. They also discuss more sophisticated versions of the same algorithm, similar to that described below.

In most non-parametric regression contexts W is diagonal and K has special banded structure which allows one to apply the smoothing operator S_W in $O(N)$ operations. This is not the case here, even though W is block diagonal and thus banded itself. K is banded if the rows are ordered with x, but then this ordering destroys the block diagonal structure of W. Thus $W + \lambda K$ is a full matrix and expensive to invert ($O(N^3)$ operations). O'Sullivan (1988) encountered exactly the same problem with the proportional hazards model and solved the system in $O(N)$ operations using a specialized preconditioned conjugate gradient technique which required 2 or 3 iterations per inversion.

Our implementation uses the diagonal approximation of W by $\overline{W}$ in order that the system be solved in $O(N)$ operations. In addition it splits each function f into linear and non-linear components, $f(x) = x\beta + g(x)$, resulting in the additive predictor, $\eta(x) = \eta_L(x) + \eta_g(x)$. This allows the linear component to be computed by projection using the full weight matrix. Thus the general orientation of the fitted functions (their slopes) is determined without Gauss-Seidel iteration. As a by-product we get the linear component of the fit which can be used to judge the amount of non-linearity. We call the result the *split backfitting algorithm* which proceeds as follows:

0: Initialize Set the non-linear component $\eta_g(\cdot) = 0$. Fit the *linear* model $\eta_L(x_{rk}) = x_{rk}^t \beta$, using the Newton-Raphson algorithm described in the appendix. Compute $\mu_{rk} = \exp(\eta(x_{rk})) / \sum_k \exp(\eta(x_{rk}))$.

1: Additive step Compute weights $\overline{w}_{rk} = \mu_{rk}(1-\mu_{rk})$ and adjusted dependent variates

$$z_{rk} = \eta_g(\boldsymbol{x}_{rk}) + \frac{y_{rk}-\mu_{rk}}{\mu_{rk}(1-\mu_{rk})}.$$

Fit the weighted additive model $\eta_g(\boldsymbol{x}_{rk}) = \sum_{j=1}^{m} g_j(\boldsymbol{x}_{rkj})$ to the z_{rk} using the backfitting algorithm. Update μ_{rk}.

2: Linear step Compute weights $w_{rk} = \mu_{rk}$ and adjusted dependent variates

$$z_{rk} = \eta_L(\boldsymbol{x}_{rk}) + \frac{y_{rk}-\mu_{rk}}{\mu_{rk}}.$$

Compute $\tilde{\boldsymbol{x}}_{rk} = (\boldsymbol{I} - \boldsymbol{1}\boldsymbol{\mu}_k^t)\boldsymbol{x}_{rk}$ and fit the linear model $\eta_L(\boldsymbol{x}_{rk}) = \tilde{\boldsymbol{x}}_{rk}^t\boldsymbol{\beta}$ to the z_{rk} by weighted least-squares. This resembles one step of the Newton-Raphson algorithm described in the appendix. Update μ_{rk}.

3: Until Repeat steps 1 and 2 until the fitted functions and coefficients do not change.

4: Finalize Compute $f_j(x_j) = \beta_j x_j + g_j(x_j)$, $j = 1, \ldots, m$.

4. Discussion.

The system of equations (8) is appropriate for a wide variety of regression estimators, not only smoothing splines. In practice this means that each variable can be associated with a particular smoothing operator. Of course, the least-squares hat matrix is a candidate as well, and is used when some of the variables are kept linear. Each time the variable has a turn in the backfitting algorithm, the smoothing operator is applied to the partial residuals resulting in a fitted function for the variable. This procedure can be further generalized to include associating a single smoothing operator with more than one variable (e.g. surface smoothing).

The point of view that the spline smoother can be replaced with any reasonable non-parametric regression estimator suggests that any of the commonly used scatterplot smoothers can be used provided they can be modified to incorporate observation weights. Amongst others, this class includes "locally weighted running lines" (Cleveland, 1979), kernel smoothers (e.g.Watson, 1964), and "supersmoother" (Friedman and Stuetzle, 1982).

Our algorithm draws only on basic tools available in most statistical packages, namely weighted least-squares and weighted scatterplot smoothing. Thus no additional specialized software is required although high quality graphical output is desirable.

We feel that the additive model should be treated as a supplement to, rather than a substitute for, the classical linear model. The non-linear contributions of covariates graphically displayed by the function plots not only indicate problems with the linear model but also indicate how to ameliorate them. Thus parametric re-expressions and innovative re-parametrizations suggest themselves rather than being left to the ingenuity of the modeler. In a more detailed version of these proceedings Hastie and Pregibon (1988) provide several examples.

References

Breslow, N.E. and Day, N.E. (1980), *The Analysis of Case-Control Studies*, IARC Scientific Publications, Lyon.

Buja, A., Hastie, T., and Tibshirani, R. (1988), "Linear Smoothers and the Additive Model", *Annals of Statistics* (to appear),

Cleveland, W.S. (1979), "Robust Locally Weighted Regression and Smoothing Scatterplots", *Journal of the American Statistical Association* , **74**, 829-836.

Friedman, J.H. and Stuetzle, W. (1982), "Smoothing of Scatterplots", *Department of Statistics Technical Report, Orion 3*, Stanford University.

Gill, P., Murray, W., and Wright, M. (1980), *Practical Optimization*, Academic Press, New York.

Hastie, T., and Pregibon, D., (1988), "A New Algorithm for Matched Case-control Studies with Applications to Additive Models", *Research Report*, AT&T Bell Laboratories.

Hastie, T. and Tibshirani, R. (1986), "Generalized Additive Models (with discussion)", *Statistical Science*, **1**, No 3., 297-318.

Hastie, T. and Tibshirani, R. (1988), "Exploring the Nature of Covariate Effects in the Proportional Hazards Model", (submitted for publication) *Division of Biostatistics Technical Report*, University of Toronto.

Jørgenson, B (1984), "The Delta Algorithm and GLIM", *International Statistical Review*, **52**, 283-300.

O'Sullivan, F. (1988), "Nonparametric Estimation of Relative Risk Using Splines and Cross-Validation", *SIAM Journal of Scientific & Statistical Computing*, **9**, 531-542.

Pregibon, D. (1982), "Score Tests in GLIM", *Proceedings of the International Conference on Generalized Linear Models, London*, Springer-Verlag lecture notes in Statistics #14, Berlin, Heidelberg.

Pregibon, D. (1984), "Data Analytic Methods for Matched Case-Control Studies", *Biometrics*, **40**, 639-651.

Rubin, R.T, Wheeler, N.C., Pregibon, D., and Poland R.E. (1988), "Neuroendocrine Aspects of Primary Endogenous Depression-V. Logistic Regression Analysis of Matched Patient-Control Hormone Data for Discrimination Between Groups", *Journal of Psychiatric Research* (to appear),

Watson, G.S. (1964), "Smooth Regression Analysis", *Sankya Series A*, **26**, 359-372.

Appendix

The special structure of the weight matrix W can be exploited to derive a diagonally weighted version of the Newton-Raphson algorithm (6) (Pregibon, 1982). Write the matrix $W_k = (I - \mathbf{1}\mu_k^t)^t U_k (I - \mathbf{1}\mu_k^t)$, where $\mathbf{1}$ is a column of R+1 ones. Now if X_k denotes the matrix of exposure variables for the kth matched set, $\tilde{X}_k = (I - \mathbf{1}\mu_k^t)X_k$ is the corresponding centered version; the centering is done by removing the μ-weighted average for each exposure variable. Thus $X_k^t W_k X_k = \tilde{X}_k^t U_k \tilde{X}_k$, and if $\tilde{X}$ denotes the entire matrix of exposure variables with each block centered by its block μ-weighted average, then $X^t W X = \tilde{X}^t U \tilde{X}$. Noting that $X^t(y - \mu) = \tilde{X}^t(y - \mu)$, substitution into (6) yields

$$\beta^{new} = \left(\tilde{X}^t U^{old} \tilde{X}\right)^{-1} \tilde{X}^t U^{old} \left(\tilde{X}\beta^{old} + (U^{old})^{-1}(y - \mu^{old})\right).$$

In algorithmic form the Newton-Raphson algorithm is as follows:

Initialize: Set $\beta = \mathbf{0}$.

Cycle: Compute $\mu_{rk} = \exp(x_{rk}^t\beta)/\sum_{r=0}^{R}\exp(x_{rk}^t\beta)$.

Center x_{rk} by its μ-weighted average, i.e. $\tilde{x}_{rk} = (I - \mathbf{1}\mu_k^t)x_{rk}$.

Compute $\tilde{z}_{rk} = \tilde{x}_{rk}^t\beta + (y_{rk} - \mu_{rk})/\mu_{rk}$.

Compute β^{new} by regressing $\tilde{z}$ on $\tilde{X}$ with weights μ.

Until: the deviance $D = -2\sum_k \log \mu_{0k}$ converges.

An Algorithm for the Approximation of N-Dimensional Distributions

J. Gordesch, Berlin

Summary: An algorithm for approximation and parameter estimation leading to L_1 and L_∞ optimization is presented, and for numerical computation Karmarkar's linear programming algorithm is adapted.
Keywords: approximation, estimation, linear programming, computer aided design, finite mixture distribution

1. Introduction

We often encounter problems like the following:

Problem 1: A (complicated) theoretical distribution function is to be approximated by a linear combination of appropriately chosen (simple) distribution functions, e. g. for drawing wire frame models.

Problem 2: An empirical distribution shall be represented by a theoretical distribution the parameters of which have to be estimated.

Problem 3: A distribution function can be decomposed as a linear combination of distributions on homogeneous strata, and the coefficients as well as the parameters of the distributions must be estimated.

Problem 4: Probability distributions and stochastic processes are often described by functional equations the solutions of which are to be approximated.

The solution of these and similar problems usually leads to optimization problems which in many cases can be reduced to L_1 or L_∞ (Chebyshev) approximation entailing the use of linear programming techniques.

Compstat 1988

2. Linear Programming Algorithm

In the worst case, the simplex algorithm is of exponential order $O(e^n)$. As the number of variables may be very large, Karmarkar's linear programming algorithm is proposed, which is approximately of order $O[(n+1)(q+\ln(n+1))]$ (q is a termination parameter connected with the desired accuracy).

Karmarkar's algorithm is a variety of the gradient method for linear objective functions which has been adapted to non-negative variables. The main steps of the algorithm are:

(1) Transform the LP problem by a projective transformation to the standard type:
minimize $c'x$
subject to $x \in \{x \mid Ax=0\} \cap$ n-dim. simplex
$c'x_{min} = 0$, center a_0 of the simplex is a feasible solution
stopping rule $c'x/c'a_0 \leq 2^{-q}$

(2) Iteration procedure: The projection onto the subspace of restrictions (to put it vaguely) is used to determine the best direction of improvement.

(3) If the minimum is not zero and not known solve the unidimensional problem:
minimize $c'x_{min} = d$
subject to $d_1 \leq d \leq d_0$, for convenient d_1, d_0

(4) Transform the solution found back to the original space.

Proper consideration may be given to statistical conditions either when defining the projective transformation in step (1) or step (2). As a full exposition would be too lengthy to be presented here, refer for some of the details to Karmarkar's original paper, and for the incorporation of statistical conditions, to Gordesch (1988).

3. Approximation and Parameter Estimation

Problem 1 (one-sided approximation on a subset of the space of continuous functions C_n) leads immediately to linear programming (if discretized, the linear programming problem is finite). For Problem 2, a combined algorithm for discrete linear L^∞ approximation and parameter estimation is developed here, which also covers some special cases of Problem 3 and Problem 4. The algorithm is based on an idea of Krabs, where a sequence of linear programming tasks leads to the desired optimal solution. Linear estimation of the parameters can be incorporated. Nevertheless, the main theorems remain valid. For a proof of the theorems of existence refer to Krabs' papers (1966, 1969), and for some of the statistical details to Gordesch (1988).

Let us assume:

(1) B is a finite subset of a metric space, $\{x_1, x_2, \dots x_m\}$, $m \geq r+s+2$

(2) C(B) is the linear space of all real-valued functions on B

(3) $\tau(f)$ is the maximum norm of f

(4) U is the n-dimensional subspace spanned by $u_j \in C(B)$, $j = 1 \dots n$

(5) $m \geq n+1$

Then a minimal solution $\hat{u} \in U$ exists for arbitrary but fixed $f \in C(B)$ such that

$$\tau(\hat{u} - f) = \inf \tau(u - f), \text{ inf taken over } u \in U.$$

Furthermore, for each minimal solution holds

$$\min\, [\hat{u}(x) - f(x)]u(x) \leq 0 \text{ for all } u \in U,$$
$$\text{min taken over } x \in E(\hat{u});$$
$$E(u) = \{x \in B \mid \text{abs}(u(x) - f(x)) = \tau(u - f)\}$$

Starting from these general theorems a practical algorithm may be developed:

(1) Detect any outliers and drop them for further computation.

(2) Find a starting solution for the minimization problem. Con-

sider numerically convenient as well as statistically meaningful procedures. The simplest technique is ordinary least squares estimation, which also often produces satisfying solutions:

$$\sum[u^o(x) - f(x)]^2 \le \sum[u(x) - f(x)]^2, \; x \in B.$$

In most cases the usual statistical tests will fail, and should as far as necessary be replaced by distribution-free tests (cf. Maritz, pp 202). Nevertheless, the distributional properties will change during the approximation procedure; so just find a statistically sound starting value. Various other starting techniques make sense, dependent on the particular problem. For instance, minimum chisquare estimates may provide a good initial solution in estimating parameters of a onedimensional probability distribution, etc.

(3) Iteration step 1: Search for a $\bar{u}^k \in E_k$ such that
$\min\, [u^k(x) - f(x)]\, \bar{u}^k(x) > 0$, or:
Maximize μ
subject to $\forall x \in E(\delta_k)\; \sum[u^k(x) - f(x)]u_j(x)a_j - \mu \ge 0$,
$\mathrm{abs}(a_j) \le 1, \quad j = 1 \ldots n$,
$E(\delta_k) = \{x \in B \mid \mathrm{abs}[u^k(x) - f(x)] \ge R_k - \delta_k\}$,
$R_k = \tau(u^k - f)$

(4) Iteration step 2:
$\varepsilon_k = \min \mathrm{abs}[\bar{u}^k(x)], \quad x \in E(\delta_k)$
$c_1 = \min \{R_k - [u^k(x) - f(x)]\}/[\varepsilon_k - \bar{u}^k(x)], \quad \varepsilon_k - \bar{u}^k(x) > 0$
$c_2 = \min \{R_k + [u^k(x) - f(x)]\}/[\varepsilon_k + \bar{u}^k(x)], \quad E_k + \bar{u}^k(x) > 0$
$\lambda^k = \min (c_1, c_2)$
$u^{k+1}(x) = u^k(x) - \lambda_k\bar{u}^k(x)$

Note: The optimization of iteration step 1 can be avoided in special cases, and be replaced by algorithms with less computational effort. If $E_k = \{x \in B \mid \mathrm{abs}(u^k(x) - f(x)) = \tau(u^k - f)\}$ consists of not more than n+1 points, then a linearly independent set of n basic functions $u_1 \ldots u_n$ can be chosen. If E_k consists of less than n points, then E_k is completed with conveniently selected points. If E_k consists of more than n+1 points, these algorithms fail. Workable algorithms exist chiefly in the case the functions $u_k(x)$ obey the Haar condition (cf. e. g. Stiefel 1959, 1960).

4. Computational Aspects

The approximations of functions of several variables is a rather delicate matter, particularly for L_1- and L_∞-norms. Much depends on the skill (and the taste) of the person who does the job. Since computational effort may be considerable, the approximation of functions of several variables by functions with a smaller number of variables is of special interest. Little is known, but one may try a polyadic approximation:

$$f(x_1 \ldots x_n) = \sum_i f_{i1}(x_1)\ldots f_{in}(x_n),\ 1 \leq i \leq m,\ m \text{ fixed.}$$

Fix all functions but one and determine an optimal approximation (function of one variable only). Then select another function, fix the rest, determine the optimal approximation, etc. Stop if the desired accuracy is reached (there is no guarantee that the minimal defect is ever reached).

For standard cases (functions of one variable) numerical software packages provide subroutines for L_2, L_1 and L_∞ approximations (e. g. IMSL Chapter I, Interpolation, Approximation, and Smoothing; and Chapter R, Regression Analysis; ICSMOU for outlier detection). Programs for various forms of the simplex algorithm and other techniques related to linear programming can be found in software packages for operations research (e. g. Hering and Scheurer 1988). Karmarkar's algorithm however is not included in the common packages, but one might follow the sample program published in BYTE (Rocket and Stevenson 1987) and prepare one's own version.

Comparing the performance of Karmarkar's algorithm to that of the simplex algorithm it can be argued that the simplex algorithm searches the best solution, while Karmarkar's algorithm stops at an answer that is a set factor better than the initial guess.

Given a good starting solution and only low requested accuracy, then Karmarkar's algorithm is much faster. It is preferable for rather large n, since it is of lower order than the simplex algorithm and also less cumbersome and more general than a decomposition algorithm. For small n the simplex algorithm behaves well, and elaborate programs are available.

References

GORDESCH, J. (1988). *Fitting n-Dimensional Distributions*. Research Paper, Free University Berlin, Institute of Sociology (to appear).

HERING, E., and SCHEURER, K. (1988). *Operations Research mit dem PC*. Braunschweig.

KARMARKAR, N. K. (1984). A new polynomial-time algorithm for linear programming. *Combinatoria* 4, 373-395.

KRABS, W. (1966). Ein Verfahren zur Lösung der diskreten rationalen Approximationsaufgabe. *ZAMM* 46, 63-66.

KRABS, W. (1969). Ein Pseudo-Gradientenverfahren zur Lösung des diskreten linearen Tschebyscheff-Problems. *Computing* 4, 216-224.

LEMPIO, F. (1971). Lineare Optimierung in unendlich-dimensionalen Vektorräumen. *Computing* 8, 284-290.

MARITZ, J. S. (1984). *Distribution-Free Statistical Methods*. London - New York.

MARSAGLIA, G. (1970). One-sided approximation by linear combinations of functions. *Approximation Theory*, ed. by A. Talbot. New York - London, 233-242.

ORTEGA, J. M., and RHEINBOLDT, W. C. (1970). *Iterative Solution of Nonlinear Equations in Several Variables*. New York - London.

ROCKET, A. M., and J. C. STEVENSON (1987): Karmarkar's algorithm. *BYTE*, Sept., 146 - 160 (Corrections: Dec. 1987, 32).

STIEFEL, E. (1959). Über diskrete und lineare Tschebyscheff-Approximationen. *Num. Math.* 1, 1-28.

STIEFEL, E. (1960). Note on Jordan elimination, linear programming and Tschebyscheff approximation. *Num. Math.* 2, 1-17.

TITTERINGTON, D. M., SMITH, A. F. M., and MAKOV, U. E. (1985). *Statistical Analysis of Finite Mixture Distributions*. Chichester (U. K.).

Further Recursive Algorithms for Multidimensional Table Computation

B. P. Murphy and G. Bartlett, Western Australia

Synopsis : We briefly review some recursive algorithms in the literature, and see that most seem to fall into two broad classes : classical branch-and-bound backtracking, and mixed-radix integer counting. New examples of the latter are shown, and it is indicated how more can be obtained from these to perform a wide range of tabular computation.

Introduction

A full review of recursive algorithms is given by Rohl[1984], using the programming language Pascal, though any cited could be simply translated to any other structured language such as C, Modula-2 and ADA. Most of Rohl's examples are of use to programming statisticians, in that they concern recursive templates for general computing techniques, such as sorting, database construction and interface parsing. Clarke[1982] and Murphy *et al.*[1983, 1986] give examples of more specific interest to statistical questions, using versions of the classic branch and bounds recursive algorithm, related to all possible model searching for 'best fit' models. Their algorithms can be summarised as

```
PROCEDURE NextTerm(NextI : integer);
VAR i : integer; fitd : boolean;

BEGIN
FOR i := 1 TO NextI DO
    IF NOT Bound THEN
        IF NOT Branch THEN begin
            DoJob(fitd);
            IF NOT fitd THEN NextTerm(i+1)
            ELSE Repair;
        END;
END;
```

where Procedure NextTerm finds the next branch of the recursive tree to attempt to fit, and the Boolean fitd, with functions Branch and Bound, effect the backtracking process. This is equivalent to the basic algorithm of the standard Knight's Tour, Eight Queens and Towers of Hanoi problems beloved of the computer scientist. The Degrees of Freedom procedure of Murphy *et al.* [1986] is of similar structure, though this is somewhat obscured by the seeming perversity of its central computation.

The procedure Sweep of Murphy *et al.* [1986] has quite different structure and uses, and it is some of these we intend to examine more closely here. This algorithm collapses a multidimensional table into a (multidimensional) subtable of marginal means, and then subtracts out the each mean from the data items which formed it. This powerful

Compstat 1988

algorithm effects the central computation of analysis of variance as indicated by Wilkinson[1961] and used in the GENSTAT (Alvey *et al.*,1986) and MASS (Murphy *et al.*,1987) statistical packages. It is however easier to understand in the form below, which clearly shows how the indices of the target cells of the summation process are easily derived through the simple recursive procedure GenerateIndexSet, leaving the summation to another simple procedure SumProcess (not given here) :

```
PROCEDURE GenerateIndexSet(d: integer);
    VAR  j: Integer;

    BEGIN
        FOR j := 1 TO NoOfLevels[d] DO BEGIN
            i[d] := J;
            IF d < NoOfFacts THEN GenerateIndexSet(d + 1)
                ELSE SumProcess(i)
        END
    END; { GenerateIndexSet }
```

GenerateIndexSet obtains in array i the index of the current (UnitNo-th) observation, converting the decimal integer UnitNo into an array of integers which can be read as the digits of a mixed-radix integer, the k-th digit having base NoOfLevels[k], the number of levels for the k-th factor of the data table. Its interest is that the process involves no computationally expensive **div** or **mod** operations, as required in other approaches.

It is important to realize that the same efeect could be obtained by a non-recursive procedure using a nested series of DO-loops. This however has the great disadvantage that since in general one must allow for run-time specification of the dimensionality of the problem, the number of nested DO's must be held at some arbitrary (largish integer) maximum, which means many usually unused extra lines of code, and more organising jumping over these unnecessary loops. A version using WHILE's or simple GOTO's would be better, and this is exactly what one gets when a recursive procedure is turned into a non-recursive one, as can always be done (Rohl, 1984).

So why bother with recursion? It is usually worthwhile because recursive procedures are (i) compact and thus (ii) easily understood and (iii) easily checked/'proved'. Moreover, the non-recursive version is often not easily obtained until the recursive one is well understood!

Example and extensions

The procedure SweepIndex is a generalisation of the Sweep algorithm of Murphy *et al.*. [1986] to handle data which has both implicit indices and indices given in a user input array. Such a situation occurs in any model in which the data has more factors than are needed to implicitly read it - e.g. Latin Square data has 3 factors, and must have at least one of these indexed to model it fully. The procedure supplies no new recursive features of itself, but shows the full basis for expansions we wish to discuss. The definitions are as in Sweep. Also note that one internal algorithm which could be made recursive has been opened out.

```
PROCEDURE SweepIndex ( Dir: Direction; Effect: FactorSet; SizeEffect: Factrange );
    TYPE FactRange =[0..8]; LevelRange =[0..127]; {allows 8 factors of 127 levels}
        IndiceArray = ARRAY [FactRange] OF LevelRange;
    VAR  i : IndiceArray;                {array holding mixed-radix index set  for data item}

    PROCEDURE SweepProcess(Index : IndiceArray); {
        VAR
            FactNo, FactCount, kk, MeanNo: integer;
            MIndex, Levels: IndiceArray;  { indices values and factor levels of effect}

        BEGIN
            UnitNo := UnitNo + 1;    {current data  item no.}
            FactCount := 0;
            FOR FactNo := 1 TO Factors DO
                IF FactNo IN Effect THEN BEGIN
                    FactCount := FactCount + 1;
                    IF Factor[FactNo].Index = NIL THEN BEGIN
                        MIndex[FactCount] := Index[FactNo];
                        Levels[FactCount] := Factor[FactNo].NoOfLevels
                    END
                    ELSE BEGIN
                        MIndex[FactCount] := Factor[FactNo].Index^.Ind[UnitNo];
                        Levels[FactCount]   := Factor[FactNo].NoOfLevels
                    END;
                END;
    {*}      MeanNo := 0;  MIndex[0] := 1; Levels[0] := 1;
                FOR FactNo := 1 TO (FactCount - 1) DO
             MeanNo := (MeanNo + (MIndex[FactNo] - 1)) * Levels[FactNo + 1];
             MeanNo := MeanNo + MIndex[FactCount];
    {**}     IF Dir = Up THEN BEGIN
                    Mean[MeanNo] := Mean[MeanNo] + Table[UnitNo];
                    Inc[Term[TermNum].MeanLoc + MeanNo - 1] :=
                            Inc[Term[TermNum].MeanLoc + MeanNo - 1] + 1;
    {***}           END ELSE Table[UnitNo] := Table[UnitNo] - Mean[MeanNo]
     END; {SweepProcess }

     PROCEDURE GenerateIndexSet(d: integer);
        VAR   j: integer;

        BEGIN
            FOR j := 1 TO Factor[d].NoOfLevels DO BEGIN
            i[d] := j;
            IF d < NoOfFacts THEN GenerateIndexSet(d + 1) ELSE  SweepProcess(i)
            END
        END; { GenerateIndexSet }

     BEGIN
     {Initialisations}
     GenerateIndexSet(1) ;
     {Cleanups}
     END {SweepIndex};
```

The procedure moves serially through the data, determining in the Up direction the position (MeanNo) of the cell in the marginal mean table (Mean) into which the current observation (Table[UnitNo]) must be accumulated and accumulating. In the Down direction, the same Mean(MeanNo) value is subtracted from the same data item.

A first modification was indicated in that paper - replacing lines {**} to {***} by

```
Table[UnitNo] := Mean[MeanNo] ;
```

which performs the propogation of a table of stratum means prior to the round of sweeps necessary to determine the effects found in that stratum. A similarly trivial modification provides the full code for iterative proportional scaling (IPS) in log-linear model fitting (Haberman, 1972). As in SweepIndex above, at the kernel of this procedure we adjust a vector of observations/expectations by a subtable of means, in this case multiplicatively by the ratio of a current total in table Mean to the corresponding total in table ModelTotal. Hence the ELSE clause becomes

```
ELSE Table[UnitNo] := Table[UnitNo] * Mean[MeanNo]/ModelTotal[MeanNo]
```

Similarly, a pass through the data to estimate missing items is performed by replacing the code section {*} to {***} by (ignoring for simplicity a few new initialisation lines)

```
MeanNo := 0; Levels[0] := 1;
FOR FactNo := 1 TO (FactCount - 1) DO
    MeanNo := (MeanNo + (MIndex[FactNo] - 1)) * Levels[FactNo + 1];
MeanNo := MeanNo + MIndex[FactCount];
Cell := Cell + MarginMean[AovModel[I].MeanLocn - 1 + MeanNo];
```

which sums the various mean estimates needed to compute the estimate in location Cell for the missing value. Within an iteration process similar to Sweep's for computing the means for the data and their location offsets AovModel[I].MeanLocn, this develops the complete Westmacott-Healy [1956] missing value estimation process.

The Yates algorithm (Gower, 1969) is obtained by an equally trivial modification and is included as a support procedure in the next illustration. Somewhat more difficult but in the same vein is the process for recombining or reordering effects produced by this in the more traditional order. Specifically, if A, B, C are factors with 2, 3, and 4 levels respectively, then a Yates algorithm produces effects in the order

$$A, B^1, B^2, AB^1, AB^2, C^1, C^2, C^3, AC^1, AC^2, AC^3, B^1C^1, B^1C^2, B^1C^3, B^2C^1 \ldots$$

One may want it reordered and rdcombined to give A, B, C, AB, AC, BC, ABC. This is related to the Entanglement algorithm of Zemroch [1984], and is seen in procedure Recombine.

```
PROCEDURE NewYates(ExptTable  : DataArray;  VAR EstTable : DataArray);
   VAR     LocTabSize, Rownum, UnitNo : DATAINDEX;
           FactNo       : FACTORINDEX;
           Idx          : CARDARRAY;
           ThisVal      : REAL;

   PROCEDURE DoYates(Fact   : FACTORINDEX ;   Mult   : REAL);
      VAR     i : LEVELINDEX;
      BEGIN
         FOR i := 1 TO Levels[Fact] DO ThisVal := Partn^[Idx[Fact]+i];
         IF Fact < nDim THEN DoYates((Fact + 1),(Mult * ThisVal))
         ELSE  BEGIN
                  UnitNo := UnitNo + 1;
                  EstTable[Row] := EstTable[Row] + ExptTable[UnitNo] * (Mult * ThisVal);
                  END;
      END {DoYates};

   BEGIN
      FOR Row := 1 TO TableSize DO BEGIN
         LocTabSize := TableSize;
         Rownum := Row-1;
         FOR FactNo := 1 TO nDim DO BEGIN
            LocTabSize := LocTabSize DIV Levels[FactNo];
            Idx[FactNo] := (Rownum DIV LocTabSize) * Levels[FactNo]
                     + POffset[FactNo];
            Rownum := rnum MOD (LocTabSize);
            END;
         Estim := 0.0;
         StDev := 0.0;
         UnitNo := 0;
         DoYates(1, 1.0);
         END;
   END {NewYates};

PROCEDURE Recombine;   { Gets the estimates in order of cardinality of effects }
   VAR     EffectNo : CARDINAL;
           Effect   : EffectType;
           TLevs , Idx      : LevelType;
           i       : FACTORINDEX;
           n, row   : DATAINDEX;

   PROCEDURE GetEstimates(FactNo : FACTORINDEX);
      VAR     k  : LEVELINDEX;
              z  : REAL;
      BEGIN
      IF FactNo IN Effect THEN st := 1  ELSE  st := 0;
      FOR k := st TO (TLevs[FactNo]) DO BEGIN
         Idx[FactNo] := k;
         IF FactNo < nDim THEN GetEstimates(FactNo + 1)
         ELSE BEGIN
```

```
            row := 0;
            n := TableSize;
            i := 1 TO nDim DO BEGIN
                n := n DIV Levels[i];
                row := row + (Idx[i] * n);
                END;
            row := row + 1;          { This is the position in the direct Yates vector }
            END;
        END;
    END     {GetEstimates};

    BEGIN       { The Hier array is a predetermined array  of terms in cardinal order
                                 i.e a b c ab ac... }
    FOR EffectNo := ModlUTerms TO 1 BY -1 DO BEGIN
        Effect := Hier[EffectNo];
        FOR FactNo := 1 TO nDim DO TLevs[FactNo] := Levels[FactNo] - 1;
        GetEstimates(1);
    END;
END {Recombine};
```

Further Extensions

It is clear that many further examples could be applied in other areas. Street (1988, personal communication) has illustrated one of the backtrack type in repeated measures design construction. We look forward to many others.

References

Alvey A.G. *et al.,* (1986) :**The GENSTAT manual**, Natl. Algorithms Group, Ch.9

Clarke MRB (1982) : A Givens algorithm for moving one model to the next without going back to the data, **Appl. Statist., 30**, 198-203.

Gower J C(1969) : Calculation of Effects **Appl. Statist., 18**, 287-290

Haberman SJ (1972) : Loglinear fit for contingency tables, **Appl. Statist., 21**, 218-225

Healy MJ and Westmacott MH (1956) : Missing values in experiments analysed in automatic computers. **Appl. Statist., 5**, 203-208

Murphy BP (1983) : Production and reproduction of statistical software for Z80 microprocessors, **Contemp. Maths., 9**, 397-400

Murphy BP, Cribb RL, Rohl JS (1986) : Recursive techniques in statistical programming. **Proc. COMPSTAT'86**, ed. Antoni *et al.,* Springer Verlag, Vienna, 338-344

Murphy, BP, Henstridge JD, Poulsen JC and Wearne KL (1987) : **The MASS Manual,** Westat Associates, Western Australia, Ch. 2.3

Rohl JS (1984) : **Recursion via Pascal,** Cambridge University Press.

Zemroch PJ : Generation of polynomial contrasts for incomplete factorial designs, **Appl. Statist., 30**, 325-333

Statistical Methods

Nonlinear Regression: Methodological and Software Aspects

A. Messéan, Paris

Abstract

Through a survey of the state of the art in nonlinear regression, this paper examines the contribution of different scientific tools: asymptotic theory, differential geometry, numerical analysis and computer science. The specific role of computer science is pointed out: hardware power, software technology, computing environnements and now, artificial intelligence. Finally, some trends for the future are presented.

1 Introduction

Nonlinear regression is now widely used in many fields of applications such as kinetics, populations dynamics, biological assays, epidemiology, ... A considerable effort of investigation has been devoted to this topic since the appearance of modern computers and the implementation of numerical algorithms able to solve optimization problems and, particularly, nonlinear least squares problems.

Originally, investigation was mainly dedicated to the extension to the nonlinear case of some classical results of the linear model theory and to the implementation of efficient numerical algorithms. Now, a wide variety of problems is considered: definition of the statistical model of regression, choice and properties of parameter estimates, statistical inference, numerical problems and also side aspects as calibration, prediction or censored regression.

Many fields of statistics and applied mathematics have been involved in nonlinear regression: statistical asymptotic theory and, particularly, higher order asymptotics, differential geometry to appreciate effects of the model nonlinearity, resampling methods to substitute amounts of computing to asymptotic theory, numerical analysis to solve specific optimization problems encountered in estimation and inference steps.

Software packages devoted to nonlinear regression have also considerably changed since the 1960's. From subroutines minimizing a least squares criterion with computation of standard confidence intervals, such as GAUSHAUS (Meeter, 1968), or libraries of routines, such as NAG or HARWELL, to general statistical packages which include nonlinear regression capabilities, such as Genstat, BMDP or SAS. There are also some more specific software products such as MLP

Compstat 1988

(Ross, 1987), which carries out an interactive study of nonlinear models through a simple interface language.

In the first part of this paper, we present the point of view of our Department of Biometry upon the state of the art of nonlinear regression both in modeling and estimation aspects, trying to identify the contribution of the different tools of applied mathematics and computer science. Then we discuss our software strategy and give some hints about a project of an Interactive System of Nonlinear Modelling based on an object-centered knowledge base management system.

2 Methodological aspects.

We consider general nonlinear regression models of the following form

$$Y_i = f(x_i, \theta) + \varepsilon_i \qquad i = 1, ..., n \tag{1}$$

where the errors (ε_i) are independent, centered and have a variance $v(x_i, \theta, \beta)$, and x_i is a set of independent variables, $\theta \in \Theta \subset R^p, \beta \in B \subset R^q$.

2.1 Definition of the statistical model.

The construction of the model, i.e. the choice of the response function f (the expectation of Y), of the variance function v and of the distribution of the ε_i's, is one of the fundamental tasks of the analysis of regression. In most of the applications, mainly in System Analysis (where the model is often defined through differential equations), most of the effort is devoted to the choice of response function f, regression analysis being reduced to curve fitting by least squares criterion. Nevertheless, although v is often considered as a nuisance component, i.e. the parameters of interest are the $\theta's$, variance modeling must be led very carefully.

Choice of the response function f. Response modeling constitutes a large field of interest we can summarize drastically by distinguishing two kinds of approaches:

1. The explicative model.

   The model structure, its behaviour and its parameters are analyzed with respect to their phenomenon significance. In this case, the parameters are of main interest and model validation is essential. Populations dynamics models described by differential equations are a good example of such a model.

2. The representative model.

   The regression model is viewed as a mathematical representation of the phenomenon and its particular structure does not play any role. Main interest is devoted to the response function

f. The calibration problem encountered in radio-immunological assays is an illustration of this situation (Huet, 1983).

Choice of the variance function v. Choice of the variance model will depend on the analysis of the variability of the raw data (variability *vs.* response f or *vs.* independent variables x_i), on the prior knowledge of the phenomenon (for example, in populations dynamics, standard error is often proportional to the response function f, so $v = \sigma^2 f^2$) and on the number and the level of hypotheses the user can or wishes to introduce in the model.

Distribution error. The user has also to decide to specify or not the distribution of the errors. And more generally, he must check that the regression framework is the adequate one for his problem. As a matter of fact, a nonlinear regression is generally used in all curve fitting situations although there could be some confusion with other statistical models, e.g. distribution fitting under multinomial likelihood.

Thus, these different steps involve different points of view whose handling supposes a great level of human expertise and the statistical model retained can carry quite different levels of hypothesis. The model can be supposed completely defined (distribution of errors, f and v known) or almost unknown (f unknown and approximated by a function g, v estimated by nonparametric methods and distribution unknown). Forthcoming analysis will be then greatly oriented by the model structure. A description of the different situations can be found in Huet *et al.* (1988).

2.2 Parameter estimation.

Suppose the regression model is well defined and consider the problem of estimation of the parameters θ. This reduces to the choice of an estimating function of the observations, $\widehat{\theta}_n = H(Y)$, the estimator of θ, which has good properties and recovers *the maximum of information* from the data. Desirable properties could be consistency, unbiasedness, first-order efficiency, asymptotic normality and second-order optimality. All of these properties involve asymptotics and their proofs rely mainly on the asymptotic theory machinery.

On the other hand, depending on the aim of the analysis, the user is interested by quantities as the bias, the precision of the estimate (through its standard deviation or its variance) and by construction of confidence intervals and statistical tests. But, in practice, if asymptotic results can be used to estimate such quantities, they are not always applicable, due to the high nonlinearity of the model.

2.2.1 Choice of the estimating criterion.

A wide class of estimating functions can be used. Their applicability and their properties will depend on the definition of the model, and in particular the error distribution. Suppose we have the standard nonlinear regression model

$$y_i = f(x_i, \theta) + \varepsilon_i \tag{2}$$

where the errors are independent with variance $\sigma^2 v(x_i, \theta)$ and f and v are known functions.

If the errors are gaussian, the Maximum Likelihood Estimate (MLE) is defined by minimizing $-2L_n(\theta)$, where L_n is the likelihood.

$$-2L_n(\theta) = \sum_{i=1}^{n} \ln(v_i(\theta)) + \sum_{i=1}^{n} \frac{(y_i - f_i(\theta))^2}{v_i(\theta)} \tag{3}$$

where $f_i(\theta) = f(x_i, \theta)$ and $v_i(\theta) = \sigma^2 v(x_i, \theta)$.

The Ordinary Least Squares Estimate (OLSE) can always be used as estimate of the expectation parameters θ and minimizes the unweighted sum of residuals squares $\sum_{i=1}^{n}(y_i - f_i(\theta))^2$. The Modified Least Squares Estimate (MLSE) does not suppose the normality of the distribution, takes into account the variance model v and is defined by a set of normal equations

$$\sum_{i=1}^{n} \frac{(\partial f_i(\theta)/\partial \theta^a)}{v_i(\theta)}(y_i - f_i(\theta)) = 0 \qquad a = 1, ..., p \tag{4}$$

The three estimates are consistent and the MLE is first-order efficient under gaussian distribution. If we suppose there are some specific parameters β in the variance and we want to estimate them, alternative procedures of estimation of θ and β can be constructed (Carroll & Ruppert, 1982; Huet *et al.*, 1988).

2.2.2 Evaluation of the estimate properties.

Consider the problem of computing some quantity as the variance of the estimator or of constructing a confidence interval $I(\widehat{\theta_n})$. The estimate variance $Var(\widehat{\theta}_n) = Var(H(Y))$ cannot be expressed analytically and has the following form

$$Var(\widehat{\theta_n}) = \frac{C_\theta}{n} + o(n^{-1}) \tag{5}$$

So we have to choose a method for estimating $\widehat{Var(\theta_n)}$. Different methods can be used in this particular context but all the methods we discuss in the following can also used in the more general framework of inference about the parameters (confidence intervals, statistical tests).

Asymptotic results. From the asymptotic normality of the estimators, $\widehat{\theta}_n \sim \mathcal{N}(0, V_\theta)$, we choose as estimate of $Var(\widehat{\theta_n})$, the asymptotic variance V_θ computed at the estimated value $\widehat{\theta_n}$, i.e. $\widehat{Var(\widehat{\theta_n})} = V_{\widehat{\theta_n}}$.

As mentioned previously, practical reliability of such asymptotic theory results can be deficient in some situations.

Beyond the asymptotic first-order results. In order to overcome the deficiency of the use of asymptotic variance, we can think at deriving some asymptotic expansions of higher order of some quantities. Edgeworth expansions of the parameter estimate distribution have been derived (Schmidt and Zwanzig, 1986) in order to produce confidence intervals or test statistics. Actual improvement of confidence intervals of second and third order based on such expansions could not be established clearly by our simulation studies and, moreover, such expansions can led to curious behaviour (Hall, 1986).

Direct computation of the higher order terms of the variance estimate can also be made through analytical arguments (for the standard gaussian model, see Clarke, 1980) or by considering geometrical representation of the nonlinear regression.

Geometry of nonlinear regression. Even before asymptotic properties of standard nonlinear regression estimators were established (Jennrich, 1969), much effort have been devoted to investigate nonlinearity through geometrical representation of the estimation problem. Beale (1960) pointed out the existence of two components in the nonlinearity: an intrinsic one, invariant by reparametrization and a parametric one reflecting the effect of the parameter system. He proposed numerical measures of these components and gave some hints about their practical use in constructing confidence regions.

Bates and Watts (1980) gave a more comprehensive and complete geometric presentation of Beale's ideas, derived some curvature measures and pointed out that the parametric component of nonlinearity is much more important than the intrinsic one.

Simulations were performed (Messéan, 1984) in order to compute the actual coverage probability of two kinds of confidence regions: the ellipsoid one, based on standard asymptotic results and the region based on the likelihood ratio test which is invariant by reparametrization. About one hundred practical examples were considered under standard nonlinear regression model (with gaussian errors and constant variance σ^2). While the likelihood-based confidence region always recovers the nominal coverage probability (95% in our study), the standard asymptotic confidence regions can present a disastrous behaviour (actual coverage probabilities of 30% or 40% in some situations). Nonlinearity measures as proposed by Bates and Watts seem to be rather good qualitative indicators of this behaviour, although it was not possible to derive any critical level of nonlinearity. Ratkowsky (1983) carried out simulations upon some classical models with different parametrizations and different data sets and compared parameter estimate distributions, bias and nonlinearity measures. He pointed out that there could be drastic differences between parametrizations but also that these discrepancies highly depend on the data set considered.

Systematic use of reparametrization was thus proposed to reduce nonlinearity and although standard parameter transformations minimizing nonlinearity measures can only be defined in the unidimensional gaussian model (Messéan, 1984), practical procedure of confidence interval construction can be proposed (Hougaard, 1982).

Following the work of Amari (1982) who proposed a differential geometry approach of exponential families problems, a general geometrical representation of general nonlinear regression model with heterogeneous variance has been described, leading to a clarification of some heuristic previous results and a very intuitive illustration of most of the asymptotic results (consistency, first-order and second-order efficiency).

Resampling methods. Due to the decrease of computational costs, computer-based methods have known a large expansion. As pointed out in the previous paragraph, Monte-Carlo simulations have been widely used in comparing procedures or algorithms and are the basic tool for anybody who wants to evaluate practical relevance of general or theoretical results.

But computer-based methods also permit to develop new procedures of inference, as those based on bootstrap methods (Efron, 1979). They substitutes amount of computation to theoretical analysis. Although Bootstrap methods are very simple in their principle and in their practical use, their reliability must be checked in each application through the study of their asymptotic properties (Bickel & Freedman, 1981).

Application of bootstrap methods in nonlinear regression has been considered in the standard model with constant variance, in order to get an identically distributed sample of residuals. Staniewsky (1984) proved the asymptotic reliability of the bootstrap and Bunke (1986) proposed different practical procedures for constructing confidence intervals.

Evaluation of estimates properties constitutes thus a large domain of investigation. In order to bring those improvements to the final user, a general strategy has to be defined and implemented in some software package. This is the aim of the general project of comparison of confidence intervals procedures we have been carrying out for two years (Huet *et al.*, 1988). All the procedures proposed up to now are compared, leading to about 30 different confidence intervals. This project involves large amount of computing (Monte-carlo comparisons, Booststrap procedures) and algorithm optimization.

2.2.3 Numerical aspects of estimation.

Many algorithms are available for solving different kinds of optimization problems as least squares minimization, nonlinear equations solving or general criterion minimization. But specific software implementation for nonlinear regression is generally devoted to standard least squares problems. Solution of nonlinear regression with heterogeneous variances is possible with packages as MLP

(Ross, 1987) or GENSTAT (Nelder, 1975) but the user has to describe the estimation criterion (for example, the likelihood function).

In order to solve this problem, we implemented a generalization of Gauss-Marquardt algorithm to general exponential families problems. As a matter of fact, nonlinear regression estimation problems reduce to solving the following system of equations

$$B_\theta^T(Z_Y - \mu_\theta) = 0 \tag{6}$$

where $Y = (Y_1, ..., Y_n)$ is the vector of observations, $\theta = (\theta_1, ..., \theta_p)$ the vector of parameters, Z is a vector of functions of the observations (maybe a subset of sufficient statistics) , μ is the expectation of Z or, more generally, the limit of Z when $n \to +\infty$ and B_θ a $m \times p$ matrix which is a function of the derivatives and which may depend on the observations, m being the dimension of Z.

This algorithm is implemented in the NL package (Bouvier *et al.*). A good behaviour has been observed in general nonlinear regression and other exponential families problems (Messéan, 1988).

3 Software aspects.

As we have seen up to now, all the new trends in statistical research have been applied to nonlinear regression: asymptotic theory, differential geometry, resampling methods. Similarly, software has been taking advantage of the evolution of computer science. From the standard subroutine able to solve nonlinear least squares problem to a knowledge base management system dedicated to general nonlinear regression problem, there have been some important changes both in programming and user interface aspects.

Although there exist a great variety of simple packages or general statistical systems including nonlinear regression capabilities, we illustrate those software aspects through the presentation of some of our products which intend to combine both methodological and software states of the art.

A first step in our software strategy was realized through the Haus82 program (Nicole *et al.*, 1984) which includes almost all methodological capabilities of standard nonlinear regression with constant variance: model construction, parameter estimation, confidence intervals and analysis of the nonlinearities. But Haus82 can be considered as a black box package.

3.1 Towards modularity and portability: the NL package.

NL stands for Non-Linear and is a collection of routines solving general nonlinear regression problems (Huet & Messéan, 1986). This product deals with nonlinear regression models with heterogeneous variances and covers all the steps of standard analysis: analysis of variability, parameter estimation (several estimating criteria), hypothesis testing, confidence intervals, prediction, calibration and comparison of curves. The following modules are available in NL.

- Study of the variability of the data (`calc_heterogeneity` and `plot_heterogeneity`)
- Parameter estimation under various statistical models (`estimation`)
- Analysis of the residuals (`calc_residuals`)
- Analysis of sensitivity (`calc_sensitivity`, `plot_sensitivity`)
- Confidence intervals for the parameters (`plot_ellipse`, `projection`)
- Likelihood contours (`calc_contours`, `plot_contours`)
- Confidence intervals for $f(x_i, \theta)$ (`suspicion`)
- Calibration problems (`calibration`)
- Confidence intervals for functions of the parameters $\phi(\theta)$ (`function`)

NL has been designed through the modularity principle, each module being devoted to a statistical, a numerical or a graphical task. All data management aspects, user interface design and input/output capabilities have been deferred to the driving program which can be a standard main program or some general software which can drive external routines. This was the Consistent System (Dawson *et al.*, 1980) in our Multics implementation (Bouvier *et al.*, 1985) and will be the S system (Becker & Chambers, 1984) under the Unix operating system. Algorithm implementation can thus be viewed independently from their system and data management environments.

3.2 Towards more intelligent and more integrated nonlinear regression software: the Chimène project.

As pointed out by Haux and Jöckel (1986), one needs more integrated and more intelligent statistical software. Chimène is a project for constructing a general system of computer-aided nonlinear regression modeling. It was designed inside the EDORA group, a club of biologists, statisticians, mathematicians and informaticians whose general aim is to improve the use of applied mathematics in biology (Pavé & Rechenmann, 1986).

A typical Chimène session contains the following steps

- The user defines the data set, specifies the variables involved in the regression and the field of application.
- The user can describe the data set in terms of geometric characteristics (asymptotes, inflexion point, curvature, growth or decay trends, ...).
- Chimène selects among a set of models defined in the knowledge base, a subset of models compatible with the data geometrics.

- The user selects one of the response model proposed, possibly specifies constraints on the parameters and the system builds the statistical model of regression by specifying the variance model through an analysis of data variability.
- The fitting context (estimating function, initial values of parameters, numerical process parameters, ...) is generated from user specification or from default values or procedures.
- Chimène fits the model, computes estimates and their asymptotic properties and checks the validity of the candidate model through a classification of the residuals.
- Chimène compares different statistical models fitted from the same data set.

All intermediate results are saved in the base and, at each step, Chimène proposes to the user some tasks to execute and the user can choose one of them or let Chimène decide.

Chimène uses an object-centered knowledge base management system, called SHIRKA (Rechenmann, 1987), whose knowledge representation is made through a hierarchy of *frames* (Minsky, 1975) without rules. In Chimène, the objects can be any of the following: regression variables, data structures, parameters with their significance, response models, variance models, numerical process characteristics, algorithms, estimation results, graphics, ... All the objects defined in the knowledge base are described in the same way, as *frames* with a tree structure and links between trees.

Different inference mechanisms are used (inheritance, classification, filtering, external procedures calls) and Chimène calls the NL routines for its statistical and graphical tasks. The arguments of these routines are also described following the same representation scheme, both with their constraints (validity of the values is thus checked before procedure calls).

A first release of the product gives some evidence to the feasibility of designing a general driving system which combines data management and external calls capabilities (as the S system for example), with some knowledge management, the knowledge covering all the statistical and numerical aspects of nonlinear regression analysis and some context information from the application field (which can help the system to determine, for example, some parameter constraints) and automatic data description.

4 Conclusion.

Nonlinear regression is a statistical topic which can benefit from both new statistical research trends and computer science new developments. Specific role of computer science must be pointed out:

- Hardware power which permits large use of Monte-Carlo simulations for comparing procedures or algorithms and the development of the resampling methods. It also allows more complex but more efficient implementation of numerical algorithms: for example confidence intervals based on likelihood ratio test are computed by an iterative process (Huet & Messéan, 1986).

- Software technology allows the design of more efficient and *robust* packages but needs a project management between the statistician and the programmer. Construction of general computing environments allows a separation between algorithmic procedures (machine- and system independent routines), general system management (including data and procedures management) and user interfaces (graphics and windows, which depend on the hardware).
- Artificial intelligence can be of considerable interest for a statistical field as nonlinear regression, especially by extending the previous concept of computing environments to knowledge base management systems. Although less developed than previous aspects, artificial intelligence will introduce some drastic changes in the user approach of nonlinear regression analysis. Rousseau (1988) makes a good survey of knowledge representation in the biometrical framework.

Future work in nonlinear regression will be dedicated to make its analysis moreintegrated with the practical problem we deal with.

From the statistical point of view, this means that more specific and complex problems, such as calibration in radioimmunological assays or censored regression in ELISA assays for example (Huet *et al.*, 1988), will be considered as a whole statistical problem and nonparametric approach will be extended to different steps of the nonlinear regression (Bunke, 1986).

From the software point of view, two major axes must be distinguished. The first one concerns the implementation of statistical research results through normalized and portable algorithms which could be included in libraries or general statistical packages. The second one will be to develop more user-specific but also more *intelligent* software products, with high graphics and interactive interface capabilities and with high level of *qualitative reasoning* that must be combined to standard statistical analysis (although there are some methodological and technical problems).

REFERENCES

AMARI S.I. (1982) Differential geometry of curved exponential families: curvatures and information loss. *The Annals of Statistics*, **10(2)**, pp 357-385.

Bates D. & Watts D. (1980) Relative curvatures measures of nonlinearity *J. R. Statist. Soc.*, **B 42(1)**, pp 1-25.

BEALE E.M.L. (1960) Confidence regions in nonlinear estimation, *J. R. Statist. Soc.*, **B22(1)**, pp 41-88.

BECKER R.A. & CHAMBERS J.M. (1984) Design of the S system for data analysis *Communications of the ACM*, Vol. 27(5), pp 486-495.

BUNKE O. (1986) Assessing the performance of regression estimators and models under nonstandard conditions *Seminabericht*, **89**, Humboldt University, pp 65-102.

BICKEL P.J. & FREEDMAN D. (1981) Some asymptotics on the bootstrap. *The Annals of Statistics*, **9**, pp 1196-1217.

BOUVIER A., GÉLIS F., HUET S., MESSÉAN A. AND NEVEU P. (1985) *CS-NL*. Reference manual, Laboratory of Biometry, Jouy-en-Josas.

CARROLL & RUPPERT (1982) Adapting for heteroscedasticity in linear models *The Annals of Statistics*, **10**, pp 1224-1233.

CLARKE G.P.Y. (1980) Moments of the least squares estimators in a nonlinear regression model, *J. R. Statist. Soc.*, **B 42(2)**, pp 227-237.

DAWSON R., KLENSIN J.C., YNTEMA D.B. (1980) The Consistent System, *The American Statistician*, Vol. 34(3), pp 169-176.

EFRON B. (1979) Bootstrap methods: another look to Jackknife, *Annals of Statistics*, **7**, pp 1-26.

HALL P. (1986) On the Bootstrap methods and confidence intervals. *The Annals of Statistics*, Vol. 14(4), pp 1431-1452.

HAUX R. & JÖCKEL K.-H. (1986) Database management and statistical data analysis: the need for integration and becoming more intelligent *COMPSTAT 86: proceedings in Computational Statistics*, Physica Verlag, pp 407-414.

HOUGAARD P. (1982) Parametrizations of nonlinear models *J. R. Statist. Soc.*, **44(2)**, pp 244-252.

HUET S. (1983) *L'analyse statistique des dosages radioimmunologiques*. Thèse de troisième cycle. Université Paris-Sud (Orsay), 174p.

HUET S. (1986) Maximum Likelihood and Least Squares Estimators for a Non-Linear Model with Heterogeneous Variances. *Statistics*, **17**, pp 517-526.

HUET S., LAPORTE J. & VAUTHEROT J.-F. (1988) Statistical methods for the comparison of antibody levels in serums assayed by enzyme linked immuno sorbent assay *Biométrie-Praximétrie*, to appear.

HUET S. & MESSÉAN A. (1986) NL: a statistical package for general nonlinear regression problems. *COMPSTAT 86: proceedings in Computational Statistics*, Physica Verlag, pp 326-331.

HUET S., JOLIVET E. & MESSÉAN A. (1988) *La régression non-linéaire: aspects statistiques*, Cahiers de Biométrie, to appear.

HUET S., JOLIVET E., MESSÉAN A. & NICOLE O. Some simulations results about confidence intervals and bootstrap methods in nonlinear regression Seminarbericht, **89**, Humboldt University, pp 1-26.

JENNRICH R.I. (1969) Asymptotic properties of nonlinear least squares estimators *The Annals of Mathematical Statistics*, **40(2)**, pp 633-643.

MEETER D. (1968) *Program GAUSHAUS*, Numerical analysis laboratory, University of Wisconsin.

MESSÉAN A. (1984) *Application de la géométrie différentielle à la statistique du modèle non-linéaire*. Thèse de Doctorat-ingénieur, Université Paris-Sud (Orsay), 170p.

MESSÉAN A. (1988) A generalization of Gauss-Marquardt algorithm for exponential families problems. *Computational Statistics Quaterly*, Vol. 4, Issue 2.

MINSKY M. (1975) A framework for representing knowledge The Psychology of Computer Vision, Winston P. ed., McGraw-Hill, pp 211-277.

NELDER J.A. (1975) *Genstat.* Interuniversity/Research council Series No 3. 3rd Edition. Program library unit, University of Edinburgh.

NICOLE O., MESSÉAN A. & VILA J.-P. (1984) Haus82. A new tool for nonlinear fitting and the study of nonlinear models *COMPSTAT 84: proceedings in computational statistics*, Physica Verlag, pp 434-439.

PAVÉ A. & RECHENMANN F. (1988) Computer aided modelling in Biology: an artificial intelligence approach *Artificial Intelligence in Simulation: State of the Art* Vansteenkiste and al. (eds.), SCS publications, Vol. 18(1).

RATKOWSKY D. (1983) *Nonlinear regression modeling: a unified practical approach*, Marcel Dekker Eds.

RECHENMANN F. (1987) *SHIRKA. Système de gestion de bases de connaissances centrées-objet*, Reference manual, Second release, INRIA, 53p.

ROSS G. (1987) *Maximum Likelihood Program. Reference manual*, Numerical Algorithms Group Limited.

ROUSSEAU B. (1988) *Vers un environnement de résolution de problèmes en Biométrie.* Thèse de doctorat, Université Claude Bernard, Lyon, 281p.

STANIEWSKY P. (1984) The bootstrap in nonlinear regression *Robustness of statitical methods and nonparametric statistics*, Rasch D. & Tiku M. (eds.), DVW, Berlin.

SCHMIDT W. & ZWANZIG S. (1986) Second-order asymptotics in nonlinear regression *J. of Multivariate Analysis*, **18**, pp 187-215.

Comparing Sensitivity of Models to Missing Data in the GMANOVA

E. P. Liski and T. Nummi, Tampere

Summary: It is important that a model has a satisfactory fit to the data we analyze. Other general acceptable features of a model are its simplicity and interpretability of parameters. Longitudinal data arising in epidemiologic or clinical studies, for example, are rarely complete. It is therefore desirable that a model is, to some extent, robust to missing data. Thus models should be compared also with respect to this property.

We investigate robustness to missing data in the generalized analysis of variance model (GMANOVA) applying a resampling approach. For this purpose we introduce a statistic which indicates the influence of deleting a set of measurements on estimated curves. For example, px100%(0<p<1) from measurements are dropped out at random and the effect on estimates is assessed using our influence measure, which can be viewed as a multivariate generalization of Cook's measure. By repeating the experiment we can determine the empirical distribution of the influence statistic, and this distribution can be analyzed further. Carrying out the analysis for different values of p and for the alternative models under consideration, robustness to missing data can be scrutinized as a function of p. Both the maximum likelihood and the least squares approach can be applied. The techniques are illustrated by analyzing data on a growth experiment for bulls. It is shown how the methods can be utilized in finding "good" design points in growth studies.

Key words: Robustness, resampling schemes, missing data, growth curves, maximum likelihood.

1. Introduction

We have studied sensitivity of models to missing data in the generalized multivariate analysis of variance. The GMANOVA model was introduced by Potthoff and Roy (1964) for investigating growth data. This model is as follows

$$\mathbf{Y} = \mathbf{XBT}' + \mathbf{E}, \tag{1.1}$$

where $\mathbf{Y}$ is an $n \times q$ matrix of responses, $\mathbf{X}$ is an $n \times m$ matrix of known constants, $\mathbf{B}$ is an $m \times p$ parameter matrix and $\mathbf{T}$ is a $q \times p$ regression matrix. We suppose that both $\mathbf{X}$ and $\mathbf{T}$ are of full rank. The rows of $\mathbf{E}$ are independently distributed with mean vector $\mathbf{0}$ and covariance matrix $\mathbf{\Sigma}$, which is supposed to be of full rank. Although model (1.1) is often called a growth curve model, it provides a general format for a variety of longitudinal analyses. Multivariate regression and MANOVA are special cases of model (1.1).

Longitudinal data arising in epidemiologic or clinical studies, for example, are rarely complete. It is therefore desirable that a model should be robust to the missing observations. Other generally desirable features are simplicity of a model, interpretability of parameters and goodness of fit to data. Often we cannot fully control the circumstances under which the measurements are taken or change the design during the course of the study. Therefore we consider also robustness to changes of the study design.

2. Estimation in the GMANOVA model

Khatri (1966) and Rao(1966) showed independently that the maximum likelihood estimator of **B** is

$$\hat{\mathbf{B}}_{ML} = (\mathbf{X}'\mathbf{X})^{-1}\mathbf{X}'\mathbf{Y}\mathbf{S}^{-1}\mathbf{T}(\mathbf{T}'\mathbf{S}^{-1}\mathbf{T})^{-1}, \tag{2.1}$$

when **Y** has the multivariate normal distribution and

$$\mathbf{S} = \mathbf{Y}'[\mathbf{I} - \mathbf{X}(\mathbf{X}'\mathbf{X})^{-1}\mathbf{X}']\mathbf{Y}\ .$$

To determine the degree of influence of a given subset of measurements we compute the estimate of **B** with the given subset deleted and compare it with the complete data estimate. If we let an arbitrary subset of measurements be deleted from the data, the estimate of **B** from the incomplete data can be obtained by applying the generalized growth curve model (GGCM) introduced by Kleinbaum (1973) or by the maximum likelihood (ML) approach introduced.

Use of Kleinbaum's method requires a consistent estimator of Σ obtained from incomplete data. Especially when plenty of data are missing, estimation of covariances may fail. These problems have been discussed by Kleinbaum (1973) and by Schwertman and Allen (1979), for example. The ML approach used to estimate parameters of (1.1) utilizes the EM algorithm (Liski 1985), which is a general iterative method of computing ML estimates when data are incomplete. We have compared sensitivity of models with respect to missing data using both the ML technique and the Kleinbaum's method. In the ML estimation normally distributed observations are assumed. As an iterative technique the EM algorithm is much heavier than the Kleinbaum's technique.

3. Comparing models with respect to robustness to missing data

The influence measure (Liski 1987a and 1987b)

$$D_{(IJ)} = \mathrm{tr}\,(S_H S_E^{-1}) \tag{3.1}$$

can be interpreted as a multivariate extension of the well-known measure proposed by Cook (1977) in linear regression for detecting influential measurements. On the right hand side of (3.1) the matrices S_H and S_E are as follows:

$$S_H = (\hat{B} - \hat{B}_{(IJ)})'X'X(\hat{B} - \hat{B}_{(IJ)}) \text{ and } S_E = (T'S^{-1}T)^{-1}, \tag{3.2}$$

where $\hat{B}$ is an estimate of B. Formally we can determine the (1-a)100% confidence set for B satisfying

$$\text{tr}\,(S_H S_E^{-1}) \le U_a^{(s)},$$

where s is the number of nonzero eigenvalues of $S_H S_E^{-1}$ and $U_a^{(s)}$ is the upper (1-a)100 percentage point of the distribution of $U^{(s)}$. We note the connection of (3.1) to the Lawley-Hotelling trace statistics (cf. Liski 1987a and 1987b). The exact distribution of $U^{(s)}$ is known and the critical values for it can be obtained using the technique as described by Seber (1984, p. 38). For example, the inequality $D_{(IJ)} > U_{0.9}^{(s)}$ means, that deleting measurements Y_{IJ} moves the estimate outside the 90% confidence set for B based on $\hat{B}$. The comparison of $D_{(IJ)}$ with the probability points of $U^{(s)}$ is not a test of significance, but only a monotonic transformation of $\hat{B} - \hat{B}_{(IJ)}$ to a more familiar scale.

Examining the values of measure (3.1) and the corresponding significance levels for different models yields interesting information concerning the relative robustness of models to missing data or to design changes. In the next section we show how the sensitivity analysis outlined above can be done in practice.

4.Results of an empirical analysis

We have examined a number of models for describing the growth of bulls tested at an experimental station in Finland in 1966 (cf. Liski 1987a). The most promising models were

$$y = \beta_0 + \beta_1 t + \beta_2 t^2 + \beta_3 t^3 + e, \tag{4.1}$$

$$y = \beta_0 + \beta_1 t + \beta_2 \ln(t) + e \tag{4.2}$$

and

$$y = \beta_0 + \beta_1 t + \beta_2 t^2 + \beta_3 \ln(t) + e. \tag{4.3}$$

The model (4.3) fitted our data best among the above three models. In order to simplify the calculations we considered only a sample of 27 Ayrshire bulls contained in data of 208 bulls tested at station in Finland in 1966. The testing covered the interval 30-365 days of age, and bulls were weighed every 30th day (12 time-points).

When applying model (1.1) the data analyst faces usually three major types of missing data problems: (i) missing observations i.e. missing individuals, (ii) deleted time-points, and (iii) other groups of missing measurements. In this example we investigate the cases (ii) and (iii). If the number of weighing times must be reduced, for example on economic grounds, the influence of deleting time points should be considered. At first we put the question: What time-points to delete if we must? In Table 1 we have displayed three least influential subsets of time-points, when the number of deleted points increases.

Table 1. The time-points at which $D_{(J)}$ attains the minimum (1), the second smallest (2) and the third smallest value (3), when deleting all one-point subsets in turn, all two-points subsets in turn and so on.

		Weighing ages											
J	Model	30	60	90	120	150	180	210	240	270	300	330	365
	(4.1)			3	2				1				
1	(4.2)	2							3			1	
	(4.3)			1		2	3						
	(4.1)			2,3	1,3				1,2				
2	(4.2)	1,2,3							1	3		2	
	(4.3)			1,2		1,3	2	3					
	(4.1)		2,3	1	1	3		2	1		2,3		
3	(4.2)	1,2,3							2,3	1,3		1,2	
	(4.3)			1,2		1,2,3	1	2,3				3	
	(4.1)		1,2,3	1,2		2,3	3	1			1,2,3		
4	(4.2)	1,2,3						2,3	1,2	1,3			1,2,3
	(4.3)					2,3		1,3	1,2,3	2		1,2	1,3
	(4.1)		1,2,3	1,2	3	1,2,3	1	2,3			1,2,3		
5	(4.2)	1,3		1,3				1,2,3	2,3	1,2,3	2	1,2	
	(4.3)			2		1,3		1,2,3	1,2,3	3		1,2,3	1,2

It can be immediately observed, that the smallest values of the influence measure are attained at different points for every model under consideration. Similarly it was found that every model also had different influential time points. However, the ages 60 and 365 days were rather influential in every model. Also other characteristics (like mean, standard deviation, min, max etc.) for the empirical distribution of D-measure can be calculated when all one-time-point subsets in turn, all two-point subsets in turn and so on, are deleted. The model (4.3) proved most robust to deleted time-points on the average when less than 4 time-points were dropped out. However, also the most influential combinations of time points were found with respect to model (4.3) when more than 3 time points were deleted.

In order to compare sensitivity of models with respect to randomly missing measurements, we carried out simulation experiments. From all measurements 10, 20, 30, 40 and 50% were deleted randomly and this experiment was repeated 100 times. For every deletion the value of $D_{(IJ)}$ was calculated using both the Kleinbaum's method and the ML approach. The results for the two best models are summarized in Table 2.

Table 2. Characteristics for the empirical distributions of $D_{(IJ)}$ when 10, 20, 30, 40 and 50 % of data are randomly deleted when estimates are calculated with the Kleinbaum's method and with the ML-method.

%	model		$\bar{D}$	s_D	min	md	max
10	(4.1)	EM	0.439	0.692	0.005	0.282	4.780
		KL	0.418	0.357	0.021	0.285	2.070
	(4.3)	EM	0.388	0.494	0.033	0.272	4.418
		KL	0.355	0.180	0.043	0.325	0.848
20	(4.1)	EM	0.860	0.969	0.052	0.584	7.818
		KL	0.666	0.650	0.022	0.485	5.037
	(4.3)	EM	0.476	0.429	0.023	0.368	3.087
		KL	0.451	0.187	0.170	0.423	1.058
30	(4.1)	EM	0.763	0.670	0.012	0.603	3.147
		KL	0.887	0.647	0.089	0.769	5.012
	(4.3)	EM	0.564	0.398	0.024	0.476	2.382
		KL	0.487	0.189	0.144	0.468	1.134
40	(4.1)	EM	1.167	1.037	0.025	0.855	4.882
		KL	1.125	0.759	0.130	1.005	5.223
	(4.3)	EM	0.631	0.411	0.075	0.566	2.334
		KL	0.517	0.216	0.203	0.486	1.203
50	(4.1)	EM	0.996	0.929	0.028	0.710	4.569
		KL	1.389	0.910	0.064	1.251	4.684
	(4.3)	EM	0.640	0.393	0.064	0.557	2.082
		KL	0.575	0.279	0.145	0.517	1.813

When measurements are randomly missing from the data, model (4.3) is clearly most robust. There was no great difference in sensitivity between the Kleinbaum's estimator and the ML estimator. For computational reasons the number of iterations in EM-algorithm was restricted to 50 iterations. Although this number of iterations is usually sufficient, in some cases the algorithm has not yet reached the solution. This causes the great maximum values of $D(IJ)$ for the EM algorithm distorting results in favor of the Kleinbaum's method. The Kleinbaum's estimator used here resembles the 2-stage least squares method. At first the parameter **B** is estimated assuming that Σ is equal to the identity matrix **I**. Thereafter missing measurements are estimated using the first stage **B**-estimate and Σ is estimated from the completed data. In the second stage **B** is obtained with the Kleinbaum's technique utilizing the estimated Σ. Although this is a 2-stage method, it is computationally much faster than the iterative EM-algorithm. When the proportion of missing data is small (e.g. 10%), each method gives estimates very near to each other. In some cases convergence of the EM algorithm is very slow, and therefore techniques to speed convergence should be adopted.

5. Concluding remarks

We have illustrated techniques for comparing sensitivity of different models to missing data and to changes in study design. The approach in this paper is based on inspecting the behaviour of our influence measure in various missing data situations. The programs are written in APL. When data are missing at random, the model (4.3) in our example was shown to be most robust. Also least influential and most influential combinations of design points in the example were identified.

REFERENCES

Cook, R.D. (1977). Detection of influential observations in linear regression. *Technometrics*, 19, 15-18.

Dempster, A.P., Laird, N.M. and Rubin, D.B. (1977). Maximum likelihood from incomplete data via the EM algorithm. *J. R. Statist. Soc.*, B. 39, 1-22.

Khatri, C.G. (1966). A note on a MANOVA model applied to problems in growth curves. *Annals of the Institute of Statistical Mathematics*, 18, 75-86.

Kleinbaum, D.G. (1973). A generalization of the growth curve model which allows missing data. *Journal of Multivariate Analysis*, 3, 117-124.

Liski, E.P. (1987a). A Growth curve analysis for bulls tested at station. *Biometrical Journal* , 29, 331-343.

Liski, E.P. (1987b). Identifying influential data in a growth curve model. Proc. Second International Tampere Conference in Statistics (Tampere, Finland, 1-4 June 1987), Ed. Tarmo Pukkila and Simo Puntanen, Dept. of Mathematical Sciences, Univ. of Tampere, Finland.

Liski, E.P. and Nummi, T. (1986). An APL-system for analyzing multivariate data. Proceedings of Compstat 1986. Physica-Verlag, Heidelberg.

Potthoff, R.F. and Roy, S.N. (1964). A generalized multivariate analysis of variance model useful especially for growth curve problems. *Biometrika*, 51, 313-326.

Rao, C.R. (1966). Covariance adjustment and related problems in multivariate analysis. Academic Press, New York. *Multivariate Analysis,* 87-103.

Seber, G.A.F. (1984). Multivariate Observations. Wiley, New York.

Ware, J.H. (1985). Linear Models for the Analysis of Longitudinal Studies. *The American Statistician*, 39, 95-101.

A Modelling Approach to Multiple Correspondence Analysis

M. Green, Lancaster

ABSTRACT

A multi-dimensional extension to the Goodman RC model is proposed as a statistical modelling form of multiple correspondence analysis. An algorithm is given for the maximum likelihood estimation of the parameters and an approximate method which provides starting values for the full ML procedure. The method is illustrated by the analysis of the Law School Admission Test. The versitility of the approach is demonstrated by the analysis of a non-standard problem in the study of Marital Endogamy.

Keywords: Statistical modelling, maximum likelihood, multiple correspondence analysis, Goodman association models.

1. INTRODUCTION

A fundamental problem in Psychometrics is the development of techniques to perform the same task of dimension reduction for categorical variables that Factor Analysis achieves for continuous variables. In general the approach has been a statistical modelling one and often in the framework of maximum likelihood or generalisations of the likelihood method. Many approaches have been proposed such as "item response models" (Andersen), "latent class analysis" (Lazarfeld and Hendry), "latent trait models" (Bock and Aitkin) and various estimation procedures used; conditional likelihood, marginal likelihood and integrated likelihood. One property that all the approaches have in common is that they are computer intensive leading to a 'batch' approach to modelling, that of fitting the most complicated model we can consider and hope we can extract enough information to make useful inferences.

In a less formal framework a similar problem arises in the analysis of survey data in which we wish to obtain some insight into the form of the interactions of categorical variables from multi-dimensional tables. A common approach, especially in France, has been the application of multiple correspondence

Compstat 1988

analysis to such problems (Lebart et al). The success of the technique is due to its ability to represent the most complicated problem in a concise and interpretable way and to the fact that the technique is computationally very efficient. In all but the most complex cases the procedure is fast enough for interactive data exploration, although present software doesn't often make this easy. The deficiency of the technique is that without the framework of a statistical model the reliability of the inferences drawn from the data cannot be assessed. The graphical representation of data is, at once, the strength and weakness of the technique as it leads to an automatic fit of two dimensions even when one would be sufficient. The desire for this representation stems from the lack of a statistical assessment of the significance of apparent effects.
This paper considers an approach which is an amalgam of the two approaches by use of a multi-dimensional form of the Goodman RC association model (Goodman). The advantage of this approach is that it retains the graphical display of Correspondence Analysis for the elucidation of associations between the variables while being in the Likelihood framework, allowing the use of statistical tests to assess the significance of such associations.

2. LINEAR SCORE PREDICTOR MODEL

Consider, for simplicity, the case of an aptitude test comprising m questions to which the respondents answers are considered as simply right or wrong. Suppose we attach a "score" of $\alpha_1^{(j)}$ to a correct answer to question j and one of $\alpha_2^{(j)}$ to an incorrect answer (taken to be $-\alpha_1^{(j)}$ in the following analysis). Consider the first l responses yield a total score S and that this is used to predict the response to question l+1 using a standard logistic regression model. Using these terms a likelihood function for the m responses is built up. Formulating this likelihood in the log-linear model form and using the exchangability of the responses leads to a model for the frequencies in the m-way contingency table indexed by $I_1,I_2,\ldots,I_m$ formed from the responses results in a the form of the log-linear model except that;

$$\log(\mu_{I_1 I_2 \ldots I_m}) = \sum_{i=1}^{m} \gamma_{I_i} + \sum_{i=1}^{m} \sum_{j>i}^{m} \alpha_{I_i}^{(i)} \alpha_{I_j}^{(j)}$$

The total score, S, for an individual with responses $(y_1,y_2,\ldots,y_m)$ is given by;

$$S = \Sigma_i \, \alpha_{y_i}$$

and is the estimated value of the latent variable for that individual. It is clear that this model is valid for multi-category variables. The model is easily generalised to more than one latent dimension by considering $\alpha^{(k,i)}$ to be the score parameters for the ith variable and the kth latent dimension. Then

$$\log(\mu_{I_1 I_2 \ldots I_m}) = \sum_{i=1}^{m} \gamma_{I_i} + \sum_{k=1}^{l} \sum_{i=1}^{m} \sum_{j>i}^{m} \alpha_{I_i}^{(k,i)} \alpha_{I_j}^{(k,j)}$$

2.1 An Estimation Algorithm

Since the model is not linear in the parameters the approach adopted here for obtaining the maximum lilelihood estimates of the parameters is based on successive relaxation on the α parameters. For simplicity take the case of a single latent variable. The rth stage of the an iteration of the algorithm estimates α_{I_r} considering the parameters $\alpha_{I_j}, j \neq r$ fixed at their current estimates. The model can then be rearranged as;

$$\log(\mu_{I_1 I_2 \ldots I_m}) = \sum_{i=1}^{m} \gamma_{I_i} + \sum_{i \neq r}^{m} \sum_{\substack{j>i \\ j \neq r}}^{m} \alpha_{I_i}^{(i)} \alpha_{I_j}^{(j)} + \alpha_{I_r}^{(r)} \sum_{i \neq r} \alpha_{I_i}^{(i)}$$

$$= C_{I_1 I_2 \ldots I_m} + \alpha_{I_r} S_{I_1 I_2 \ldots I_m}$$

The "constants" S are the total scores ommitting scores α_{I_r}.
This is now in the form of a standard GLM with S taken as an independent covariate and could be fitted by any GLM software, such as GLIM, or more efficiently by directly solving the maximum likelihood equations which result in updating formulae of the form:

$$\alpha = \alpha + \delta$$
$$\delta = \Sigma(y - \hat{\mu})S \, / \, \Sigma \, \hat{\mu}S^2$$
$$\hat{\mu} = \hat{\mu} \exp(\delta S)$$

Which is a generalised form of the IPS algorithm.

2.2 The Law School Admission Test

The data consist of the responses of 1000 individuals to 5 questions. The responses recorded whether the individual gave a correct or incorrect answer. These data have been extensively analysed by various methods (Bock and

Aitkin). The latent variable is taken to be aptitude and the object often taken to be to attach an aptitude score to each individual. However, a designer of such tests may well be interested in identifying those questions with the greater discriminating power. These are merely two aspects of the same analysis. Applying the above analysis to the data gives results:

	Deviance	df
independence model	74.80	26
linear score model	21.23	21

with estimates of the score parameters (for correct answers):

Question	1	2	3	4	5
Score	0.338	0.297	0.372	0.284	0.269

showing that question 5 has considerably less discriminating power than question 3.

2.3 Approximate Estimation Procedure

With high dimensional tables this fitting procedure can be expensive in cpu time and storage. Since iterative procedures benefit from having good starting values an approximate method that is economical is useful both as an exploratory tool and as providing good starting values for the full maximum likelihood procedure.

It is clear from the likelihood that the two-way margins of the table are sufficient statistics for the parameters and we may look for a technique that is based on these margins. Borrowing an approach from multiple correspondence analysis we consider using the Burt table (Greenacre). The likelihood for the Goodman RC model applied to the Burt table is very similar to that of the full likelihood method and fitting this model gives a good approximation to the parameter estimates. For example the estimates for the Law School Admission data are:

Question	1	2	3	4	5
Score	0.257	0.241	0.306	0.232	0.221

The deflation of parameter estimates when using the Burt table is well understood in correspondence analysis.

3. NON-STANDARD ANALYSES

The advantage of this approach over multiple correspondence analysis is not only that the likelihood framework provides the ability to assess the significance of apparent effects but that the generality of the framework allows a much greater versitility. Models can include ordinary log-linear interaction terms and can specify the form of the association terms. This versitility is demonstrated by the following example.

3.1 Marital Endogamy Study

It is a persistent feature of British society that people tend to choose marriage partners of the same or similar social class (Penn and Dawkins). Two way tables of groom's and bride's social class are well modelled by the Goodman RC model which can be formulated as:

$$\log(\mu_{ij}) = \alpha_i + \beta_j - \Phi(u_i - v_j)^2$$

where u_i and v_j are the positions of the groom's social class, i, and bride's social class, j, on a continuous class scale and the last term is the negative effect of class difference. The situation in the data considered here is complicated by the attempt to assess the effect of parental class on such marriages. In the following example only father's social class was incorporated due to the problem of coping with the high proportion of economically inactive mothers resulting in a 4 dimensional table based on 498 marriages. Generalising the previous approach we consider a bride score and groom score, in each case formed as a sum of a score for their own social class and the score of their father's social class. The proposed model is:

$$\log(\mu_{ijkl}) = (\alpha\beta)_{ij} + (\gamma\delta)_{kl} - \Phi\{(u_i + u_j^f) - (v_k + v_l^f)\}^2$$

where
- i - groom's social class
- j - groom's father's social class
- k - bride's social class
- l - bride's father's social class

and the prefix, f, denotes father's score.

The $(\alpha\beta)_{ij}$ and $(\gamma\delta)_{kl}$ terms allow for the social mobility effects.

The endogamy association term is now

$$2\Phi\{(u_i+u_j^f)(v_k+v_l^f)\}$$

which is a restricted form of the model in the previous example. Adding this term to the model produces a reduction in deviance of 61.5 on 19 df. The estimated scores are

	groom		bride	
class	self	father	self	father
self-employed	-0.248	0.202	-0.824	-0.022
professional	0.801	0.431	0.724	0.094
supervisors	0.371	-0.417	0.749	0.162
clerical	0.533	0.640	0.244	0.920
skilled manual	-0.169	-0.418	-0.570	-0.690
non-skilled manual	-1.288	-0.438	-0.324	-0.464

It is noticeable that the fathers' contributions to the combined score are not only large but often considerably different in form from the self score and that there are some important differences between the groom's and bride's parameters.

The approximate procedure for this case is modified to the fitting of the Goodman RC model to a table formed from a subset of the two way tables. viz

$$\left[\begin{array}{c|c} G \times B & G \times BF \\ \hline GF \times B & GF \times BF \end{array}\right]$$

G - groom GF - groom's father
B - bride BF - bride's father

References

Andersen E.B. (1973) Conditional Inference for Multiple-choice Questionnaires, Br.J.Math.Statist.Psychol., 26, pp 31-44.

Bock D. and Aitkin M. (1981) Marginal Maximum Likelihood Estimation of Item Parameters : An Application of an E.M. Algorithm, Psychometrica, 46, pp 443-459

Goodman L.A. (1986) Some Useful Extensions of the Usual Correspondence Analysis Approach and the Usual Log-linear Models Approach in the Analysis of Contingency Tables, Int. Statist. Review, 54, 243-309

Lazarfeld P. and Hendry N. (1968) Latent Structure Analysis, Houghton Mufflin co., Boston.

Lebart L., Morineau A. and Warwick K.M. (1984) Multivariate Descriptive Statistical Analysis, New York, John Wiley.

Penn R.D. and Dawkins D.C. (1983) Structural Transformations in the British Class Structure: A Log-linear Analysis of Marital Endogamy in Rochdale 1856-1964, Sociology, 17, 4, 506-523

Multidimensional Scaling on the Sphere

M. A. A. Cox and T. F. Cox, Newcastle upon Tyne

INTRODUCTION

Nonmetric multidimensional scaling (MDS) has become a popular method for graphical representation of objects or individuals based on dissimilarity measures. Since the pioneering work of Shepard (1962a,b) and Kruskal (1964a,b), various MDS models have been produced together with associated computer programs such as MDS(X) which is described in Coxon and Davies (1982).

The aim of MDS is to find a configuration of points in p dimensional Euclidean space (p is usually taken as 2 or 3) where each point represents an object or individual under study. The configuration of points is found such that the distances between pairs of points match as well as possible dissimilarities calculated for the pairs of objects. The usual distance between points is taken as Euclidean distance but other distance measures are possible, such as the city block metric. Dissimilarities between objects can be calculated in various ways such as $1.0-s_{ij}$ where s_{ij} is the simple matching coefficient between objects i and j.

One disadvantage of using points in a Euclidean space to represent the objects under study is that some objects have to be at the 'edge' of the configuration of points. However, if the objects are represented by points on a hypersphere then none of the objects need be at the 'edge' of the configuration. If it is the case that some objects <u>ought</u> to be 'edge' objects this would be manifest on the hypersphere by the configuration being restricted to a local region of the hypersphere.

Compstat 1988

MDS IN EUCLIDEAN SPACE

The discrepancy between a configuration of points in a p dimensional space with distances between pairs of objects $\{d_{ij}\}$ and the original dissimilarities $\{\delta_{ij}\}$ between the pairs of objects is measured by the stress

$$S = \{ \sum_{i<j} (d_{ij} - \hat{d}_{ij})^2 / \sum_{i<j} d_{ij}^2 \}^{\frac{1}{2}}$$

Here $\{\hat{d}_{ij}\}$ is the primary monotone least-squares regression of $\{d_{ij}\}$ on $\{\delta_{ij}\}$, found by minimising $\sum_{i<j} (d_{ij}-\hat{d}_{ij})^2$ and ensuring that if $\delta_{ij} < \delta_{kl}$ then $\hat{d}_{ij} \leq \hat{d}_{kl}$ for any i,j,k,l. The configuration is sought that minimises the stress S.

Further details can be found in textbooks on multivariate analysis, see for example Chatfield and Collins (1980).

SPHERICAL MDS

The points within a configuration on the sphere have spherical coordinates $(1, \theta_i, \phi_i)$, for individual i. In Cartesian coordinates these become $(\cos\theta_i \sin\phi_i, \sin\theta_i \sin\phi_i, \cos\phi_i)$. The distance between two points on the sphere will be defined as the length of the shortest arc of the great circle passing through the two points. It is these distances between each pair of points that are to represent the dissimilarities between the pairs of objects.

Because there is a one to one relationship between the distance around the sphere between two points as described above, and the Euclidean distance between the two points (i.e. the distance of the straight line joining the two points, through the sphere), the multidimensional scaling for the sphere can be carried out on these Euclidean distances. These are given by

$$d_{ij} = \{2 - 2 \sin\phi_i \sin\phi_j \cos(\theta_i - \theta_j) - 2 \cos\phi_i \cos\phi_j\}^{\frac{1}{2}}.$$

The stress S is then minimised with respect to $\{(\theta_i, \phi_i)\}$. This is done by following Kruskal (1964) using the method of steepest descent. Details are not given here but will be published elsewhere.

The configuration obtained as giving the minimum stress is not a unique solution, as negating one of the coordinates θ or ϕ will also be a solution. An arbitrary rotation of the sphere will give another solution. Now d_{ij} is invariant to addition of an arbitrary angle α to each θ_i, but is not invariant to addition of an arbitrary angle β to each ϕ_i. However arbitrary rotations of the sphere can be achieved by first rotating about the z-axis and then rotating about the y-axis. This leads to the mapping

$$(\cos\theta \sin\phi,\ \sin\theta \sin\phi,\ \cos\phi) \longmapsto (\cos(\theta+\alpha) \sin\phi \cos\beta - \cos\phi \sin\beta,$$
$$\sin(\theta+\alpha) \sin\phi,\ \cos(\theta+\alpha) \sin\phi \sin\beta + \cos\phi \cos\beta),$$

and it can be shown that d_{ij} is invariant under this mapping.

SPHERICAL MDS IN PRACTICE

MDS in Euclidean space is usually carried out in two dimensions for ease of displaying derived configurations. Configurations from MDS on the sphere are not so easy to display. Possible methods for doing so are: i) a Mecator projection ii) a physical sphere with points annotated; iii) three dimensional plots of the configuration at several rotations; iv) a spinning sphere depicted on the computer screen using animated graphics. At present the authors have only adopted iii).

Cities around the world

The longitude and latitude of several cities around the world were used to calculate Euclidean distances between cities. This data set was chosen as a basic test of the method. The data were subjected to MDS on the sphere and MDS in two dimensional Euclidean space. The stresses of the final configurations found for the two methods were 0% and 13% respectively.

For the spherical case two different starting configurations were used. The first used a configuration extremely close to the true positions of the cities and the second used a completely random starting configuration. Both gave zero stress for the final configuration. After reversing the sign of θ and rotating the sphere as described above, the two configurations were shown to agree.

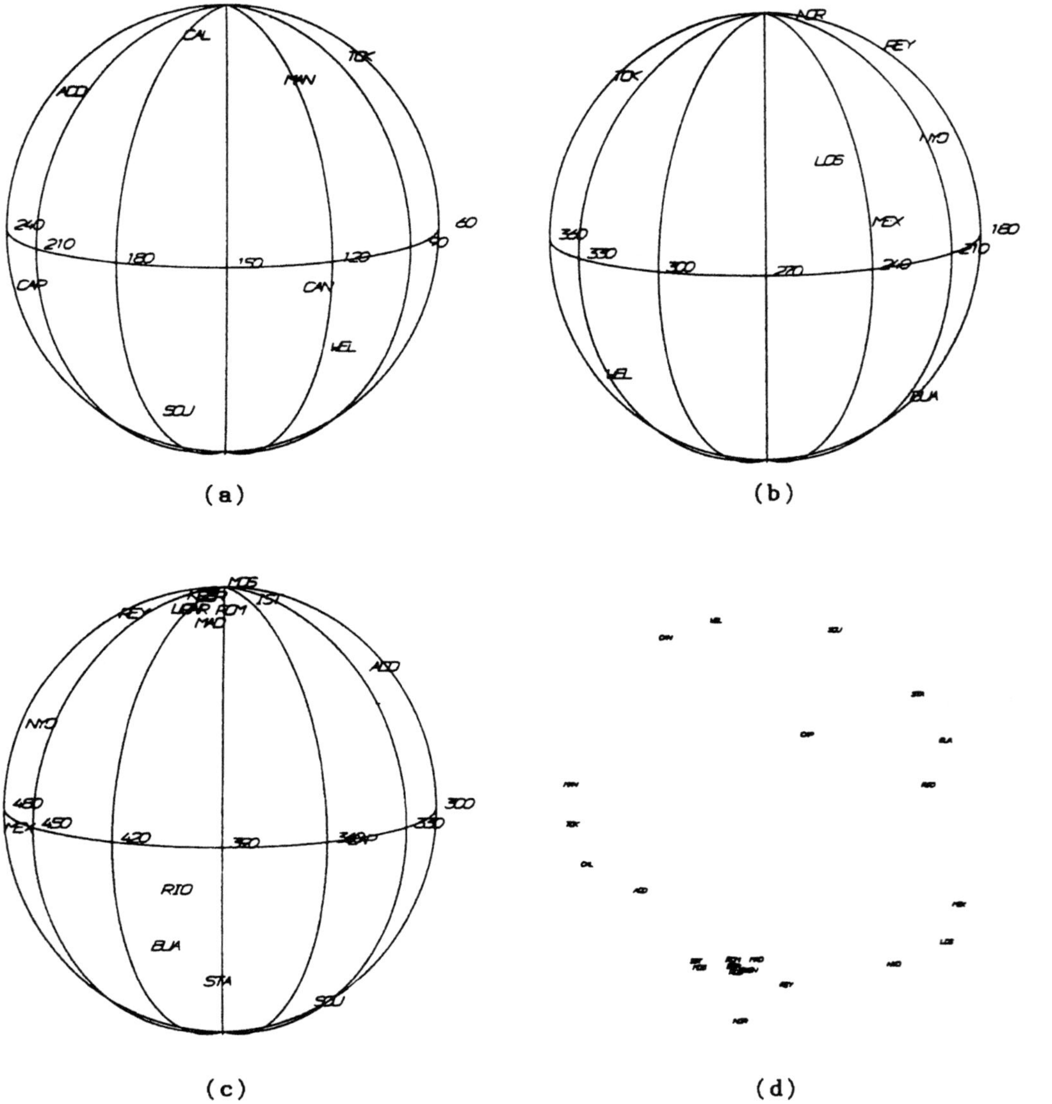

Figure 1.Spherical and planar MDS for world cities.
(Addis Ababa, Berlin, Buenos Aires, Calcutta, Canberra, Cape Town, Istanbul, København, London, Los Angeles, Madrid, Manila, Mexico City, Moscow, New York, Paris, Reykjavik, Rio de Janeiro, Rome, Stanley, Tokyo, Wellington, North pole, South pole)

Figures 1(a), 1(b), 1(c) show the results for spherical MDS giving three rotations about the z-axis in order to bring all the points into view. MDS will never reproduce exactly a configuration of points even when starting with dissimilarities which <u>are</u> the distances of an actual configuration of points. Figure 1(d) shows the results for planar MDS.

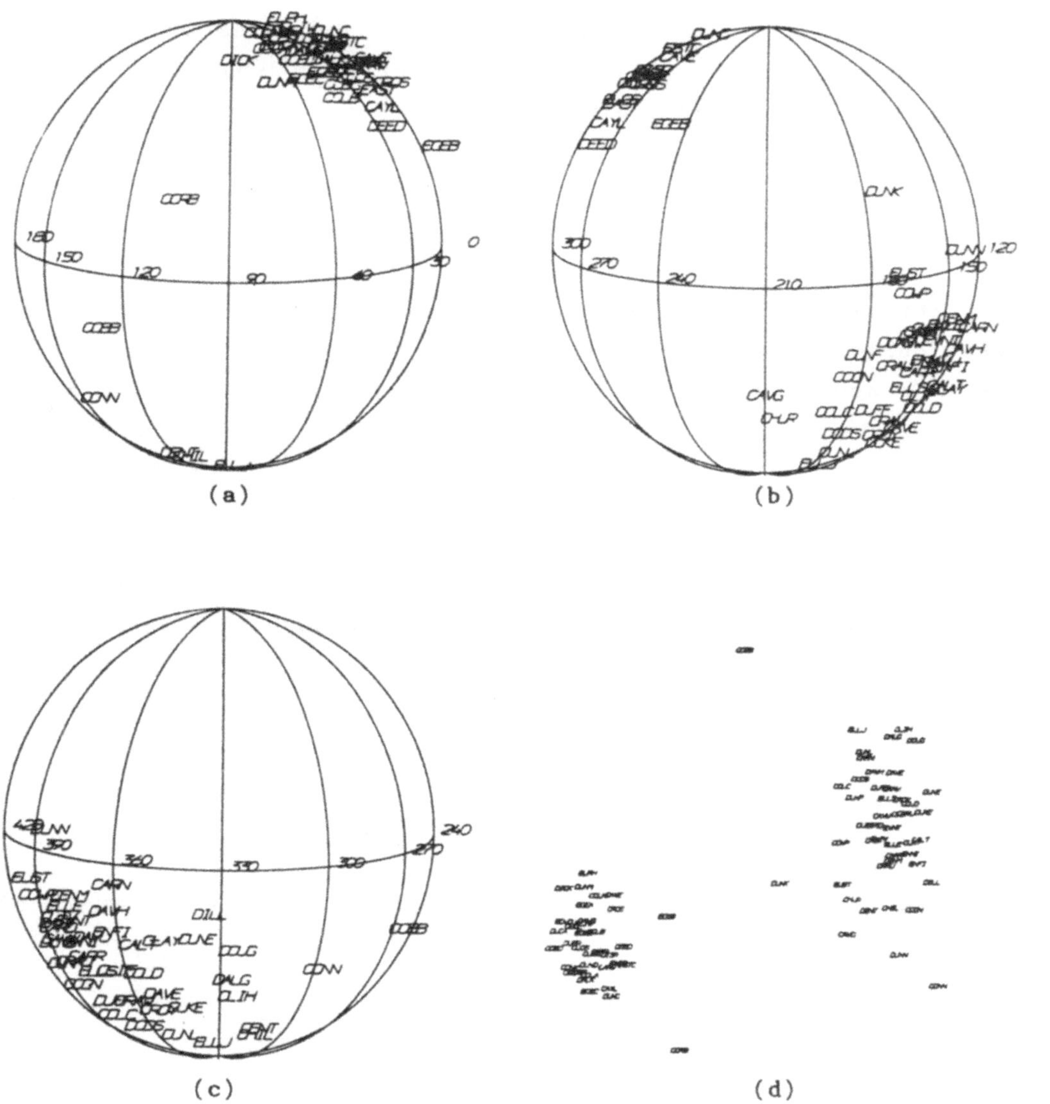

Figure 2. MDS for the voting behaviour of members of the British Parliament in 1861.

Spherical MDS was tried on several other sets of data, and was shown to give comparable stress to planar MDS in many situations. Often the difference in the two stress levels was less than 1%, indicating that the stresses for the two methods can be compared numerically even though their definitions are not precisely the same. The discrepancy occurs in the scaling of the coordinates of the points in the two situations. In planar MDS, the coordinates are scaled so that the coordinates on each axis have mean zero and variance unity, while in spherical MDS no scaling is required as the points will always sit on the unit sphere.

Historical voting data

The second set of data illustrated here concerns the voting behaviour of 81 members of the British Parliament in 1861. These are a random subset of those members voting more than twenty times in the whole year. The dissimilarity between a pair of members is calculated as the number of votes in which they differed divided by the number of times they simultaneously voted.

Figures 2(a), 2(b), 2(c) show the results obtained using spherical MDS, giving three rotations of the sphere. Figure 2(d) shows the result for planar MDS. The stress for spherical MDS was 11%, and that for planar MDS 15%. It can be seen that there are two main groups of members who are 'poles' apart.

Acknowledgement

The historical voting data were provided by V. Cromwell of Sussex University.

References

Chatfield C, and Collins AJ (1980) Introduction to Multivariate Analysis. Chapman and Hall. London.

Coxon APM, and Davies PM (1982) The User's Guide to Multidimensional Scaling. Heinemann. Exeter.

Kruskal JB (1964a) Multidimensional scaling by optimising goodness of fit to a nonmetric hypothesis. Psychometrika 29, 1-27.

Kruskal JB (1964b) Nonmetric multidimensional scaling: a numerical method. Psychometrika 29, 115-129.

Shepard RN (1962a,b) The analysis of proximities: multidimensional scaling with an unknown distance function. Psychometrika 27, 125-139, 219-246.

A Monte Carlo Evaluation of the Methods for Estimating the Parameters of the Generalized Lambda Distribution

M. C. Okur, Izmir

SUMMARY: This paper presents the results of a Monte Carlo study concerning the problems of parameter estimation in the generalized lambda distribution. Statistical properties of estimates obtained by using four available methods are contrasted and it is concluded that the exact least squares procedure performs better than the other methods.
Key Words: Generalized lambda distribution, nonlinear least squares, method of moments, probability weighted moments, quantile estimation.

1. INTRODUCTION

Families of multiparameter probability distributions have proven to be useful for the representation of data when the underlying model is unknown. An interesting example of such distributions is the generalized lambda distribution (GLD) which was introduced by Ramberg and Schmeiser(1974). Because of its generality and flexibility, the GLD can approximate a wide range of distributions and provides a simple algorithm for generating random variates.

A major difficulty involved in using the GLD in applications is the determination of its four parameters. Ramberg et al.(1979) used the method of moments(MOM) and provided a table to obtain approximate values for the parameters. Öztürk and Dale(1985) developed a nonlinear least squares(LS) procedure for estimating the parameters of the GLD. Öztürk(1988) has formulated another procedure based on so called probability weighted moments.

All these estimation methods require nonlinear function minimization and as such their computational complexities are comparable. Öztürk and Dale(1985) indicated tha neither the MOM nor the LS method offered any major computational advantages over the other. Dudewicz and Mishra(1988), on the other hand, notes that the MOM is a powerful method in applications and is in wide use. Okur(1988) fitted the GLD to air pollution data by the method of LS and obtained satisfactory results.

Compstat 1988

The objective of this study is to investigate some statistical properties of the parameter estimation methods by using computer simulation. In Section 2, the methods are briefly described and in Section 3 the simulation results are presented.

2. ESTIMATION OF THE PARAMETERS OF THE GLD

Let P be a uniformly distributed random variable on the interval zero to one. The GLD is defined by its four-parameter quantile function

$$x_p = Q(p) = \lambda_1 + [p^{\lambda_3} - (1-p)^{1-\lambda_4}] / \lambda_2 \ . \tag{1}$$

Ramberg and Schmeiser(1974) determined the regions of parameters where the GLD is legitimate and derived expressions for the central moments of the distribution.

Since the GLD has no distribution function in simple closed form, the maximum likelihood estimates of the parameters are difficult to obtain. The MOM estimates may be obtained by equating the first four sample moments to the corresponding moments of the distribution. However, the sensitivity of the sample moments to extreme observations is a major problem.

The LS procedure proposed by Öztürk and Dale(1988) is based on minimising the squared distance between the quantile function in (1) and the empirical quantile function. In this procedure the ith sample order statistic is expressed as

$$X_{(i)} = E\,[X_{(i)}\,] + e_i \tag{2}$$

which is the Model 1 of Öztürk and Dale. A different version of this model is Model 2 which uses first order approximations for the expected values. In either case, the ordinary LS estimates of the parameters are obtained by employing a function minimization routine such as Nelder and Mead procedure.

The probability weighted moments (PWM) can also be used for estimating the parameters of the GLD. Öztürk(1988) obtained the rth PWM for the GLD as

$$M_r = \lambda_1/(r+1)+[1/(\lambda_3+r+1)-B(\lambda_4+1,\ k+1)]/\ \lambda_2 \tag{3}$$

where B(.) is the usual Beta function. By using the unbiased estimators $\hat{M}_r$, one may obtain the four PWM equations as usual. From these, the solutions for

$\underline{\lambda}$ are obtained by an appropriate numerical procedure.

Since the GLD is a four-parameter system, the parameter estimates are expected to be sensitive to the estimation method being used and to the peculiarities of particular data sets. One way of evaluating the consequences of estimation according to various methods is to investigate the accurracy and precision of the parameter as well as quantile estimates.

3. MONTE CARLO RESULTS

Monte Carlo experiments were performed to investigate some statistical properties of parameter and quantile estimates. Given samples of size n = 50, 100, 200; the parameters of three particular GLD's and the corresponding quantiles Q(p); where p=.001, .01, .05, .50, .95, .99, .999, were estimated by 1.The MOM, 2.The method of LS based on Model 1(LS1), 3.The method of LS based on Model 2(LS2), and 4.The method of PWM. The parameter values and corresponding skewness(α_3) and kurtosis(α_4) coefficients of the selected GLD's are given in Table 1.

Table 1. Parameter values and corresponding coefficients of α_3 and α_4.

Distribution	α_3	α_4	λ_1	λ_2	λ_3	λ_4
GLD1	0	3	0	.1974	.1349	.1349
GLD2	0	6	0	−.1686	−.0802	−.0802
GLD3	1.5	10	−.382	−.1803	−.0524	−.1104

All these distributions have mean zero and variance one.The exact quantile values Q(p) for the distributions are given in Table 2.

Table 2. Exact quantile values for the distributions.

Distribution	.001	.01	.05	.50	.95	.99	.999
GLD1	−3.07	−2.34	−1.65	.00	1.65	2.34	3.07
GLD2	−4.39	−2.65	−1.59	.00	1.59	2.65	4.39
GLD3	−2.80	−1.89	−1.29	−1.5	1.78	3.29	5.96

By performing N=500 experiments, the mean and standard deviation of the estimates $\hat{\theta}_i$, corresponding to a parameter or quantile were determined for each sample size and GLD combination. The mean $\mu(\hat{\theta})$ and standard devation $\sigma(\hat{\theta})$ were then used to determine the empirical bias and mean square error of an estimator as follows:

$B(\hat{\theta}) = \mu(\hat{\theta}) - \theta$

$$MSE(\theta) = \sigma^2(\hat{\theta}) + [\,B(\hat{\theta})\,]^2 \qquad (4)$$

A Kolmogorov-Smirnov type statistic denoted by DMAX and defined as

$$DMAX = \text{Max}\ \left|\ i/(n+1) - Q^{-1}(\,\underline{\hat{\lambda}},\ x_{(i)}\,)\ \right| \qquad (5)$$

was also computed in each case.

The simulation results for estimating the three parameters of the GLD are summarized in Table 3 and Table 4.

Table 3. Bias in estimates of the GLD parameters

n	Method	GLD1				GLD2				GLD3			
		λ_1	λ_2	λ_3	λ_4	λ_1	λ_2	λ_3	λ_4	λ_1	λ_2	λ_3	λ_4
50	MOM	.01	.03	.05	.05	-.01	.29	.16	.17	-.23	.31	.09	.24
	LS1	.00	-.05	.00	.00	.00	.02	.04	.04	-.07	.04	.02	.08
	LS2	.01	-.15	-.05	-.07	.00	-.27	-.06	-.06	-.02	-.27	-.05	-.04
	PWM	.01	-.14	-.05	-.05	.00	-.47	-.08	-.08	.00	-.43	-.07	-.07
100	MOM	-.02	.02	.03	.03	.02	.23	.12	.12	-.21	.24	.07	.17
	LS1	.00	-.02	-.01	.00	.00	.00	.01	.01	-.05	.02	.01	.05
	LS2	.00	-.08	-.04	-.04	.00	-.10	-.06	-.06	.02	-.22	-.05	-.07
	PWM	.00	-.07	-.04	-.04	.00	-.22	-.05	-.05	.00	-.17	-.03	-.03
200	MOM	-.02	.01	.01	.02	.00	.15	.08	.08	-.18	.19	.06	.12
	LS1	.00	-.01	.00	.00	.00	.00	.01	.01	-.02	.03	.01	.03
	LS2	-.01	-.05	-.03	-.03	.00	-.14	-.05	-.04	.01	-.13	-.03	-.04
	PWM	.01	-.05	-.03	-.04	-.01	-.19	-.06	-.06	.01	-.14	-.04	-.06

Table 4. Standard Deviation ofThe Parameter Estimates

n	Method	GLD1				GLD2				GLD3			
		λ_1	λ_2	λ_3	λ_4	λ_1	λ_2	λ_3	λ_4	λ_1	λ_2	λ_3	λ_4
50	MOM	.53	.09	.15	.14	.40	.12	.10	.10	.34	.12	.07	.13
	LS1	.49	.17	.17	.17	.28	.39	.16	.16	.25	.37	.09	.20
	LS2	.45	.30	.19	.17	.24	.55	.19	.19	.24	.74	.15	.26
	PWM	.45	.27	.18	.18	.24	.98	.20	.19	.23	.79	.15	.25
100	MOM	.37	.07	.10	.10	.25	.12	.07	.07	.25	.11	.03	.10
	LS1	.34	.10	.10	.11	.17	.25	.10	.10	.17	.30	.07	.15
	LS2	.30	.13	.11	.11	.15	.36	.13	.13	.15	.46	.10	.17
	PWM	.31	.14	.11	.11	.16	.46	.14	.14	.15	.45	.10	.16
200	MOM	.28	.06	.07	.09	.16	.13	.07	.07	.18	.10	.20	.08
	LS1	.22	.07	.07	.06	.12	.18	.07	.08	.12	.20	.05	.10
	LS2	.20	.08	.07	.07	.12	.27	.10	.10	.13	.28	.07	.12
	PWM	.18	.09	.08	.07	.11	.27	.09	.09	.11	.32	.07	.12

The empirical bias values indicate that the LS1 procedure is clearly superior to the others for all sample sizes. The MOM is comparable to the LS1 for only GLD1. When the α_3 or α_4 values increase, the MOM estimates usually exhibit considerable positive biases. The LS2 and PWM estimates are also poor and they have considerable negative biases. The biases tend to decrease as n increases for all methods. The standard deviations of the estimates are

Table 5. Bias in Estimates of the GLD Quantiles

n	Meth.	GLD1							GLD2							GLD3						
		.001	.01	.05	.50	.95	.99	.999	.001	.01	.05	.50	.95	.99	.999	.001	.01	.05	.50	.95	.99	.999
50	MOM	.25	.12	.03	.00	-.05	-.16	-.31	1.18	.35	.02	.00	-.04	-.37	-1.21	.75	.28	.04	.00	-.03	-.55	-2.03
	LS1	-.09	.01	.02	.00	.00	.01	.11	-.04	.06	.02	.00	-.02	-.07	-.03	.07	.04	.00	.00	.00	-.10	-.08
	LS2	-.74	-.25	-.08	.00	.08	.27	.84	-2.33	-.55	-.14	.00	.14	.56	2.42	-.58	-.27	-.08	.00	.18	.82	2.10
	PWM	-.47	-.09	.01	.01	.00	.10	.47	-2.00	-.26	.04	.00	-.05	.23	1.86	-.62	-.06	.04	-.01	-.13	.12	1.92
100	MOM	.15	.07	.01	.00	.00	-.06	-.13	.84	.21	-.01	.00	.01	.22	-.87	.68	.26	.06	-.01	.04	-.30	-1.40
	LS1	-.02	.02	.02	.00	.00	.02	.07	-.05	.03	.02	.00	.01	.01	.11	.07	.03	.00	.00	-.03	-.09	-.05
	LS2	-.38	-.15	-.05	.00	.05	.15	.39	-1.31	-.34	-.08	.00	.08	.35	1.36	-.70	-.20	-.05	.00	.11	.57	2.18
	PWM	-.27	-.07	.00	.00	.00	.08	.28	-.88	-.13	.02	.00	-.02	.15	1.00	-.32	-.05	.00	.00	-.03	.16	1.32
200	MOM	.09	.05	.02	.00	.00	-.04	-.08	.61	.17	.01	.00	.01	-.12	-.50	.60	.24	.05	-.02	.06	-.15	-.93
	LS1	.02	.02	.02	.00	-.01	-.01	.00	.07	.05	.02	.00	-.02	-.05	-.06	.05	.04	.01	.00	-.03	-.06	-.11
	LS2	-.18	-.06	-.01	.00	.01	.07	.20	-.86	-.21	-.04	.00	.05	.22	.84	-.39	-.10	-.02	.00	.06	.31	1.14
	PWM	-.20	-.06	.00	.00	.00	.06	.20	-.74	-.13	.02	.00	.00	.17	.86	-.29	-.05	.02	-.01	-.01	.25	1.47

Table 6. Standard Deviation of Quantile Estimates

n	Meth.	GLD1							GLD2							GLD3						
		.001	.01	.05	.50	.95	.99	.999	.001	.01	.05	.50	.95	.99	.999	.001	.01	.05	.50	.95	.99	.999
50	MOM	.60	.38	.25	.15	.25	.37	.56	1.13	.63	.34	.14	.34	.62	1.10	.61	.36	.20	.13	.42	.75	1.37
	LS1	.92	.44	.24	.15	.26	.43	.86	2.41	.77	.33	.13	.33	.72	2.07	1.00	.43	.21	.12	.39	.98	3.49
	LS2	1.51	.54	.26	.15	.26	.60	1.85	4.62	1.11	.37	.13	.38	1.16	4.84	2.28	.60	.21	.12	.50	2.08	9.68
	PWM	1.25	.48	.26	.17	.26	.48	1.23	4.54	.93	.31	.14	.31	.89	4.41	1.99	.54	.23	.13	.48	1.54	7.46
100	MOM	.45	.29	.19	.14	.18	.27	.43	1.01	.50	.25	.10	.24	.48	.98	.56	.31	.16	.10	.33	.65	1.35
	LS1	.54	.29	.18	.11	.18	.30	.57	1.51	.52	.22	.10	.24	.54	1.60	.69	.29	.14	.08	.31	.81	2.89
	LS2	.72	.33	.18	.11	.18	.34	.76	2.42	.69	.25	.09	.25	.70	2.54	1.30	.39	.15	.09	.36	1.23	6.09
	PWM	.71	.32	.18	.11	.17	.30	.67	2.21	.60	.23	.10	.22	.63	2.30	1.02	.34	.14	.09	.29	.87	3.78
200	MOM	.34	.20	.13	.08	.13	.20	.34	.82	.35	.16	.07	.18	.43	1.04	.43	.25	.13	.08	.25	.57	1.36
	LS1	.33	.18	.12	.09	.12	.19	.35	1.00	.36	.15	.07	.15	.37	1.05	.43	.19	.10	.07	.20	.53	1.81
	LS2	.42	.20	.12	.09	.12	.21	.43	1.67	.50	.17	.07	.18	.48	1.54	.72	.26	.11	.06	.23	.72	2.75
	PWM	.43	.21	.12	.08	.12	.21	.44	1.50	.44	.16	.07	.15	.43	1.55	.68	.24	.10	.06	.20	.64	2.83

generally lower for the MOM estimators. The LS1 estimators are better than the other two and close to the MOM estimators in most cases.

The simulation results concerning quantile estimation for the three GLD's are presented in Table 5 and Table 6. The results indicate that the LS1 is the best procedure for estimating the quantiles of the GLD. In many cases the empirical biases turn out to be negligible i.e. close to zero. The MOM tends to overestimate the left tail and underestimate the right tail quantiles. The LS2 and PWM estimators of the quantiles are also severly biased. The increase in sample size, although reduces the biases, does not seem to be sufficient for obtaining acceptable quantile estimates for the methods other than the LS1. The standard deviations of the MOM estimates are generally lower than the other methods. However, in terms of the MSE of the estimators the LS1 appears to be the most efficient one among the four procedures.

In many studies the value of DMAX is used as an indicator for goodness of fit.In order to give an idea about the behavior of the DMAX in the case of the GLD, the mean values obtained for GLD2 and GLD3 are presented in Table 7.

Table 7. Mean DMAX values in estimating the GLD2 and GLD3.

	GLD2				GLD 3			
n	MOM	LS1	LS2	PWM	MOM	LS1	LS2	PWM
50	.073	.068	.065	.077	.075	.067	.065	.085
100	.057	.055	.053	.058	.065	.053	.053	.058
200	.044	.041	.040	.040	.053	.041	.041	.044

According to the mean DMAX values the LS1 and LS2 are almost the same and all four methods are quite similar. These and previous results demonstrate that measures like DMAX could be misleading in judging the performance of an estimation method or in assessing the goodness of fit. Apparently this was the main reason why Öztürk and Dale(1985) failed to discover the advantages of LS1 over the other methods.

REFERENCES

Dudewicz,E.J. and M.S.Mishra(1988), Modern Mathematical Statistics. Wiley, New York.

Okur,M.C.(1988), On fitting the generalized lambda distribution to air pollution data. Atmospheric Environment (To appear).

Öztürk,A. and R.F.Dale(1985), Least squares estimation of the parameters of the Generalized lambda distribution. Technometrics, 27, 81-84.

Öztürk,A.(1988), Probability-weighted moments estimation of the generalized lambda distribution. Technical Report. Department of Computer Engineering, Ege University, İZMİR.

Ramberg,J.S. and B.W.Schmeiser(1974), An approximate method for generating asymmetric random variables. Comm. ACM., 17, 78-82.

Ramberg,J.S., E.J.Dudewicz, P.R.Tadikamalla and E.F.Mykytka(1979), A probability distribution and its uses in fitting data. Technometrics, 21, 201-214.

Statistical Guidance for Model Modification in Covariance Structure Analysis

T. C. Luijben and A. Boomsma, Groningen

SUMMARY

Four topics on model modification in covariance structure analysis are covered. 1. A comparison by Monte Carlo methods of statistical criteria used to detect the correct model under conditions of misspecification. 2. The effect of a number of factors on the behaviour of these criteria. 3. The scale dependency of the criteria. 4. An alternative model test procedure.

Key words: model modification, model test, specification errors, covariance structure analysis, scale dependency.

INTRODUCTION

In a process of model modification for covariance structure analysis consider the situation where a researcher starts with some baseline model. Often this initial model does not properly fit the data, which may lead to a decision to relax a fixed parameter in order to improve the fit. Under conditions of proper fit the model may be changed in a way to diminish its complexity. Although the justification of model modification should be guided mainly by theoretical considerations, here the behaviour is studied of three statistical criteria, used to expand a model by relaxing a parameter.

1. The first order derivative (FD) of the maximum likelihood fitting function for a fixed parameter is, to some extent, indicative for the change in fit when that parameter would be relaxed. In a model modification process the parameter with the largest absolute value of its FD could be freed.

2. The modification index (MI) of a fixed parameter is an estimate of the decrease in the χ^2 test statistic, if just that parameter would be relaxed (Jöreskog & Sörbom, 1984; Sörbom, in press). A large value of MI suggests a substantial improvement of overall goodness-of-fit. Therefore, the parameter with the largest MI could be freed. When the fitted model is correct, asymptotically each MI has a χ^2 distribution with 1 degree of freedom (df).

3. The expected parameter change (EPC) of a fixed parameter is an estimate of its changed value, when it would be freed (Saris, Satorra & Sörbom, 1987). The EPC equals -MI/[(N-1)FD], where N is the sample size. The fixed parameter with the largest absolute EPC could be freed. For MI and EPC, the accuracy of the estimate depends on the size of the misspecification: it should be small.

Compstat 1988

A FACTOR ANALYSIS MODEL AND THE BASIC MONTE CARLO DESIGN

The behaviour of the criteria for model modification is studied by Monte Carlo methods. To that purpose, as an example, consider an oblique two-factor model of simple structure. The model is defined by $\mathbf{x} = \mathbf{\Lambda\xi} + \mathbf{\delta}$, where $\mathbf{x}(6x1)$ is a vector of observed variables, $\Lambda(6x2)$ a matrix of loadings on two common factors ξ_1 and ξ_2, and $\delta(6x1)$ a vector of measurement errors. Under usual assumptions the population covariance matrix of $\mathbf{x}$ has the form $\mathbf{\Sigma} = \mathbf{\Lambda\Phi\Lambda}' + \mathbf{\Psi}$, with $\mathbf{\Phi} = E(\xi\xi')$, and $\mathbf{\Psi} = E(\delta\delta')$ diagonal with the variances of the unique scores as elements. The first three variables have a positive loading on ξ_1 $[\lambda_{11}=\lambda_{21}=0.4,\ \lambda_{31}=0.6]$ and a zero loading on ξ_2, the other three variables have a positive loading on ξ_2 $[\lambda_{42}=\lambda_{52}=0.4,\ \lambda_{62}=0.6]$ and a zero loading on ξ_1. The variances of x_i are normalized: $\mathbf{\Psi} = I - \mathrm{diag}(\mathbf{\Lambda\Phi\Lambda}')$. The correlation $\phi=\phi_{21}$ between ξ_1 and ξ_2 equals 0.3. This population model M_1 is specified by $\theta_1'=[\lambda_{11},\lambda_{21},\lambda_{31},\lambda_{42},\lambda_{52},\lambda_{62},\phi,\psi_{11},\psi_{22},\psi_{33},\psi_{44},\psi_{55},\psi_{66}]$, the vector of parameters to be estimated if M_1 would be fitted to a sample covariance matrix **S**.

In our basic Monte Carlo design, pseudo-random samples of size N=200 were taken from the multivariate normal population $N_6(0,\Sigma)$, using IMSL-routines. The resulting samples of covariance S were fitted against a misspecified baseline model M_0, specified by θ_0 , which differs from θ_1 by omitting parameter ϕ. This misspecification implies that it is falsely assumed that the two factors are orthogonal. Maximum likelihood estimation procedures of LISREL VI (Jöreskog & Sörbom, 1984) were used. Under these modeling conditions the likelihood-ratio test statistic can be approximated as a noncentral χ^2_9 variable, with noncentrality parameter τ=4.27. The number of replications is 300.

A COMPARISON OF MODIFICATION CRITERIA

1. **Empirical power**. A total of 309 samples S had to be generated for a required number of 300 converging LISREL solutions; 9 samples showed problems of convergence. When these 300 samples were analyzed, for 27% of them model M_0 was rejected in favour of the saturated model using the likelihood-ratio test criterion (α=0.05); for 220 samples model M_0 was falsely accepted. This empirical power of 27% is well approximated by the 24% given by the estimation procedure of Satorra & Saris (1985). For 10% of the 300 samples improper solutions (negative estimates of variances) occurred.

2. **The behaviour of FD, MI, and EPC**. Which parameter relaxation for each sample S for which M_0 was rejected, is suggested by the modification criteria FD, MI, and EPC? More specifically, given our design, which statistic most often selects ϕ as a misspecified parameter? The results are summarized as

follows. In modifying towards M_1, using FD is not very helpful. The percentage of absolute largest FD is rather homogeneously spread over all parameters; for ϕ it was 4% out of 80, only. High percentages were acquired for ψ_{63}, ψ_{42}, and ψ_{52}: 16%, 10%, and 10%, respectively. Using MI is rather promising: in 46% (37) of the samples rejecting M_0, the MI of ϕ was largest. The remainder was equally spread over the other parameters. Relaxing the parameter with the largest MI only if it exceeds a critical value of 3.84, instead of doing so unconditionally, has no effect for this model: the percentages remained the same. The EPC performs better than the MI: in 74% (59) of the samples rejecting M_0, ϕ was selected to be relaxed.

3. **Discussion**. The results of this study are discouraging. For a simple baseline model, which differs from the population model in one parameter only, a one step modification procedure leads just for 37 (MI), or 59 (EPC), out of 300 samples to the selection of the correct model with an acceptable fit. This result is mainly due to the low power of 27%. If model M_0 is rejected both MI and EPC appear to be useful in finding the misspecified parameter; the value of FD as a modification criterion for this model is not impressive. Thus, in a one step modification process the correct modification of model M_0 was not always obtained, when this model was rejected. With additional modification steps an even worse result can be expected (see, for example, MacCallum, 1986). Also, notice that when M_1 was accepted, the estimate of ϕ appears to be biased (Luijben, Boomsma & Molenaar, in press).

FACTORS THAT AFFECT THE BEHAVIOUR OF MODIFICATION CRITERIA

Here, the results are summarized of a Monte Carlo study on factors which influence the power of the model test, and hence the behaviour of statistical modification criteria. The following factors were studied: the population value of ϕ [symbolized as M_2, M_3], the sample size N [M_4], the α-level of the model test [M_5], the reliability of the indicators [M_6], and the number of indicators [M_7]. If model M_1 is compacted as ϕ=0.3, N=200, α=0.05, the other factors under study are summarized by their deviance from M_1. Therefore, they are distinguished as follows: M_2: ϕ=0.4 (i.e. the same design as M_1, except that ϕ equals 0.4 instead of 0.3); M_3: ϕ=0.5; M_4: N=400 (i.e. model M_1, but a sample size of 400); M_5: α=0.20; M_6: $\lambda_{11}=\lambda_{21}=\lambda_{42}=\lambda_{52}=0.6$, and $\lambda_{31}=\lambda_{62}=0.8$; and finally M_7: $\lambda_{11}=\lambda_{21}=\lambda_{31}=\lambda_{41}=\lambda_{62}=\lambda_{72}=\lambda_{82}=\lambda_{92}=0.4$, and $\lambda_{51}=\lambda_{10,2}=0.6$.

The number of converging replications is 300 in all cases. The results from the Monte Carlo studies are summarized in Table 1; for the estimated power the method of Satorra & Saris (1985) was used. It is shown that the behaviour of FD improves substantially only when the factor loadings are larger (M_6). As for MI and EPC, all factors under study tend to improve their behaviour, except for a larger α-level (M_5). Notice, that the increase of the number of indicators (changing from M_1 to M_7), and the consequential decrease in empirical power, still results in enhanced MI and EPC behaviour.

Table 1. The effect of a number of factors on the power and on the behaviour of statistical modification criteria. The meaning of the increased empirical power (column 6) is discussed in the last section.

model	df	τ	estim. power	empirical power	increased emp. power	correct modification FD	MI	EPC
M_1	9	4.27	24%	27%	3.7%	4%	46%	74%
M_2	9	7.65	43	41	7.3	4	66	90
M_3	9	12.08	66	70	5.7	9	72	92
M_4	9	4.27	50	52	6.3	5	66	95
M_5	9	8.55	49	50	4.0	3	40	72
M_6	9	10.16	57	53	7.3	22	78	95
M_7	35	5.90	18	19	11.0	5	50	81

THE SCALE DEPENDENCY OF STATISTICAL MODIFICATION CRITERIA

Confining ourselves to a change of scale in the latent variables, consider model M_0, where ϕ_{11} and ϕ_{22} are fixed at one, and the equivalent model M_{0*} with λ_{31} and λ_{62} fixed at one, instead. Thus, in M_0 the scale of ξ_1 and ξ_2 is fixed directly by fixing their variances, in model M_{0*} this is established indirectly by fixing λ_{31} and λ_{62} at one. The values of the population parameters for model M_{1*} (compare with model M_1) are: $\lambda_{11}=\lambda_{21}=\lambda_{42}=\lambda_{52}=0.667$, $\lambda_{31}=\lambda_{62}=1.000$, $\phi_{11}=\phi_{22}=0.360$, $\phi_{21}=0.108$, $\psi_{11}=\psi_{22}=\psi_{44}=\psi_{55}=0.840$, and $\psi_{33}=\psi_{66}=0.640$. When M_0 and M_{0*} are fitted to a sample covariance matrix S from Σ, both the total proportion of explained variance and the χ^2 goodness-of-fit are the same for the two models.

The question now is whether the behaviour of the MI and the EPC depends on the scaling of the latent variables. More specifically: does the behaviour of MI and EPC change when samples S are fitted against M_{0*}, instead of against M_0? Luijben (1988; Theorem 1) proves that when two models are equivalent, the values of MI are equal for all parameters of both models. It thus follows that the MI is scale independent: whether M_0 or M_{0*} is fitted, ϕ or any other parameter is detected equally often as the parameter to be relaxed. In contrast, from the proof of the same theorem and the definition of the EPC, it can be shown that the EPC is scale dependent. By similar arguments it also follows that a standardized version of the EPC, the SEPC, introduced by Kaplan (1987), depends on how latent variables are being scaled.

To illustrate these analytical results, consider the model fit of M_0 and M_{0*}, respectively, to the population covariance matrix $\Sigma(\theta_1)=\Sigma(\theta_{1*})$. In this way the so-called theoretical MI and EPC values are obtained, under the two scaling conditions: shortly, MI-0, MI-0*, EPC-0, and EPC-0*. (From above it follows that MI-0 equals MI-0*.) These theoretical values can be compared with Monte Carlo results based on 300 replications (N=200): see Table 2.

Table 2. Theoretical MI and EPC under M_0 and M_{0*}, and the percentage of their largest (absolute) value among all parameters in 300 samples.

	λ_{12}	λ_{22}	λ_{32}	λ_{41}	λ_{51}	λ_{61}	ϕ	ψ_{41}	ψ_{42}	ψ_{43}	ψ_{51}	ψ_{52}	ψ_{53}	ψ_{61}	ψ_{62}	ψ_{63}
MI-0	.49	.49	1.82	.49	.49	1.82	**4.22**	.06	.06	21	.06	.06	.21	.21	.21	.79
%	5	5	6	4	3	9	**35**	4	5	6	2	4	4	3	3	4
EPC-0	.07	.07	.13	.07	.07	.13	**.30**	.02	.02	.03	.02	.02	.03	.03	.03	.06
%	5	4	6	5	5	4	**62**	0	1	1	1	1	1	1	1	0
EPC-0*	.11	.11	.22	.11	.11	.22	**.11**	.02	.02	.03	.02	.02	.03	.03	.03	.06
%	13	13	21	10	10	28	**0**	0	0	0	0	0	1	1	1	0

As expected from the theoretical values EPC-0 and EPC-0*, fitting M_{0*} results in a substantial drop in detecting ϕ as the misspecified parameter, compared to fitting the equivalent model M_0: in none of the 300 samples ϕ had the largest absolute EPC value! This does not necessarily imply that the EPC cannot be used. Careful inspection of the population parameter values of M_{1*} learns, for example, that the EPC value of 0.22 for λ_{61} should be interpreted as an increase in the true score variance of x_6 of 2% [(0.22)(0.22)(0.36)]; and the EPC value of 0.108 for ϕ suggests that the correlation between the two factors equals 0.108/{(0.60)(0.60)}=0.30. Under such considerations the decision could still be that ϕ, rather than λ_{61}, should be relaxed. However, can it be expected to argue properly, when complex models are applied in practice? Surely, an unreflecting, automatic use of the EPC (or SEPC) in the process of model modification is out of the question. Its use can only be recommended, when its values are interpreted in relation to the variances of the observed and latent variables, as well as to the range of the parameters.

It is known that the MI suffers from a differential sensitivity of parameter misspecification (Saris, Satorra & Sörbom, 1987); the EPC suffers from the nuisance of scale dependency. Since the statistical approach, and interpretation of both modification criteria differs, it is not quite easy to give general practical recommendations with respect to the balance of their use.

AN ALTERNATIVE MODEL TEST PROCEDURE

As an alternative model test the following procedure is suggested. Expand the model by the parameter with the largest MI: (a) when the model is rejected based on the familiar χ^2 likelihood-ratio test, and (b) when the model is not rejected by that test, but one of the MI exceeds the critical χ^2_1 value of size $\alpha_B=\alpha/$(the total number of MI considered in the model); i.e. a Bonferroni approach. As an example, for M_0 the procedure implies that $\alpha_B=0.05/16=0.0031$, which means that when M_0 would be accepted initially, it could still be modified if one of the MI exceeds the value of 8.75.

If this procedure should be followed, the size of α increases with the probability that the model is accepted given a value larger than 8.75 for at least one MI, under the assumption that the population model has been tested. Monte Carlo studies show that this increase is limited; for example, for model M_0 ($\alpha=0.05$, N=200) the empirical increase in the size of α equals 1.7%.

The proposed test procedure appears to be more powerful than the usual one. In the first Monte Carlo study (model M_1) 11 of the 220 samples accepting M_0 had a parameter with a MI larger than 8.75, in 6 cases of which (correctly) associated with ϕ. So, here is a small increase in empirical power of 3.7%. The percentage of increase in power for other models is given in column 6 of Table 1: under circumstances, the power raises substantially.

REFERENCES

Jöreskog, K.G., & Sörbom, D. (1984). LISREL VI: Analysis of linear structural relationships by the method of maximum likelihood, instrumental variables, and least squares methods: User's guide (3rd ed.). Mooresville, IN: Scientific Software.

Kaplan, D.W. (1987). On specification error issues in structural equation models. Unpublished doctoral dissertation, University of California, Los Angeles.

Luijben, T.C. (1988) Equivalent models in covariance structure analysis. Heymans Bulletins, HB-88-883-EX. University of Groningen, Psychologische Instituten.

Luijben, T.C., Boomsma, A., & Molenaar, I.W. (in press). Modification of factor analysis models in covariance structure analysis: A Monte Carlo study. In T.K. Dijkstra (Ed.). Capitalization on chance and sample re-use models. Berlin: Springer.

MacCallum, R. (1986). Specification searches in covariance structure modeling. Psychological Bulletin, 100, 107-120.

Saris, W.E., Satorra, A., & Sörbom, D. (1987). The detection and correction of specification errors in structural equation models. In C.C. Clogg (Ed.). Sociological Methodology 1987. San Francisco: Jossey-Bass, 105-129.

Satorra, A., & Saris, W.E. (1985). The power of the likelihood ratio test in covariance structure analysis. Psychometrika, 50, 83-90.

Sörbom, D. (in press). Model modification. Psychometrika.

Similarities Functions

G. le Calvé, Rennes

INTRODUCTION

The main problem in data analysis is the problem of the representation of this data by a visuable display understandable by every body ; such are, for example, hierarchical classification, additive tree, Euclidean representation. With this end in view, what is really analysed is not exactly the data but relations between them, namely similarities or dissimilarities. It is interesting to quote that the representations are mostly read in terms of dissimilarities, while they are constructed in terms of similarities. It is then very important to study the relations between these two kinds of association coefficients.

We listed below some of the usual similarities indexes, defined as decreasing functions of a dissimilarity index D. In the first paragraph we study the really important case of Euclidean representation, whose study is based on the positive definiteness of a matrix. For this purpose we give a generalisation of the Schur's lemma on the Hadamard product of two matrices and its applications. In the second section we are more specially concerned by a generalisation of the linear similarity function which seems to be very useful. The last section is devoted to applications.

KEYWORDS : Similarities, dissimilarities, quadratics forms, additive trees, L_1 norm.

I - NOTATIONS AND DEFINITIONS

We consider a finite set, of cardinality n.

- D a dissimilarity index on I is a square symmetric matrix with null diagonal and positive elements.
- D is a distance if, in addition, the triangular inequality $D_{ij} \leqslant D_{ik} \leqslant D_{kj}$ is fulfilled for every i, j, k.
- S a similarity index on I is a square symmetric matrix such that $S_{ii} > S_{ij}$.
- f, a similarity function, is a real decreasing function. Writing $S = f(D)$ instead of $S_{ij} = f(D_{ij})$, it will be noted that S is a similarity index iff D is a dissimilarity index.

Let D be a dissimilarity index. We consider the following similarity function :

- Linear : $S_{ij} = 1-D_{ij}$. This function is very often used for qualitative variables. Well adapted to the hierarchical classification.
- Homographic : $H_{ij} = \frac{2}{1+D_{ij}} - 1$; used in the same case.
- Quadratic : $W^M(D^2)_{ij} = \frac{1}{2}[D^2_{Mi} + D^2_{Mj} - D^2_{ij}]$, well adapted to the Euclidean case.
- The Exponential : $E_{ij} = \exp(-D_{ij})$ used, for example, by the Nei index in genetric
- The "Circular" : $C_{ij} = \cos D_{ij}$.

II - POSITIVE DEFINITENESS OF THIS FUNCTIONS

In order to answer the question of the Euclidean representation of such indexes, we study the positive definiteness of similarity functions. Indeed, it is well known that a dissimilarity D admits a Euclidean representation iff the quadratic similarity function $W^M(D^2)$ is positive definite (see for example, Schöenberg).

We first establish a generalisation of the Schur's lemma, saying that the Hadamard product $A * B$ of two non negative definite matrices is non negative definite.

THEOREM 1

Let $t \to f(t)$ a real function, admitting the development :

$$f(t) = \sum_{r=0}^{\infty} a_r \, t^r \quad \forall \; t < R$$

Let $M = (M_{ij})$ a nxn symmetric matrix, non negative definite.

a) the matrix $F = (F_{ij})$ with $F_{ij} = f(M_{ij})$ is n.n.d

b) F is positive definite iff the following condition is fulfilled.

CONDITION (C)

$\forall X \in \mathbb{R}^n$, there exists $r \in \mathbb{N}$ such that $X^t A^{*^r} X > 0$

PROOF. Use the Schur's lemma if $M_{ij} < R$, and a continuity argument if there exists i and j such that $M_{ij} \leqslant R$. (For more details see Joly-Le Calve).

THEOREM 2

Let S be a similarity index. Then S verify the condition (C).

PROOF. The proof lies on two remarks :

- If $\lim_{r \to \infty} (A^{*^r}) = I$, then A verify the condition (C)
- If there exists a matrix M of rank 1 such that $\lim_{r \to \infty} (M^{*^r} * A^{*^r}) = I$, then A verify the condition (C).

Now let us define $R_{ij} = \dfrac{S_{ij}}{\sqrt{S_{ii}\ S_{jj}}}$. Clearly R has the property that $\lim_{r \to \infty} R^{*^r} = I$, define $M_{ij} = \sqrt{S_{ii}\ S_{jj}}$, S can be written as $S = M * R$ so that S verify condition (C).

This theorem shows us that, in particular every correlation matrix such that $|R_{ij}| \neq 1$ for $i \neq j$ verify condition (C).

THEOREM 3

Let D be a dissimilarity index and consider the following assertions :

a) $W^M(D^2)$ is n.n.d

b) $E(\lambda^2 D^2)$ is n.n.d $\forall \lambda$

c) C(D) is n.n.d

d) S(D) is n.n.d

e) H(D) is n.n.d

and the assertions a', b', c', d', e' deduced from the preceedings by "p.d" instead of "n.n.d". Then

$a \Longleftrightarrow b \qquad b \Longrightarrow d' \qquad c \Longrightarrow d' \qquad d \Longrightarrow e'$

PROOF. $a \Longleftrightarrow b$ is a Schoenberg's result.

The others assertions derive from the relations :

$$S = 1 - (2-2W^M(D^2))^{1/2}$$
$$S = 1 - \text{Arcos } C$$
$$H = \frac{S}{2 - S}$$

and the application of Theorem 1.

III - THE LINEAR SIMILARITY FUNCTION

We now consider the linear generalised similarity function defined by :

$$W^k(D)_{ij} = \frac{1}{2} [D_{ki} + D_{kj} - D_{ij}]$$

This is exactly the definition of $W^k(D^2)$ with D instead of D^2. We list below some properties of this similarity function. The proofs are omitted, being only rewriting of known results.

THEOREM 4

- D is a distance iff $W^k(D)_{ij} \geqslant 0$
- D is a chain implies $W^k_{ij} = 0$ (within a permutation)
- D is a symmetric difference iff $W^k(D)_{ij}$ is an integer and $W^\varphi(D)_{ij} = |X_i \cap X_j|$
- D is a quasi-hypermetric iff $W^k(D)$ is n.n.d
- D is an ultrametric implies $W^i_{jk} = W^j_{ik} = \frac{1}{2} D_{ik}$
- D is an additive tree implies that $\forall$ i,j,k $\exists$ M such that $W^M_{ij} = W^M_{ik} = W^M_{jk} = 0$.

FOOT OF A POINT

For every point i, let us define i^* by $D_{ii^*} = \inf_{k,\ell} W^i(D)_{k,\ell}$. D_{ii^*} is positive if D is a distance. Define also $\delta_{ij} = D_{ij} - D_{ii^*} - D_{jj^*}$ for $i \neq j$, and $\delta_{ii} = 0$; and finally define iRj iff $\delta_{ij} = 0$.

THEOREM 5

i) δ_{ij} is a distance

ii) R is an equivalence

iii) iRj iff $W^k_{ij} = W^k_{jx} = W^k_{ix} \quad \forall k, \ \forall x$.

PROOF

- The first assertion is a consequence of the inequality :

$$2 D_{jj^*} \leqslant D_{ij} + D_{jk} - D_{ki} \quad ;$$

so that

$$D_{ki} - D_{kk^*} - D_{ii^*} \leqslant D_{ij} + D_{jk} - D_{kk^*} - D_{ii^*} - D_{jj^*}$$

- The second is a consequence of the fact that δ is a distance

- The third is based on the relation

$$2 W^k_{ix} - 2 W^k_{jx} = \delta_{ki} - \delta_{ix} - \delta_{kj} + \delta_{jx}$$

It is interesting to note that if the third point is verify for one k, it is verify for every k.

This notion of "foot of a point" play an important role in many situations as it will be seen in the last section. For instance, a distance is a star distance iff all the points are equivalent in the preceeding sense. In an additive tree, the foot of a point is the branching point of the point in the tree.

ADDITIVE TREE DISTANCE

Let us examine the case of additive tree which is a very useful representation in data analysis. Recall that D is an additive tree distance iff, for every i,j,k,ℓ we have

$$D_{ij} + D_{k\ell} \leqslant \text{Max}(D_{ik} + D_{j\ell}, D_{i\ell} + D_{jk})$$

We can prove the following.

THEOREM 6

A dissimilarity D is an additive tree distance iff for every point M there exists a partition in M_r subsets such that $W^M(D)_{ij} = D_{MM^*}$ iff i and j belongs to two different subsets.

PROOF. It can be easily proved that in an additive tree, $W^M(D)_{ij} = W^{M^*}(D)_{ij} + D_{MM^*}$, M^* is the branching point of M in the tree, and D is an additive tree iff $W^{M^*}(D)_{ij} > 0$ iff i and j belongs to different edges issued from M^*.

IV - APPLICATIONS

• *ADDITIVE TREE DISTANCE*

As an application, Theorem 6 can be used for obtaining an algorithm for the construction of an additive tree. This algorithm works in the following manner.

Let us take a point M and compute the matrix $W^M(D)_{ij}$ and denote by α the $\inf_{i,j} W^M(D)_{ij}$. Then $W^M(D)$ can be, easily, written as shown in figure 1, by

$>\alpha$	α	α
α	$>\alpha$	α
α	α	$>\alpha$

FIGURE 1

the use of the M_k decomposition of Theorem 6 (the joint figure is obtained with $k = 3$).

Then the additive tree is decomposed in k additive trees and the same operation is done again in each tree. It will be noted that this algorithm is of order n^2 while the classical ADDTREE is in n^4. This algorithm can also be extended to the case in which D is not exactly an additive tree distance.

• *PINS-AND-HOLES REPRESENTATION*

Let D be a dissimilarity. The problem of finding an (approximate) Euclidean representation of this data is not easy to solve. An interesting solution consist to find the smallest constant C such that $E_{ij} = D_{ij} + C$ admits a Euclidean representation. A few years ago, Cailliez gives the solution to this problem. (Namely the first eigenvalue of an 2nx2n matrix). It seems, however, that every body represent only the Euclidean E_{ij}, missing the representation of C. But this representation is easy : every point is represented by a circle of radius C/2, and the distance between two points must be read as the distance of the edges of the circles. This representation can be called the hole-representation.

On the other hand, another possibility consists in finding C' such that $D_{ij} = E'_{ij} + C'$, E'_{ij} being Euclidean. (In our knowledge, this problem is not yet solved). In this case the representation, I propose to call pin-representation, consist by representing each point by a pin of length C/2, such that the distance between two points must be read as the distance between the two feet of the pin plus the length of the pin.

In each case the "deformation" of the dissimilarity which is measured by $\frac{n(n-1)}{2}$ C, is uniformly distributed over all points. Or, in many situations, only a few points do not agree with a Euclidean representation, the magnitude of this disaccord being extremely various. In this condition we think that the following result is of some interest.

THEOREM 7 : PINS AND HOLES

Each dissimilarity D admits the following decomposition $D_{ij} = X_i + E_{ij} + X_j$, X_i being a pin or a hole (positive or negative) and E_{ij} being Euclidean.

PROOF. Theorem 6 asserts that D_{ij} can be written as $D_{ij} = D_{ii^*}+\delta_{ij}+D_{jj^*}$. Now, let C be the additive constant of δ_{ij}, computed by the Cailliez's formula and let us write $X_i = D_{ii^*} - \frac{C}{2}$, $\delta_{ij}+C$ is Euclidean and $D_{ij} = X_i+E_{ij}+X_j$.

The interest of this decomposition is that we have an exact representation of the dissimilarity. E_{ij} has the same interpretation as in factor analysis. In addition $|X_i|$ can be interpreted as the "personality" or the "personal dimension" of i, E_{ij} being the distance of i and j in the community, apart of this individual variation. In this point of view, greater is the variance of X_i, better is this representation, on the opposite of the additive constant representation.

REFERENCES

GOWER J.C., LEGENDRE P. (1986) : "Metric and Euclidean properties of dissimilarity coefficients".
Journal of Classification 3, pp. 5-48.

FICHET, LE CALVE G. (1984) :"Structure des principaux indices de dissimilarité sur signes de présence-absence".
Statistiques et analyse des données, Vol. 9, n° 3, pp. 11-44.

JOLY S., LE CALVE G. (1986) : "Etude des puissances d'une distance".
Statistiques et analyse des données, Vol. 11, n° 3, pp. 30-50.

SATTAH S., TVERSKY A. (1977) : "Additive similarity trees".
Psychometrika 42, pp. 319-345.

SCHOENBERG I.J. (1973) : "On certain metrice spaces arising from Euclidean spaces by a change of metric and their imbedding in Hilbert Space".
Annals of Mathematics, Vol. 38, n° 4, pp. 787-793.

Robust Bayesian Regression Analysis with HPD-Regions

K. Felsenstein and K. Pötzelberger, Wien

1. Introduction

Robust methods for inference in the linear regression model have been discussed from different points of view. In the Bayesian analysis robustness with respect to the prior distribution of the parameter plays an important role. Various papers (Chamberlain and Leamer (1976), Leamer (1982), Polasek (1984)) deal with the sensitivity of the posterior mean to changes in the prior variance. This concept has been extended to credible regions by Pōtzelberger (1986) and Polasek and Pōtzelberger (1987).

In this paper we consider the case of prior independent parameters. Let $y \in \mathbb{R}^n$ be the vector of observations and $X \in \mathbb{R}^{n \times p}$ the design matrix. We assume that

$$y = X\beta + \epsilon \tag{1}$$

for a $\beta \in \mathbb{R}^p$ and ϵ normally distributed with mean 0 and known (at least up to a constant) covariance Σ. The conjugate prior distribution for β is a p-dimensional normal with precision matrix P,

$$\beta \sim N(b_0, P^{-1}). \tag{2}$$

Since β is a location parameter the noninformative prior density is constant. This case can be identified with $P = 0$.

In this paper we focus on a problem which could be formulated in two ways. For fixed prior mean b_0 we analyze the sensitivity of the posterior distribution to changes of the prior distribution. The second more practical formulation of the problem is the specification of P. It is easier to assess the prior location b_0 than to report the complete covariance. Therefore we try to use methods of inference which do not depend on or are insensitive to changes in the prior covariance matrix.

2. Bayesian Inference - Feasible Sets

Throughout the paper we assume $\Sigma = \sigma^2 I$. The prior distribution (2) leads to a normal posterior distribution with mean

$$b(P) = (X^\top X + \sigma^2 P)^{-1}(X^\top y + \sigma^2 P b_0) \qquad (3)$$

and precision $\sigma^{-2}X^\top X + P$. $b(P)$ is a weighted matrix average of the prior mean b_0 and the m.l. estimate $\hat{\beta} = (X^\top X)^{-1}X^\top y$. $b(P) = b_0$ and $b(P) = \hat{\beta}$ correspond to the extreme situations of complete and totally missing prior information. Any other prior precision gives a $b(P)$ "between" b_0 and $\hat{\beta}$. Given b_0, $\hat{\beta}$ and the information matrix $X^\top X$, F, the closure of the set of possible posterior means forms an ellipsoid.

$$\beta \in F \Leftrightarrow (\beta - m)^\top X^\top X(\beta - m) \leq (m - b_0)^\top X^\top X(m - b_0) \qquad (4)$$

with $m = \frac{1}{2}(\hat{\beta}+b_0)$. The size of F depends upon the distance between $\hat{\beta}$ and b_0 and upon $X^\top X$. Each point in the interior of F is a reasonable estimate for β if P is unspecified. Especially for large p the size of F tends to increase considerably. If partial prior information is available then we replace F by some smaller feasible set. Assume P to lie between two positive definite matrices $P_1 \leq P_2$ the feasible set of posterior means is again an ellipsoid contained in F.In the case $P = \lambda P_0$, P_0 fixed, $\lambda > 0$ the feasible set reduces to a curve, the so called "curve decolletage".

A-priori independent parameters are described by a diagonal prior precision. The feasible set is contained in the convex hull of the 2^p constrained least-square estimates. Each $I \subseteq \{1, \ldots, p\}$ gives a constraint of the form $\beta_i = (b_0)_i$ for $i \in I$.(For details see Chamberlain and Leamer (1976).)

3. Robust Credible Sets - Hi-Fi Regions

Credible sets reflect more information about the parameter than a single estimate does. Bayesians consider regions of highest posterior density (HPD-regions). Let $\pi(\beta \mid y)$ denote the posterior density of β and $\alpha \in [0, 1]$ then a HPD-region with level α is

$$\Omega_\alpha = \{\beta \mid \pi(\beta \mid y) \geq c\}$$

with c such that

$$\int_{\Omega_\alpha} \pi(\beta \mid y) d\beta = \alpha \qquad (5)$$

For the linear model (1) and (2) this HPD-region is given by

$$\Omega_\alpha = \{\beta \mid (\beta - b(P))^\top (\sigma^{-2} X^\top X + P)(\beta - b(P)) \leq \chi^2_{p,\alpha}\}. \tag{6}$$

For $p = 1$, $\Omega_\alpha(P = 0)$, the HPD-region generated by the noninformative prior coincides with the union of all HPD-intervalls with varying P, in contrast to $p \geq 2$. $\Omega_\alpha(P = 0)$ is then a proper subset of the so called high fiducuary (Hi-Fi) region H_α where

$$H_\alpha := \bigcup_{P>0} \Omega_\alpha(P). \tag{7}$$

H_α can be regarded as a robust HPD-region with robustness concerning the variation of P. For $p \geq 2$ H_α is no more a genuine HPD-region.

Without loss of generality we set $b_0 = 0$. A point $z \in \mathbb{R}^p$ is in the boundary of H_α, if $z \neq 0$, $\chi^2_{p,\alpha} z^\top X^\top X z \geq 4 \mid z^\top X^\top y \mid^2$ and

$$\| z^\top X^\top X(z - \hat{\beta}) \| = \sqrt{\chi^2_{p,\,\alpha}} \, \| Xz \| \, \sigma \tag{8}$$

or $\chi^2_{p,\alpha} z^\top X^\top X z < 4 \mid z^\top X^\top y \mid^2$ and

$$z^\top X^\top y = -\frac{1}{4} \chi^2_{p,\alpha} \sigma. \tag{9}$$

We remark that the inequality preceeding (9) cannot hold for any z if $\chi^2_{p,\alpha} > 4\hat{\beta}^\top X^\top X \hat{\beta}$.

For large α, the difference between the sizes of H_α and $\Omega_\alpha(P = 0)$ becomes negligable, the quotient of the volumes of $\Omega_\alpha(P = 0)$ and H_α tends to 1 for $\alpha \to 1$ (see Pötzelberger (1986)).

4. Hi-Fi Regions for Diagonal Matrices

Robust analysis with the knowledge of prior independence of the parameter demands the variation of P in the set of diagonal matrices . If the information matrix $X^\top X$ is not diagonal the parameters given the observation are no longer independent. Therefore the assumption $X^\top X = diag(m_1, \ldots, m_p)$ seems reasonable. Diagonalization of $X^\top X$ is equivalent to change to orthogonal regressors. If the controlled factors can be adjusted diagonal information matrices frequently result from optimal continuous designs .(See Felsenstein (1988).)

We are now going to describe the Hi-Fi region H_α for diagonal prior precision matrices and diagonal information matrix. For a point $\beta = (\beta_1, \ldots, \beta_p)^\top$ in the boundary of H_α we express β_1 as a function of $\beta_2, \ldots, \beta_p$. Let $\phi(\beta_2, \ldots, \beta_p)$ denote the function given by

$$\phi(\beta_2, \ldots, \beta_p) = \frac{\chi^2_{p,\alpha}}{m_1} - 2 \sum_{|\beta_i| \leq |\hat{\beta}_i|} \frac{m_i}{m_1}(|\beta_i \hat{\beta}_i| - \beta_i \hat{\beta}_i)$$

$$- \sum_{|\beta_i| > |\hat{\beta}_i|} \frac{m_i}{m_1}(\beta_i - \hat{\beta}_i)^2. \tag{10}$$

Let $B = \{(\beta_2, \ldots, \beta_p)^\top \in \mathbb{R}^{p-1} \mid \phi(\beta_2, \ldots, \beta_p) \geq 0\}$, $B^+ = \{(\beta_2, \ldots, \beta_p)^\top \in B \mid \phi(\beta_2, \ldots, \beta_p) \leq 4\hat{\beta}_1^2\}$ and $B^- = B \backslash B^+$. A β_1 exists such that $\beta = (\beta_1, \beta_2, \ldots, \beta_p)^\top$ lies in the boundary of H_α, iff $(\beta_2, \ldots, \beta_p) \in B$. In fact there exist two β_1 for each $(\beta_2, \ldots, \beta_p)^\top \in B$. One point is given by $\beta_1 = \hat{\beta}_1 + \sqrt{\phi}$ if $\hat{\beta}_1 \geq 0$ and $\beta_1 = \hat{\beta}_1 - \sqrt{\phi}$ if $\hat{\beta}_1 < 0$. If $(\beta_2, \ldots, \beta_p)^\top \in B^+$, then the second point is given by $\beta_1 = -\phi/4\hat{\beta}_1$. If $(\beta_2, \ldots, \beta_p)^\top \in B^-$, then the first component of the other point is $\beta_1 = \hat{\beta}_1(1 - \frac{\sqrt{\phi}}{|\hat{\beta}_1|})$. If in addition $(\beta_2, \ldots, \beta_p)^\top$ lies in the boundary of B , $\phi(\beta_2, \ldots, \beta_p) = 0$, then the line between both points belongs to the boundary of H_α.

5. Example

For purposes of illustration, we apply the concept of robust regions to a simple linear regression model. The yearly data (1955-84) of the real consumption C and the real income x are used for estimation of the simple Austrian consumption function $C = \beta_1 + \beta_2 x$. After a transformation of the data to the first differences of the logarithms (growth rates) the m.l. estimate is $\hat{\beta}^\top = (1.43; 0.67)$ and the information matrix is

$$\sigma^{-2} X^\top X = \begin{pmatrix} 7.98 & 31.39 \\ 31.39 & 192.90 \end{pmatrix}. \tag{11}$$

The prior mean vector is supposed to be $b_0^\top = (0, 0.7)$. We change to orthogonal regressors by means of an orthogonal matrix B with $\sigma^{-2} B^\top X^\top X B = diag(2.79, 198.08)$ and the new parameter $\beta^* = B^\top \beta$. The prior assumption of independent components of β^* means that we only consider diagonal precision matrices Λ for β^* and therefore precision matrices of the form $B^\top \Lambda B$ for β. The feasible set $\tilde{F}$ consists of all posterior means $b(P)$ where $P = B^\top \Lambda B$ and Λ an arbitrary positive definite diagonal matrix. $\tilde{F}$ is a rectangle inside the feasible ellipsoid (4) with diagonal $\overline{b_0\hat{\beta}}$. The corresponding Hi-Fi region is defined by

$$\tilde{H}_\alpha := \bigcup_{\Lambda \in D^+} \Omega_\alpha(B^\top \Lambda B) \tag{12}$$

where D^+ denotes the set of positive definite diagonal matrices. In our example both $\tilde{H}_\alpha$ and the unrestricted Hi-Fi region H_α have "edges" for common values of α. If H_α contains points that fulfill equation (9) and the preceeding inequality then H_α has "edges".

Figure 1 shows the feasible set $\tilde{F}$, the Hi-Fi region $\tilde{H}_\alpha$ and the feasible ellipsoid F for $\alpha = 0.95$.

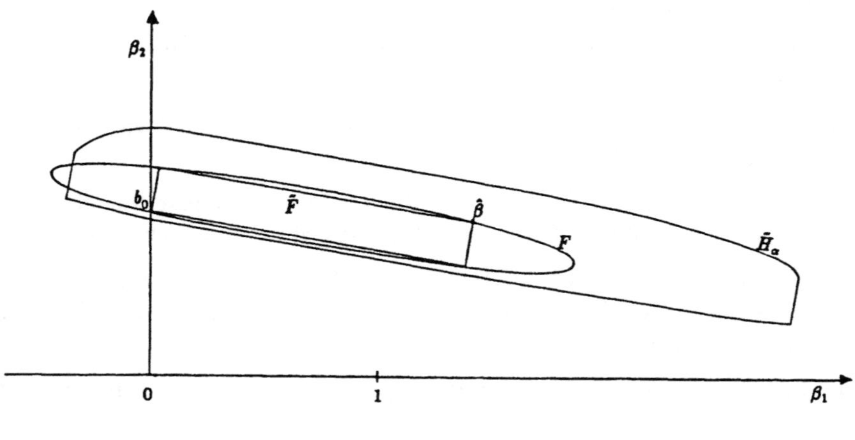

Figure 1

The size of $\tilde{F}$ is 63.5% of the size of the feasible ellipsoid F whereas the restricted Hi-Fi region $\tilde{H}_\alpha$ covers 83.8% of H_α.

6. Conclusions

The assumption of independent components of the parameter seems reasonable after orthogonalization in many practical situations. Whereas the theoretical restriction is of minor importance the afford of computation decreases essentially. The calculation of $\tilde{H}_\alpha$ is much simpler than the solution of (8) and (9) which describe H_α.

The idea of feasible sets and Hi-Fi regions can be extended to robust Bayesian decisions as Bayes tests of parameter hypothesis. Pötzelberger and Polasek (1988) used Hi-Fi regions for the Bayesian analysis of the structural change in a linear regression model. The concepts of feasible sets and Hi-Fi regions can be applied to models with autocorrelated error terms.

References

CHAMBERLAIN G. and LEAMER E. (1976), Matrix weighted Averages and Posterior Bounds, J. R. Stat. Soc. B 38, 73-84.

FELSENSTEIN K. (1988), Statistische Versuchsplanung, Research report RIS-1988-2.

LEAMER E. (1982), Sets of Posterior Means with Bounded Variance Priors, Econometrica 50, 725-36.

PÖTZELBERGER K. (1986), HPD-Regions for the Linear Model, in R. Viertl (ed.), Probability Theory and Bayesian Statistics, Plenum Press.

PÖTZELBERGER K. and POLASEK W. (1988), Robust Bayesian Analysis of a Parameter Change in Linear Regression, to appear in W. Kramer (ed.), Empirical Economics.

POLASEK W. (1984), Multivariate Regression Analysis: Estimation and Sensitivity Analysis for Two-Dimensional Data, in J. Kadane (ed.), Robustness in Bayesian Statistics, 229-309, North Holland.

POLASEK W. and PÖTZELBERGER (1987), Robust HPD-Regions in Bayesian Regression Models, submitted for publication.

Time Series

Estimation of ARMA Process Parameters and Noise Variance by Means of a Non Linear Filtering Algorithm

M. P. Muñoz-Gracia, J. Pages-Fita and M. Marti-Recober, Barcelona

KEYWORDS: Non linear filtering, ARMA(P,Q) process estimation, recursive parameter estimation.

1. INTRODUCTION

The estimation of the parameters associated with an ARMA(p,q) process can be formulated as a non linear filtering problem. Optimal filtering allows to follow the evolution of the a-posteriori probability density; all desired information about parameters can be obtained from their a-posteriori density. First of all, optimal estimators can be obtained directly from it. Secondly, this is a good test for sub-optimal algorithms, and finally, a lower bound of the simple size to get a fixed accuracy can be obtained.

The objective of the work that we present is to compute recursively the a-posteriori probability density for the parameters vector of an ARMA(p,q) process, incluiding the case where the noise variance is unknouwn. This problem was previously solved for an AR(p) process with simultaneous estimation of the unknown noise variance throught non linear filtering [1]. To determine recursive estimators of the AR(p,q) parameters and the noise variance an extended state vector is defined including as state variables the parameters to be estimated.

Through a bayesian approach to the problem, an estimate of this extended state can be

Compstat 1988

determined from the a posteriori distribution, making a grid over the extended state and computing sequentially the a-posteriori probability density for each point of the grid, by the means of numerical methods. The starting a-priori probability distribution is stablished as non-informative, say the a-priori density function constant over the parameter admissible range.

At the present, the computer power needs for the optimal non linear filtering is quite high in CPU time and in memory capacity when the number of parameters to estimate grows.

The filter's formulation and the estimation algorithm are described in section 2. In section 3 the cases of ARMA(1,1) and ARMA(2,1) models are studied and comparative results presented. In section 4 some conclussions are discussed.

2. MODEL AND ALGORITHM DESCRIPTION

Given an ARMA(p,q) process

$$y_t = a_1 y_{t-1} + \dots + a_p y_{t-p} + e_t + c_1 e_{t-1} + \dots + c_q e_{t-q} \tag{1}$$

where $\{e_t\}$ is gaussian white noise with noise variance σ^2 unknown. We define the unknown parameters vector $\emptyset' = (a_1,\dots,a_p,c_1,\dots,c_q)$, and $y^t=\{y_1,\dots,y_t\}$ is the set of all observations until time t. The a-posteriori probability density is:

$$p(\emptyset,\sigma^2/ y^t) \propto p(y^t/ \emptyset,\sigma^2)\, p(\emptyset,\sigma^2) \tag{2}$$

where $p(\emptyset,\sigma^2)$ is the prior density and

$$p(y^t/\emptyset,\sigma^2) = \frac{1}{(2\pi)^{t/2}(\det R)^{1/2}} \exp \{- \frac{1}{2} (y^t)' R^{-1} y^t \} \tag{3}$$

is the joint observations density, where R is the covariance matrix of the observations.

If we take the prior distribution non informative, then

$$p(\emptyset,\sigma^2/y^t) = k\, p(y^t/\emptyset,\sigma^2) \tag{4}$$

wher k is a normalization constant.

The a-posteriori probability density is easy to calculate using recurrent expressions that we obtained for det R using its Toeplitz structure (the well known Levinson recurrence) and the recurrence for the quadratic form $(y^t)'R^{-1}y^t$ [2], [3] is the following:

If we have the quadratic form at time t-1, we obtain the quadratic form at time t as the quadratic form at time t-1 plus a linear combination between the set of the observations until time t and the last row of the inverse covariance matrix of the observations, i.e. if we define the quadratic form at time t as f_t, then

$$f_t = f_{t-1} + \left(\sum_{i=1}^{t} y_i b_i \right)^2 \qquad (5)$$

where $[b_1,\ldots,b_t]$ is the last row of the inverse covariance matrix of the observations R^{-1} and $\{y_1,\ldots,y_t\}$ is the set of all observations until time t.

With this two recurrences for the computation of the determinant and the quadratic form, we obtain a third recurrence to the computation of the a posteriori density function at time t, $p_t(Ø,\sigma^2/y^t)$ as a function of the posteriori density obtained at time t-1, $p_{t-1}(Ø,\sigma^2/y^{t-1})$ by means of the Levinson recurrence to the determinant:

$$\det R_t = \det R_{t-1}\, D_t \qquad (6)$$

where the coeficients D_t are obtained by means of Levinson recurrence.The coeficients D_t are the diagonal elements of the Cholesky factoritzation of the covariance matrix R_t. Then

$$p_t(Ø,\sigma^2/y^t) = k'\, p_{t-1}(Ø,\sigma^2/y^{t-1})\, D_t^{-1/2} \exp\{ -\frac{1}{2} \left(\sum_{i=1}^{t} y_i b_i \right)^2 \} \qquad (7)$$

The algorithm implementation is as follows:

To obtain the value of the probability density, a (p+q+1) dimensional grid over the space of poles, zeros and noise variance of the ARMA process is considered. For each point of the grid, the value of the a posteriori probability density is numerically sequentially computed and also marginal probability density functions are computed and ploted. The grid is at each time adapted to obtain better accuracy and occupy a few memory capacity as possible.The points, that have associated a probability lower than a certain tolerance value, are eleminated. Also the total number of points in each dimension must be higher to a prefixed value, therefore if this total number is little than this prefixed value, we reduce the interval of the discretisation.

3. SOME RESULTS

This algorithm presents an advance with respect to the algorithm used to obtain a non linear filtering by means the state space representation. The memory requirement and the CPU time are reduced.

Severels models have been tested and we present here the results obtained for the parameters of an ARMA(1,1) model with variance noise unknown . In this case the dimension of the grid is three. Several time series have been simulated with the coeficients shown in Table 1, where the notations are:

a_1^*= Conditional mean after N observations, calculated from marginal density function of coefficient a_1

sa_1^*= Standard deviation of coefficient a_1 after N observations

c_1^*= Conditional mean after N observations, calculated from marginal density function of coefficient c_1

sc_1^*= Standard deviation of coefficient c_1 after N observations

s^2= Conditional mean after N observations, calculated from marginal density function of σ^2.

s_s2= Standard deviation of coefficient σ^2 after N observations.

In the following table, we denote by (1) the subtable that has the results obtained by means of the a-posteriori density (7) and by (2) the results obtained through the statistical package BMDP2T.

MODEL	N=50 (1)		N=100 (1)		N=100 (2)	
$y_t=0.5y_{t-1}+e_t+0.8e_{t-1}$	$a_1^*=0.56$	$sa_1^*=0.14$	$a_1^*=0.48$	$sa_1^*=0.11$	$a_1^*=0.43$	$sa_1^*=0.10$
	$c_1^*=0.52$	$sc_1^*=0.13$	$c_1^*=0.65$	$sc_1^*=0.09$	$c_1^*=0.75$	$sc_1^*=0.07$
	$s^2 = 0.56$	$s_s2= 0.14$	$s^2 = 0.48$	$s_s2 = 0.11$	$s^2 =1.03$	
$y_t=0.5y_{t-1}+e_t+0.1e_{t-1}$	$a_1^*=0.46$	$sa_1^*=0.22$	$a_1^*=0.50$	$sa_1^*=0.15$	$a_1^*= 0.48$	$sa_1^*=0.17$
	$c_1^*=0.05$	$sc_1^*=0.20$	$c_1^*=0.06$	$sc_1^*=0.18$	$c_1^*= 0.04$	$sc_1^*=0.19$
	$s^2 = 1.51$	$s_s2= 0.31$	$s^2 = 1.12$	$s_s2 = 0.13$	$s^2 = 1.04$	
$y_t=0.5y_{t-1}+e_t-0.6e_{t-1}$	$a_1^*=-0.21$	$sa_1^*=0.40$	$a_1^*=-0.06$	$sa_1^*=0.36$	$a_1^*= 0.14$	$sa_1^*=0.38$
	$c_1^*=-0.1$	$sc_1^*=0.40$	$c_1^*=-0.2$	$sc_1^*=0.37$	$c_1^*=-0.39$	$sc_1^*=0.35$
	$s^2 = 1.12$	$s_s2= 0.22$	$s^2 = 1.13$	$s_s2 = 0.13$	$s^2 = 1.02$	
$y_t=0.1y_{t-1}+e_t+0.8e_{t-1}$	$a_1^*=0.24$	$sa_1^*=0.17$	$a_1^*=0.20$	$sa_1^*=0.13$	$a_1^*= 0.14$	$sa_1^*=0.11$
	$c_1^*=0.83$	$sc_1^*=0.09$	$c_1^*=0.81$	$sc_1^*=0.08$	$c_1^*= 0.93$	$sc_1^*=0.05$
	$s^2 = 1.12$	$s_s2= 0.25$	$s^2 = 1.13$	$s_s2 = 0.12$	$s^2 = 0.99$	

TABLE 1

Several ARMA(2,1) time series have been simulated and the numeric results for the coefficients are presented in table 2. In this case the dimension of the grid is three, one for each coefficient. The dimensions of the autoregressive coefficients are not independent because they must satisfy the following restrictions in the stationary models:
$-1<a_1<1$, $a_2+a_1<1$ and $a_2-a_1<1$. The notations in Table 2 are:

a_1^*= Conditional mean after N observations, calculated from marginal density function of coefficient a_1

sa_1^*= Standard deviation of coefficient a_1 after N observations

a_2^*= Conditional mean after N observations, calculated from marginal density function of coefficient a_2

sa_2*= Standard deviation of coefficient a_2 after N observations

c_1*= Conditional mean after N observations, calculated from marginal density function of coefficient c_1

sc_1*= Standard deviation of coefficient c_1 after N observations

Also, in the following table, we denote by (1) the subtable that has the results obtained by means the a-posteriori density (7) and by (2) the results obtained by means the statistic package BMDP2T.

MODEL	N=50 (1)		N=100 (1)		N=100 (2)	
y_t=-0.4y_{t-1}+0.32y_{t-2}+ e_t-0.8e_{t-1}						
(one pole=-0.8,	a_1*=-0.37	sa_1*=0.27	a_1*=-0.22	sa_1*=0.14	a_1*=-1.12	sa_1*=0.06
the other=0.4,	a_2*= 0.20	sa_2*=0.22	a_2*= 0.34	sa_2*=0.13	a_2*=-0.14	sa_2*=0.11
the zero=0.8)	c_1*=-0.69	sc_1*=0.27	c_1*=-0.85	sc_1*=0.12	c_1*=-1.02	sc_1*=0.00
y_t=-0.3y_{t-1}+0.4y_{t-2}+ e_t+0.7e_{t-1}						
(one pole=-0.8,	a_1*= 0.34	sa_1*=0.47	a_1*= 0.45	sa_1*=0.46	a_1*= 0.78	sa_1*=0.86
the other=0.5,	a_2*= 0.14	sa_2*=0.27	a_2*=-0.01	sa_2*=0.26	a_2*=-0.21	sa_2*=0.45
the zero=-0.7)	c_1*= 0.13	sc_1*=0.46	c_1*= 0.11	sc_1*=0.44	c_1*=-0.20	sc_1*=0.88
y_t=-0.3y_{t-1}+0.4y_{t-2}+ e_t+0.1e_{t-1}						
(one pole=-0.8,	a_1*=-0.12	sa_1*=0.15	a_1*=-0.33	sa_1*=0.10	a_1*=-0.40	sa_1*=0.14
the other=0.5,	a_2*= 0.66	sa_2*=0.11	a_2*= 0.52	sa_2*=0.08	a_2*=0.50	sa_2*=0.11
the zero=-0.1)	c_1*=-0.05	sc_1*=0.23	c_1*= 0.20	sc_1*=0.14	c_1*= 0.30	sc_1*=0.16
y_t=0.4y_{t-1}+0.05y_{t-2}+ e_t+0.8e_{t-1}						
(one pole=-0.1,	a_1*= 0.71	sa_1*=0.24	a_1*= 0.60	sa_1*=0.14	a_1*= 0.59	sa_1*=0.13
the other=0.5,	a_2*=-0.14	sa_2*=0.22	a_2*=-0.01	sa_2*=0.14	a_2*= 0.00	sa_2*=0.13
the zero=-0.8)	c_1*= 0.70	sc_1*=0.21	c_1*= 0.75	sc_1*=0.11	c_1*= 0.80	sc_1*=0.09
y_t=0.9y_{t-1}-0.2y_{t-2}+ e_t+0.8e_{t-1}						
(one pole=-0.4,	a_1*= 1.03	sa_1*=0.18	a_1*= 0.99	sa_1*=0.11	a_1*= 1.01	sa_1*=0.13
the other=0.5,	a_2*=-0.26	sa_2*=0.17	a_2*=-0.27	sa_2*=0.11	a_2*=-0.27	sa_2*=0.13
the zero=-0.8)	c_1*= 0.64	sc_1*=0.15	c_1*= 0.63	sc_1*=0.09	c_1*= 0.64	sc_1*=0.10

TABLE 2

4. CONCLUSIONS

Right now, the optimal non linear filtering must be regarded as a theoretical tool, because the CPU time is proportional to the product between the square of the number of observations to treat, by the number of the points of discretitzation of the a-posteriori density elevated to the number of parameters to be estimated.

The practical implementation of the optimal estimation algorithms of ARMA(P.Q) parameters has been done in the VAX-8600 computer . This computer performs 4.6 Mips. The CPU time spent in the estimation of the ARMA(1,1) parameters with 25 observations is 0.40 seconds and that used for the same problem with 100 observations is 8 seconds.

The experience in optimal estimation for the ARMA(1,1) problem with unknown noise variance shows that the precision essentialy depends on the absolute value of the pole and zero and on the distance between them. In this case, we also see that the estimation of the noise variance is little influenced by the estimation of the autoregressive and moving average coefficients of the model.

It has been also observed that for the ARMA(2,1) problem, the precision essentialy depends on the absolute value of the poles and zeros and the distance between them.

REFERENCES

[1] MUÑOZ-GRACIA, M.P., EGOZCUE-RUBI, J.J., MARTI-RECOBER, M. (1987): "Estimació del pol i de la variància del soroll d'un model AR(1) utilitzant un filtre no lineal".QUESTIÓ (pending publication).
[2] MUÑOZ-GRACIA, M.P., PAGES-FITA, J., MARTI-RECOBER, M. (1987): "An exact algorithm for the recursive calcul of a normalized likelihood function". 18th European Meeting of Statisticians. Berlin. August 1988, accepted for presentation.
[3] MUÑOZ-GRACiA, M.P. (1988): "Estimacó dels paràmetres de models ARMA(p,q) mitjançant algorismes de filtratge óptim" Tesi Doctoral. Universitat Politècnia de Catalunya. Spain.

Autoregressive Models with Latent Variables

P. H. C. Eilers, Schiedam

Abstract. Observations of time series often are corrupted by noise. This has a negative influence on the quality of estimated parameters of autoregressive models. A new technique is presented here. The individual, not directly observable, samples of the corrupted time series are modeled and estimated as latent variables. The estimation-process leads to the iterative solution of two non-linearly coupled systems of linear equations. Adaptations for missing values, additive and innovative outliers are described. Some applications are presented.

1 Introduction

The autoregressive (AR) model is used frequently for time series of dependent observations. For the estimation of it's parameters the methodology of Box and Jenkins (1976), based on autocorrelations, or direct multiple regression are commonly used. It is well known that the resulting estimates are sensitive to additive noise, missing observations and outliers.

In this paper an alternative approach is presented. It is assumed that a given series consists of corrupted observations of a latent series, generated by an AR-process. Both the latent series and it's AR-parameters are estimated. At first sight this might seem paradoxical: from a series of, say, m observations $m + n$ unknowns are estimated, if n is the order of the AR-model. However, with a suitably chosen objective function the ill-conditioned problem is regularized. The estimated latent variables are not independent — but they should not be, as they follow an autoregressive pattern.

The theoretical foundation is described in section 2. From the optimization of a quadratic objective function follow two systems of nonlinearly coupled linear equations. These can be solved iteratively for AR-parameters and latent variables. The amount of computation is roughly proportional to mn^2.

By introducing (adaptive) weights, provisions can be made easily for missing observations, extrapolation and additive or innovative outliers. In section 3 some applications to simulated and empirical time series are presented.

Compstat 1988

2 Theoretical foundation

2.1 The objective function of an LAR(2)-model

Consider as an example the AR(2) process $\mathbf{x}$ and it's observations $\mathbf{y}$:

$$\mathbf{x}_i = b_1\mathbf{x}_{i-1} + b_2\mathbf{x}_{i-2} + \mathbf{e}_i; \quad \mathbf{y}_i = \mathbf{x}_i + \mathbf{u}_i.$$

Expansion to a higher order process is straightforward, but complicates the algebra. The innovations $\mathbf{e}_i$ and the additive noise $\mathbf{u}_i$ are independent with expectation zero. The length of the observed series is m.

We wish to obtain estimates of x_i, denoted by z_i, and estimates of b_1 and b_2, denoted by a_1 and a_2. The series z has to conform well to an AR(2) process with parameters a_1 and a_2. It also should differ as little as possible from the observed series y. Both objectives can be combined in one function to be minimized:

$$S = \sum_{i=1}^{m}(y_i - z_i)^2 + \beta\sum_{i=3}^{m}(z_i - a_1z_{i-1} - a_2z_{i-2})^2. \tag{1}$$

How to choose β will be discussed below; for the moment we assume it to have a suitable (positive) value. Note that (1) is a weighted sum of the variances of estimated additive and innovative noise. A model with this objective function will be called a latent autoregressive (LAR) model.

Instead of optimizing (1), a more general formulation will be considered, by giving weights to the individual terms. The purpose of these weights will be explained below. Thus (1) changes to

$$S = \sum_{i=1}^{m} w_i(y_i - z_i)^2 + \beta\sum_{i=3}^{m} v_i(z_i - a_1z_{i-1} - a_2z_{i-2})^2. \tag{2}$$

The minimization of S leads to the following system of equations for a_1 and a_2:

$$a_1\sum_{i=3}^{m} v_iz_{i-1}^2 + a_2\sum_{i=3}^{m} v_iz_{i-1}z_{i-2} = \sum_{i=3}^{m} v_iz_iz_{i-1}; \tag{3}$$

$$a_1\sum_{i=3}^{m} v_iz_{i-1}z_{i-2} + a_2\sum_{i=3}^{m} v_iz_{i-2}^2 = \sum_{i=3}^{m} v_iz_iz_{i-2}. \tag{4}$$

These are seen to be the normal equations for regression of z_i on z_{i-1} and z_{i-2}. For the series z we find the following equations:

$$c_i''z_{i-2} + c_i'z_{i-1} + c_0z_i + c_{i+1}'z_{i+1} + c_{i+2}''z_{i+2} = w_i(y_i - z_i)/\beta, \tag{5}$$

where

$$c_i'' = -a_2v_i; \quad c_i' = -a_1v_i + a_1a_2v_{i+1}; \quad c_i = v_i + a_1^2v_{i+1} + a_2^2v_{i+2}. \tag{6}$$

For given values of a_1 and a_2 this is a banded system of linear equations. Several efficient solution methods exist for such systems, see e.g. Pissanetzky (1984).

The equations (3–4) and (5) are solved iteratively. Generally it works well to start by solving (5) with $a_1 = 2$ and $a_2 = -1$. The number of iterations needed is generally from 10 to 20.

It is not necessary that the observed series is — or seems to be — stationary. No prior manipulations with first or second differences are needed. The parameters a_1 and a_2 can take on values that give a nonstationary AR-model, but this is no problem for the description of a time series of finite length.

2.2 Missing values and outliers

The weights w_i in (2) can be used for missing observations: if observation i is missing, put $w_i = 0$ and give y_i an arbitrary value. These weights can also be made adaptive, to reduce the influence of outliers. A useful choice is

$$w_i = \left(\frac{p^2}{p^2 + (y_i - z_i)^2}\right)^q. \qquad (7)$$

with $q = 1$ (Kappenman, 1986) or $q = 2$ (Eilers, 1987). The value of p determines the range of influence of extreme observations.

One can use $w_i = 0$ with great flexibility. Even large gaps in the data can be interpolated. To generate forecasts or "back"-casts one can introduce "missing observations" at both ends of the series.

Normally $v_i \equiv 1$ for $i = 3 \ldots m$ and zero elsewhere, but where innovation-outliers are suspected, one can put $v_i = 0$. This allows $z_i - a_1 z_{i-1} - a_2 z_{i-2}$, the estimate of the innovation e_i, to take on large values locally.

2.3 An optimal value for β

The value of parameter β has a strong influence on the results of the LAR-model: if β is too high, a latent series will be found that conforms excellently to an AR-model, but diverges strongly from the observed series, while too low a value of β makes no essential difference with direct estimation of the AR-parameters. One way to determine an "optimal" value, is to measure the predictive performance of the model.

A systematic procedure is cross-validation. A part of the observations is set apart and the model is fitted to the rest of them. Then values for the left-out observations are predicted and compared to their real values. This is repeated for a number of patterns of left-out observations. An extreme case is to leave out each of the m observations in turn. It is clear that a large amount of computation can be required. It was found however that a simple procedure performs reasonably.

From the latent variables z_{i-1} and z_{i-2} follows a "prediction" $\hat{y} = a_1 z_{i-1} + a_2 z_{i-2}$. It is not a real prediction, because all y_i are used in the estimation of the model. Nevertheless we treat it as such and compare it to the observed value y_i. We do this for each i and use squared differences to form

$$T = \sqrt{\sum_{i=3}^{m} v_i w_i (y_i - a_1 z_{i-1} - a_2 z_{i-2})^2 / \sum_{i=3}^{m} v_i w_i}. \qquad (8)$$

The weights w_i and v_i are introduced here because a prediction cannot be judged if y_i is missing ($w_i = 0$) or an innovation-outlier is suspected ($v_i = 0$). The idea is to compute T for a number of values of β and choose the one that gives the lowest value of T. In practice this works reasonably well, especially if the amount of additive noise is relatively high. Of course it should not be considered to be more than a useful heuristic.

2.4 Extension to a higher order model

The LAR(2)-model was used as an illustration, but extension to a higher order is straightforward. In (2) $z_{i-3} \ldots z_{i-n}$ are introduced for models with $n > 2$. The system (3–4) will

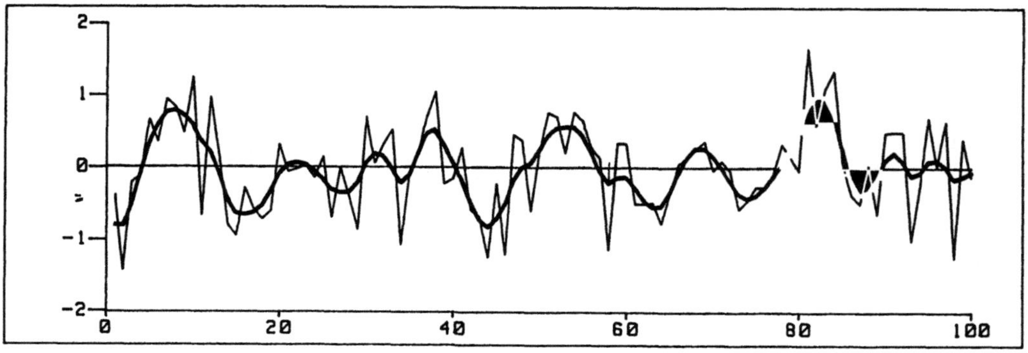

Figure 1: Simulated series (thin line) and estimated latent series (thick line).

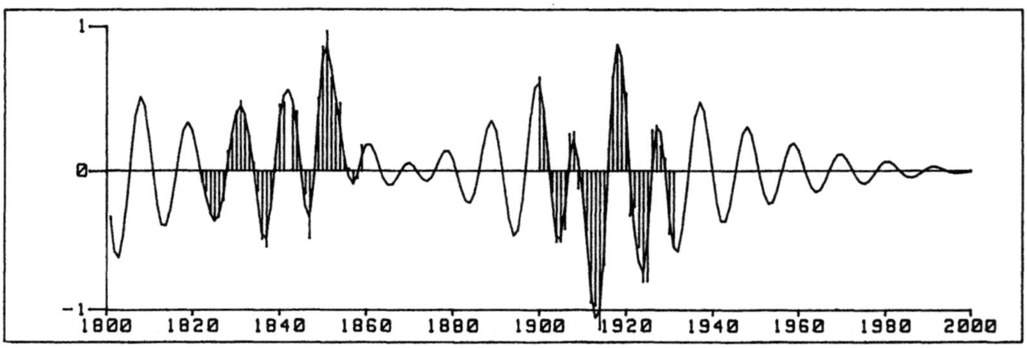

Figure 2: Logarithms of counts of Lynx-furs. Observed series as thin line, estimated latent series as thick line.

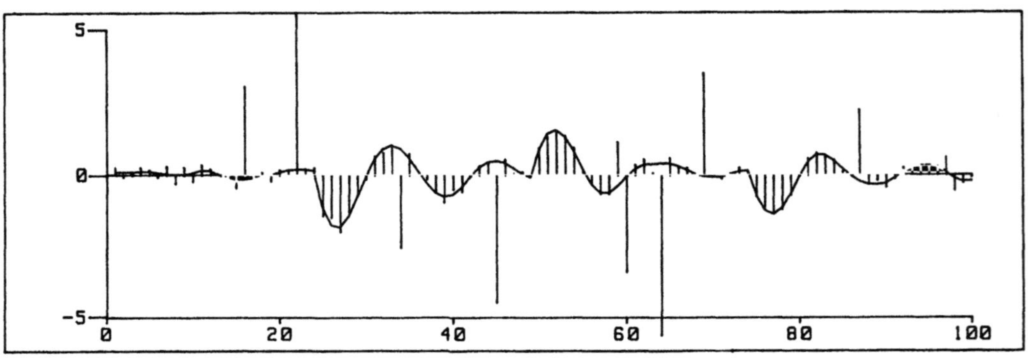

Figure 3: Simulated series with additive and innovative outliers (thin line) and estimated latent series (thick line).

consist of n equations and the system (5) has bandwidth $2n+1$. The storage requirements are of the order of mn locations for floating-point numbers; the number of arithmetical operations is of the order mn^2. The equation for T is changed in an obvious way.

3 Applications

First we consider a simulated series, as shown in figure 1. A series was simulated with an AR(2) process, with $b_1 = 1.7$, $b_2 = -0.85$ and $m = 100$. The innovations come from a normal distribution with mean zero and standard-deviation 0.1, denoted as $N(0, 0.1)$. Pseudo-random numbers from $N(0, 0.5)$ were added, to give the series that is plotted as a thin line in figure 1. Five values of β were tried (0.1, 0.3, 1, 3, 10). The lowest value of T occured at $\beta = 3$ with $a_1 = 1.639$ and $a_2 = -0.850$. After 6 iterations the values of a_1 and a_2 changed no more than 0.001.

An example of inter- and extrapolation is given in figure 2. The data, plotted as a thin line, are based on the number of Lynx-furs traded in the West district of the Hudson Company (Elton and Nicholson, 1942). Logarithms (base 10) were taken of the original counts and then the mean value was subtracted. Data are given for 1821–1934 with a gap from 1857 to 1896. To illustrate extrapolation, "missing values" were introduced for 1801–1820 and for 1935–2000. These are extreme extrapolations, but they serve well to illustrate the behaviour of the LAR-model. The value of β is 1; $a_1 = 1.608$ and $a_2 = -0.924$.

Finally, figure 3 presents additive and innovative outliers. The data were generated by an AR(2) process, with $b_1 = 1.6$ and $b_2 = -0.85$. The innovations come from $N(0, 0.1)$, but the values 1, -1 and 1 were used at $i = 25, 50, 75$. The additive noise is from $N(0, 0.2)$, but 10 additive outliers were placed randomly, their values chosen from $N(0, 5)$. The positions of the innovation-outliers were assumed to be known, thus $v_{25} = v_{50} = v_{75} = 0$ was used. Adaptive weights were used as specified by (7) with $p = 2$ and $q = 2$. The optimal value of β is 3; $a_1 = 1.539$ and $a_2 = -0.800$.

4 Further developments

Because of the limited space, only a rough sketch can be given of some further developments of the LAR-model.

The first part of S, that measures the fit to the time series, can be generalized in several ways. Any robust alternative to the sum of squared differences might be introduced. Also the deviance of a generalized linear model can be used (McCullagh and Nelder, 1983). This opens a broad perspective on the modeling of series of counts, binary or otherwise non-normally distributed observations. Consider as an example a series of Poisson-distributed observations y_i. Let the expectation of y_i be $\mu_i = \exp(z_i + t)$, where both z_i and t are to be estimated. Then S becomes

$$S = 2\sum_{i=1}^{m} w_i(y_i z_i - e^{z_i+t}) + \beta \sum_{i=3}^{m} v_i(z_i - a_1 z_{i-1} - a_2 z_{i-2})^2. \tag{9}$$

This also sheds another light on the LAR-model: it is a special case of penalized maximum likelihood estimation.

The LAR-model can be extended to input-output modeling. If an input series x_i is given, the second part of S can be formulated as $\sum v_i(z_i - a_1 z_{i-1} - a_2 z_{i-2} - c_0 x_i + c_1 x_{i-1} - \ldots)^2$, where $c_0, c_1, \ldots$ are parameters to be estimated. A similar extension leads to the latent autoregressive moving average (LARMA) model:

$$S = \sum_{i=3}^{m} w_i (y_i - z_i - c_1 z_{i-1} - c_2 z_{i-2})^2 + \beta \sum_{i=3}^{m} v_i (z_i - a_1 z_{i-1} - a_2 z_{i-2})^2. \tag{10}$$

The parameters c_1 and c_2 account for the moving average part of the model.

The foundations of the model need further investigation. It seems probable that S can have local minima. If that is the case, what are save starting values for the iterations? In some cases convergence is slow and one might like to have faster algorithms for the minimization of S.

The ad-hoc approach to choosing an optimal value of β needs improvement. Straightforward cross-validation is expensive, but it seems possible that some of the fast algorithms from the literature (Akaike, 1984; Hutchinson and de Hoog, 1985) can be adapted to this problem.

References

- Akaike H. (1984). On the Use of Bayesian Models in Time Series Analysis. In: *Robust and Nonlinear Times Series Analysis.* Franke J., Härdle W. and Martin D. (eds.). Springer.
- Box G. E. P. and Jenkins G. M. (1976). *Time Series Analysis, Forecasting and Control.* Holden-Day.
- Eilers P. H. C. (1987). Adaptive Weights, an Explorative Technique for Outliers and Mixtures of Regression Models. (In Dutch) *Kwantitatieve Methoden* **23**, 63-84.
- Elton C. and Nicholson M. (1942). The Ten-Year Cycle In Numbers of the Lynx in Canada. *J. Animal Ecology* **1**, 215-244.
- Hutchinson M. F. and de Hoog F. R. (1985). Smoothing Noisy Data with Spline Functions. *Numerische Mathematik* **47**, 99-106.
- Kappenman R. F. (1986). Adaptive M-Estimation of Symmetric Distribution Location. *Commun. Statist.—Theor. Meth.* **15**, 2935-2951.
- Kitagawa G. (1983). State Space Modeling of Nonstationary Time Series and Smoothing of Unequally Spaced Data. In: *Time Series Analysis of Irregularly Observed Data*, Parzen E. (ed.). Springer.
- McCullagh P. and Nelder J. A. (1983). *Generalized Linear Models.* Chapman and Hall.
- Pissanetzky S. (1984). *Sparse Matrix Technology.* Academic Press.

An Algorithm for Time Series Decomposition Using State-Space Models with Singular Transition Matrix

E. Alvoni, Bologna

Key words: time series, seasonal adjustment, likelihood, AIC, optimal filtering, Kalman filters, smoothers.

Introduction

Recently, the problem of seasonal adjustment of time series has received considerable attention by several authors, see e.g. Bell and Hillmer (1984) for a historical review with analysis and discussion of current issues.

Among others, Kitagawa and Gersch (1981,1984) formulated a method for the decomposition of time series in trend and seasonality by introducing a state-space representation. In their procedure, the trend and seasonal components are each characterized by stocastically perturbed difference equations. The perturbations are independent of each other, have zero means and unknown variances. These variances, with the variance of the noise on the observation , (the hyperparameters of the model), are estimated over a discrete likelihood parameter space using Akaike's AIC criterion to select the best of the alternative models fitted to the data. The problem was numerically solved using models with a non singular state transition matrix and a Kalman filter-fixed-interval smoother for the evaluation of the likelihood of the models and state estimations were employed.

Lately, much interest has been taken in filtering algorithms that don't require the inversion of the transition matrix, with particular reference to the solution of time-delay systems. In this area, Watanabe (1986) developed a new U-D factorized smoother which is proved to be numerically stable, efficient and reliable.

Compstat 1988

In this paper we use a model class with a singular transition matrix and we discuss the applicability of Watanabe's smoother.

The convergence of the algorithm is investigated in the light of recent papers by Chan, Goodwin, Sin (1984) and De Souza, Gevers, Goodwin (1986), and a few implementational details are given.

A model class with singular transition matrix

An observed time series (y_h), $h=1,\dots,N$ is to be decomposed into

$$y_h = t_h + s_h + i_h \tag{1}$$

where t_h denotes the trend, s_h the seasonal and i_h the irregular component, and it is assumed that $i_h \sim NID(0,\xi^2)$.

Alternative time series model classes formulated by Kitagawa are characterized by alternative subsets of the constraint equations on the components.

Now, let's assume that the trend satisfies a kth-order stocastically perturbed difference equation

$$\nabla^k t_h = w_{1h}\ , \qquad w_{1h} \sim NID(0,\tau_1^2) \tag{2}$$

where the difference operator ∇ is defined by $\nabla t_h = t_h - t_{h-1}$ and that the seasonal component satisfies the following relations

$$\sum_{i=0}^{p-1} s_{h-i} = w_{2h}\ , \qquad w_{2h} \sim NID(0,\tau_2^2) \tag{3}$$

$$s_h - s_{h-p} = w_{3h}, \qquad w_{3h} \sim NID(0,\tau_3^2) \tag{4}$$

where p is the seasonal period, usually 4 or 12.

Then, the model can be written compactly as

$$x_h = F\, x_{h-1} + G\, w_h \tag{5}$$

where $x_h=(t_h,t_{h-1},\dots,t_{h-k+1},s_h,s_{h-1},\dots,s_{h-p+1})^T$, $w_h=(w_{1h},w_{2h},w_{3h})^T$,

$$F=\begin{array}{c}\begin{bmatrix} F_t & \\ & F_s \end{bmatrix}\\ \underbrace{\hphantom{F_tF}}_{k}\underbrace{\hphantom{F_s}}_{p}\end{array},\ F_t=\begin{bmatrix} c_1 & c_2 & \dots & c_k \\ 1 & & & \\ & 1 & & \\ & & \ddots & \\ & & 1 & 0 \end{bmatrix},\ F_s=\begin{bmatrix} 0 & 0 & 0 & \dots & 1 \\ 0 & -1 & -1 & \dots & -1 \\ 0 & 1 & & & \\ & & 1 & & \\ & & & \ddots & \\ & & & 1 & 0 \end{bmatrix},\ G=\begin{bmatrix} 1 & & \\ & 1 & \\ & & 1 \end{bmatrix}\begin{array}{l}\left.\vphantom{\begin{matrix}1\\1\end{matrix}}\right\} k\\ \left.\vphantom{\begin{matrix}1\\1\end{matrix}}\right\} p\end{array}$$

with $c_1 + c_2 + \dots c_k = 1$, while the equation (1) can be written as

$$y_h = H\, x_h + i_h\ , \tag{6}$$

with $H=[1\ 0 \ldots 1\ 0 \ldots 0]$; $\{w_h, i_h\}$ are zero-mean processes with covariances $E\{[{w_h \atop i_h}][w_l^T\ i_l^T]\} = \begin{bmatrix} Q & 0 \\ 0 & R \end{bmatrix} \delta_{hl}$, in which E denotes the expectation, δ_{hl} denotes Kronecker delta, $Q = \mathrm{diag}(\tau_1^2, \tau_2^2, \tau_3^2)$, $R = [\xi^2]$.

Numerical solution

The transition matrix F of the system (5) turns out to be singular and has eigenvalues on the unit circle. Indeed, the spectrum of F is given by $\sigma(F) = \sigma(F_t) \cup \sigma(F_s) = \{1\} \cup \{0, -1, \mu_1, \mu_2, \ldots, \mu_{p-2}\}$, where the eigenvalue 1 of F_t has algebraic multiplicity k and the p-2 eigenvalues μ_i, i = 1.., p-2, are the p-2th roots of -1.

We note that the system, or the pair (F, H^T), we are considering, is detectable , that is, $u^*H=0$ and $u^*F^T=\lambda u^*$ for some constant λ implies $|\lambda| < 1$ or $u=0$, (u^* denotes the conjugate traspose of u).

In optimal filtering, the algorithm converges if there exists a limiting solution P_s of the solutions of matrix Riccati difference equation (RDE). If this solution exists, then it satisfies the algebraic Riccati equation (ARE) and P_s is a strong solution.

Chan, Goodwin, Sin (1984) studied the convergence of the solutions of RDE for systems that, like our trend and seasonal models, have zeros on the unit circle. Later, De Souza et al.(1986) generalized the above-mentioned results for the case of singular state transition matrix. They proved, in paricular, that if the system is detectable then there exists a unique solution of ARE [see their theorem 3.2, pg.833] and, under a suitable condition on the initial covariance, the solutions of RDE converge to the strong solution of the ARE [see their theorem 4.2 , pg.834]. An algorithm for the solution of the ARE for system with a singular transition matrix can be found in Pappas et al.(1980) [see "real Shur vector approach", pg.638].

These results provide the theoretical basis for the applicability of filters. The numerical solution to our problem needs algorithms which don't require the inversion of the transition matrix.

Watanabe (1986) developed a new forward-pass fixed-interval smoother that utilizes bacward-pass information filter outputs which can be described as in the following recursions (7)-(15);

The subscripts "B" and "F" are used to indicate estimations and associated estimation-error covariances computated by backward or forward algorithm, respectively; we denote by $\hat{x}_{h|h-1} := E[x_h \mid y_1,\dots,y_{h-1}]$, $\hat{x}_{h|h} := E[x_h \mid y_1,\dots,y_h]$, $\hat{x}_{h|N} := E[x_h \mid y_1,\dots,y_N]$, $P_{h|h-1} := E\{[x_h - \hat{x}_{h|h-1}][x_h - \hat{x}_{h|h-1}]^T\}$, $P_{h|h} := E\{[x_h - \hat{x}_{h|h}][x_h - \hat{x}_{h|h}]^T\}$ $P_{h|N} := E\{[x_h - \hat{x}_{h|N}][x_h - \hat{x}_{h|N}]^T\}$, $S := P^{-1}$.

Forward smoother

For $h = 0,N-1$

$$_F\bar{x}_{h+1|N} = F\ _F\hat{x}_{h|N} \tag{7}$$

$$_F\hat{x}_{0|N} = [I + P_0\ _BS_0]^{-1}\ [P_0\ _Bq_0 + \hat{x}_0]$$

$$_F\hat{x}_{h+1|N} = [I - C\ _BK^T_{h+1}]\ _F\bar{x}_{h+1|N} + C\ [I + C^T\ _BS_{h+1}\ C]^{-1}\ C^T\ _Bq_{h+1} \tag{8}$$

where C is a nxr matrix satisfying $C\,C^T := G\,Q\,G^T$

$$_F\bar{P}_{h+1|N} = F\ _FP_{h|N}\ F^T \tag{9}$$

$$_FP_{0|N} = [I + P_0\ _BS_0]^{-1}\ P_0$$

$$_FP_{h+1|N} = [I - C\ _BK^T_{h+1}]\ _F\bar{P}_{h+1|N}\ [I - C\ _BK^T_{h+1}]^T + C\ [I + C^T\ _BS_{h+1}C]^{-1}\ C^T \tag{10}$$

The initial state x_0 is assumed to be normally distributed with mean $\hat{x}_0$ and covariance P_0, $E\,[w_h, x_0^T] = E\,[i_h, x_0^T] = 0$. The quantities $_BK_{h+1}$, $_BS_{h+1}$, S_0, $_Bq_0$, $_Bq_{h+1}$, are provided by the following

Backword information filter

For $h = N-1,0$

$$_BK_{h+1} = {}_BS_{h+1}\ C\ [I + C^T\ _BS_{h+1}\ C]^{-1} \tag{11}$$

$$_B\bar{S}_{h+1} = [I - {}_BK_{h+1}\ C^T]\ _BS_{h+1} \tag{12}$$

$$_B\bar{q}_{h+1} = [I - {}_BK_{h+1}\ C^T]\ _Bq_{h+1} \tag{13}$$

$${}_B S_h = F^T \; {}_B \bar{S}_{h+1} \; F \; + H^T \; R^{-1} \; H \tag{14}$$

$${}_B q_h = F^T \; {}_B \bar{q}_{h+1} \; + \; H^T \; R^{-1} \; y_h \tag{15}$$

with the initial conditions ${}_B S_{N+1} = 0$, ${}_B q_{N+1} = 0$.

It is interesting to note that for any change of the initial condition x_0 ($\hat{x}_0$, P_0) , the smoothed estimates can be computed from the obtained estimates without reprocessing of the data. An algorithm to evaluate the effect of changes in the initial statistic on the smoothed estimates is obtained in Watanabe (1985) see corollary 4.2 , pg. 400 .

The present smoother is in a numerically stable form. In this form one can apply U-D factorization algorithms, as described in Thorton and Bierman (1977,1980) , that are efficient and numerically reliable.

The storage requirement of the smoother is (n+2)rN, where n is the dimension of the state vector x, r is the dimension of the vector w.

The computational cost of a step of Watanabe's smoother, with U-D implementations, as reported in Thorton and Bierman(1977,1980),is approximately $(1.5\, n^3 + 4.5\, n^2 + 9n)r + 3n^3 + 6n^2$.

In our model class the matrices F_{nxn} (here, n=k+p), G_{nxr} (here,r=3), H_{1xn} are constant and sparse; if the presence of zeros and ones is taken into account the number of operations of a step of U-D-Watanabe smoother can be reduced, approximately, to $(1.5\, n^3 + 3n^2 + 2n)r + 3n^2(n-k+1)$.

For example, in the case N = 180, p = 12, k = 2, we have $8.6\; 10^3$ locations and $4.\; 10^6$ operations.

Numerical experiments are in progress; their results will be collected and documented in a forthcoming contribution.

Acknowledgment

The author would like to thank Prof.Corrado Corradi for helpful suggestions and encouragement.

This research was supported in part by M.P.I.(60%)

References

BELL W.R. and HILLMER S.C.(1984): Issues involved with the seasonal adjustment of economic time series. J.of Bus.Econ.Stat.,2,4,291-349.

BIERMAN G.J. and THORTON C.L.(1977): Numerical comparison of Kalman filter algorithms: orbit determination case study. Automatica, 13, 23-35.

CHAN S.W., GOODWIN G.C. and SIN K.S.(1984): Convergence properties of the Riccati difference equation in optimal filtering of nonstabilizable systems. IEEE Trans.Automat.Contr., AC-29, 110-118.

DE SOUZA C.E., GEVERS M.R. and GOODWIN G.C.(1986): Riccati equations in optimal filtering of nonstabilizable systems having singular state transition matrices. IEEE Trans.Automat.Contr., AC-31, 831-838.

KITAGAWA G.(1981): A nonstationary time series model and its fitting by a recursive filter. J.Time Series Anal., 2, 103-116.

KITAGAWA G. and GERSCH W.(1984): A smoothness priors state space modelling of time series with trend and seasonality. JASA, 79, 378-389.

PAPPAS T., LAUB A.J. and SANDELL N.R.Jr.(1980): On the numerical solution of the discrete time algebraic Riccati equation. IEEE Trans.Automat.Contr. AC-26,4,631-641.

THORTON C.L. and BIERMAN G.J.(1980):UDU^T covariance factorization for Kalman filtering. In C.T.Leondes(Ed.) Control and Dynamic Systems, 16, 177--247, Academic Press, New York.

WATANABE K.(1985): On the relationship between the Lagrange multiplier method and the two-filter smoother. Int.J.Control, 42, 2, 391-410.

WATANABE K.(1986): A new forward-pass fixed-interval a smoother using the U-D information matrix factorization. Automatica, 22, 4, 465-475.

Statistical Data Bases and Survey Processing

New Perspectives in Computer Assisted Survey Processing

W. J. Keller and J. G. Bethlehem, Voorburg

Abstract

National statistical offices process huge quantities of data. The use of recent developments in the fields of statistics and computer technology are a prerequisite for efficient production of high quality statistics. At the Netherlands Central Bureau of Statistics an increasing use is made of (personal) computers in all steps of the statistical production process. This paper discusses the role of the computer in data collection, data editing, tabulation, and analysis. Particularly, attention is payed to the Blaise System for computer assisted survey processing, a system which controls various steps in the statistical production process.

1. The statistical production process

The production of statistical information is a complex process, in which data on persons, households and businesses, collected by means of surveys, have to be transformed into accurate statistics. The first step in this process is data collection, usually by means of a questionnaire. This ensures that all respondents are presented with the same questions. Data collected in this way will always contain errors. Therefore, in the second step a data editing process is necessary. The data are checked for errors, and detected errors are corrected as much as possible. After editing the data are still not ready for analysis and tabulation. Nearly always the data are not representative for the population from which they originate. To repair this a weighting procedure is carried out, which assigns adjustment weights to the sampled elements. Finally, a clean and representative data set is obtained, and then tabulation and analysis form the final step in the production process.

The role of the computer in statistical information processing has always been an important one. Already in 1880 Herman Hollerith constucted his famous punched card machine, which was succesfully used in the U.S. censuses of 1880 and 1890. And also one of the first electronical computers

Compstat 1988

was built for the Census Bureau. In the beginning the computer was only used for sorting, counting and tabulation, but in the seventies a new generation of computers emerged, and statisticians obtained the possibility to carry out statistical analyses with packages like SPSS, SAS, P-STAT, and BMDP. Now the computer is also increasingly used in other parts of the production process: data editing, Computer Assisted Telephone Interviewing (CATI), and Computer Assisted Personal Interviewing with handheld computers (CAPI).

Presently, the computer is fully integrated in the statistical production process. The Netherlands Central Bureau is equipped with hundreds of microcomputers. Each department has its own local area network (LAN), and all LAN's are connected with each other and with the mainframe and the mini's in the Automation Department. Simple and straightforward electronic data processing is now carried out by the subject matter departments themselves, leaving design and maintenance of complex information systems to be carried out by the computer specialists of the Automation Department.

The intensive and more sophisticated use of computers increases on the one hand efficieny and data quality, but on the other hand it also creats new problems. Due to the decentralisation of computer activities there is a need for standardization of automation tools. Furthermore, the various steps in the process are carried out by different departments, on different computer systems, using different computer programs. Therefore, there is also a need for integration. In this paper we will describe the different steps in the production process and the role played by the computer. Next, the Blaise System is presented as an attempt towards more integration and coordination within the production process.

2. Data collection

Statistical Agencies collect most of their data by means of (sample) surveys. A questionnaire is defined, containing the questions to be asked of respondents. Traditionally the questionnaires are completed in face-to-face interviews: interviewers visit the respondents, ask the questions, and fill in the answers on the (paper) questionnaire. The quality of the collected data tends to be good. However, face-to-face interviewing is expensive. It requires a large number of interviewers, whom all have to do a lot of travelling. Therefore telephone interviewing has become popular. Interviewer calls respondents from a statistical office, and no more travelling is necessary. Still, telephone interviewing is not always feasible: only people having a telephone can be contacted, and the questionnaire may not be too long and too complicated. Still cheaper is a mail survey: no interviewers at all are necessary. Questionnaires are mailed to potential respondents with the request to return completed forms. Although reminders can be send, the pursuasive power of the interviewer is lacking, and therefore response tends to be lower in this type of survey.

In the last decade the computer is increasingly used in the data collection stage. This occured first in telephone interviewing, and thus CATI-interviewing was born. The computer schedules the calls to make and controls the interview. It determines which questions to be asked and checks the ansers for consistency. This computer assistence has three mayor advantages. In the first place, it releases the interviewer of the burden of selecting the next question based on the answer of the current one and possibly present skip instructions. In the second place, this technology makes it possible to carry out checks during the interview. Detected errors can be discussed with the respondent, and hence corrected. This is not possible if error correction must carried out afterwards, at the office. In the third place, the data is already entered into the computer during the interview, thus making a separate data entry fase superfluous. CATI-interviewing is now a well-established data collection technique. See for a full discussion of CATI-programs e.g. Nichols and Groves (1986a, 1986b).

More recently, the advent of the small laptop computers made it possible for interviewers to take the computer with them to the homes of the respondents. CAPI-interviewing combines the advantages of face-to-face interviewing and computer assisted interviewing. Experiments carried out by the Netherlands Central Bureau of Statistics with the predecessor of the Blaise System (see Bemelmans-Spork and Sikkel, 1985[a] and 1985[b]), showed that interviewers were able to work with laptops, and that its use in face-to-face interviews did not cause any anxiety. Due to the success of these experiments, the Netherlands Central Bureau of Statistics started in 1987 with full scale use of CAPI in a regular survey, i.e. the labout force survey (see Van Bastelaer et. al, 1987a). Each month about 300 interviewers, equipped with laptops, visit 10,000 addresses. After a day of interviewing, the batteries of the hand-held computer are recharged at night, and the information collected that day is transmitted by telephone to the central office. The next morning, the interviewer finds a recharged machine with a clean workspace, ready for new interviews.

3. Data editing

Currently, the mayor part of data is collected in a traditional way, on paper forms, and without the use of a computer. Completed questionnaires have to undergo extensive treatment. For producing high quality statistics, it is vital to remove errors. Three types of errors are distinguished: A range error occurs if a given answer is outside the valid set of answers, e.g. an age of 348 years. A consistency error is caused by an inconsistency in the answers to a set of questions. An age of 11 years may be valid, a marital status 'married' is also not uncommon, but if both ansers are given by the same person, there is definitely something wrong. The third type of error is the routing error. If the interviewer, or the respondent fails to follow the specified branch or skip instructions, i.e. the route through the questionnaire is incorrect: irrelevant questions are answerd, and relevant questions are left unanswered.

Detected errors have to be corrected, but this can be very difficult if it has to be done afterwards, in the office. Particularly for household surveys, the respondent can not be contacted anymore, so other ways have to be found to something about the problem. Sometimes it is possible determine a reasonable approximation of a correct value by means of an imputation technique, but more often an incorrect value is replaced by the special code indicating the value is 'unknown'.

Beside data editing, sometimes also another activity is carried out during this stage of the production process: coding of open answers. A typical example is the question which asks about the occupation of the respondent. Questions are most easy to process if a respondent selects one possibility from a list of pre-coded answers. However, for a question like occupation, this set of pre-coded answers will be very long, and it will be vary hard for the respondent to select the proper answer. This problem is avoided by letting the respondent formulate his own answer, and the answer is literally copied on the form. To be able to analyse this type of information, the answers must be classified afterwards, and this is a time consuming job, to be carried out by experienced subject-matter specialists. Presently, ways are explored to incorporate the computer in this activities. For the labour survey a computer assisted coding system was developed, see Van Bastelaer et.al. (1987b). This system is indeed called 'computer assisted' and not 'automatic'. The mean reason is that a lot of the coding is not context-free, e.g. the code assigned to a particular occupation also depends on the description of the level of education and the height of the income. Furthermore, automatic text recognition techniques are not 100% succesful.

4. Weighting

After data editing, and possibly coding of open answers, the result is a 'clean' file, i.e. file without errors. However, this file is not yet ready for further analysis. In the first place, the sample is sometimes selected with unequal probabilities, i.e. businesses are selected with probabilities proportional to their size. A clever choice of selection probabilities makes it possible to produce more accurate estimates of population parameters, but only in combination with an estimation procedure which corrects for this inequality. In the second place, representativity may be affected by nonresponse, i.e. for some elements in the sample the required information is not obtained. If nonrespondents behave differently with respect to the population characteristcs to be investigated, the results will be baised.

In order to correct for unequal selection probabilities and nonresponse, often a weighting procedure is carried out. Every record is assigned some weight, and this weights are computed in such a way, that the weighted distribution in the sample of characteristics like sex, age, marital status, and area reflects the known distribution of these characteristics in the population. Research has been carried out at the Netherlands Central Bureau of Statistics in order to improve weighting techniques, see Bethlehem and Keller (1987). The increased power if the computer makes it possible to really implement these new techniques, see Bethlehem (1987).

5. Tabulation and analysis

Finally, we have a clean file which is ready for analysis. The first step in the analysis phase will nearly always be tabulation of the basis characteristics. A table looks like a simple thing, but in daily practice there is more to it. The composition of rows and columns (often built from more variables), the quantities displayed in cells (counts, means, percentages), the way in which percentages are computed, treatment of multiple response variables, the position of totals and subtotals, and many other things, can make life very difficult. Good tabulation software packages can help a lot, but the usefulness of packages depend on their possibilities, the user-friendliness of the control language, and the possibility to produce camera-ready output.

Many statistical agencies also carry out analyses on their data, in order to reveal underlying structures, and thus to gain insight in the data. Information obtained in this way, may improve a next survey, and thus improve quality or reduce costs. In many cases the data are categorical, and so many well-known statistical analysis techniques for interval data can not be used. Of course, there are techniques for categorical data, e.g. loglinear analysis. Still, there is a need for other techniques which can also be applied on on very large data sets. This his lead to the development and application of new techniques, like ANOTA or Correspondence Analysis. ANOTA (see Keller et.al. (1987)) is a simple alternative to logit analysis to investigate the relationship between a dependent categorical variabele and a set of explanatory variables. Correspondence Analysis, see Bethlehem and Sikkel (1987), is a technique to explore relations between categorical variables in a graphical way. Both techniques use bivariate tables as input instead of the raw data matrix, and therefore analysis on large data set can even be carried out on a microcomputer.

6. The Blaise System

BLAISE is the name of a new system for survey data processing. BLAISE controls a large part of the survey process: design of the questionnaire, data collection, data editing and data analysis. The basis of the BLAISE system is a powerful, structured language (also called BLAISE) which describes the questionnaire: questions, possible answers, routing, range checks and consistency checks. With the thus specified questionnaire as input, the system automatically generates software modules for data processing. In the first place, a CADI-program (Computer Assisted Data Input) can be produced. This program provides an intelligent and interactive environment for data entry and data editing, of data collected by means of questionnaires on paper forms. Another important module is the CATI-program. The BLAISE System automatically produces the software needed for telephone interviewing, including call management. A module which resembles the CATI-program, is the CAPI-program. This program for face-to-face interviewing with hand-held computers can also be generated by the BLAISE System. Furthermore, BLAISE automatically generates setups and system files for the statistical package SPSS. Interfaces to other statistical packages (e.g. PARADOX and STATA) are scheduled in the near future.

Traditional questionnaires are controled by goto's, but this jumping makes the questionnaire unreadable. The BLAISE language supports structured questionnaire design. This makes the structure of the questionnaire very clear. In almost one glance one is able to distinguish in which cases which questions are asked. Figure 1 gives an example of a simple questionnaire in BLAISE. A BLAISE questionnaire is like a recipe for an apple pie. Such a recipe contains three parts:

Part 1: Preparing the ingredients

For the preparation of an apple pie first all ingredients must be layed ready. Likewise, in a questionnaire first all questions must be specified. This takes place in the quest-paragraph. A question consists of an identifying name, the text of the question and a specification of valid answers. Three types of answers can be distinguished in the example. Marstat and Job have user defined answer categories. Note that such answers consist of a short name, used for checking and routing purposes in other parts of the questionnaire, and an extended description, used for explication purposes in generated software modules. The question Age expects an integer answer which is restricted to the range 0, 1, 2, ..., 120. The question KindJob expects an open answer, i.e. any text of atmost 80 characters.

Figure 1. A simple BLAISE questionnaire

```
QUESTIONNAIRE Demo;
QUEST
   Seq (KEY) "Sequence number of interview": 1..10000;
   Age       "What is the age of the respondent": 0..120;
   MarStat   "What is the marital status of the respondent":
              NotMar  "Never has been married",
              Married "Married",
              Divorce "Divorced"'
              Widowed "Widowed");
   Job       "Does the respondent have a job?":
             (Yes     "Yes, has a job",
              No      "No, does not have a job");
   KindJob  "What kind of job does respondent have?": STRING[80];

ROUTE
   Seq; Age; MarStat;
   IF Age > 15 THEN
      Job;
      IF Job = Yes THEN
         KindJob;
      ENDIF;
   ENDIF;

CHECK
   IF Age < 15 THEN MarStat = NotMar ENDIF;
ENDQUESTIONNAIRE.
```

Part 2: Mixing the ingredients

In the preparation phase all ingredients are mixed in the proper order. The analogue for the BLAISE questionnaire is the route-paragraph. It describes in which situations which questions must be answered and the order in which the questions must be answered. There is a simple rule: writing down the name of a questions means asking it. In the example it is obvious that the questions Age and MarStat are asked of every respondent. Only persons older than 15 are asked whether they have a job. The question KindJob is only asked if the respondent is older than 15 and has a job.

Part 3: Tasting the result

After finishing the pie, it is tryed to establish whether the result is tasteful. Likewise, a questionnaire should contain consistency checks. In a BLAISE questionnaire checks are specified in the check-paragraph. In particular, consistency checks are specified (range checks are automatically generated from the quest-paragraph and route checks are implied from the route-paragraph). The interpretation of this check-paragraph is: someone under 15 may not be married.

The description above does not exhaust the power of the BLAISE language. In this general discussion, we have ignored the possibility of the block concept to build large and complex questionnaires from simple sub-questionnaires. We also did not mention the treatment of tables of questions in a spreadsheet-like way, local variables, user-defined answer-types, the use of external files, and variable texts, to name just a few. Details on these matters can be found in the BLAISE reference manual (Bethlehem et. al, 1987[b]).

The Blaise System will also contain a module for interactive coding, which should be capable of handling the mayor part of the cases, leaving only a small number of hard cases, to be dealt with by a specialist. The module contains two different tools. The first tool implements a hierarchical approach to coding. Coding of an answer starts by entering the first digit of the code by selecting the proper category from a menu. After entering a digit, a subsequent menu is presented containing a refinement of the previously selected category. So, the description becomes more and more detailed until the final digit is obtained. The second tool contains a dictionary approach to coding. It tries to locate an entered description in an alfabetically ordered list. If the description is not found, the list is displayed, starting at the point as close as possible to the entered description. The list can be made so that almost any description, including permutations, is present. The advantage of this method is that it is simple, fast and controlable. Both coding tools can be used simultaneously.

An important data processing applications which can be produced by the BLAISE System, is the CADI-machine. The CADI machine is an intelligent and interactive system for data input and data editing. Data can be processed in two ways. In the first approach the subject-matter specialist sits behind a microcomputer with a pile of forms and processes the forms one by one. He enters the data at the proper places and after completion of the form he activates the check option, to test routing and consistency (range errors are checked during data input). Detected errors are highlighted and explained on the screen. Errors can be corrected, by consulting the form or calling the supplier of the information. After elimination of all errors, a clean record is written to file. If the specialist does not succeed in producing a clean record he can write the record to a separate file of problem records. These hard cases can be dealt with later on, also with the CADI-machine. In the second approach the data have already been entered beforehand by data typists, without error checking. With the CADI machine the records of the file can be edited. One by one a record is read from file, the check option is activated and, detected errors can be inspected and corrected.

The CADI-machine concentrates on processing data which have been collected on paper forms previously. However, the BLAISE System can also be used for data collection. The system is able to produce the software, required to carry out CATI-interviewing or CAPI-interviewing. A CATI-program applies dynamic routing, i.e. the software takes control of the routing. After an answer has been entered, the CATI machine decides which question is to be asked next, and automatically jumps to that question. Note the difference with the static routing as supported by the CADI-machine. This difference arises because of the completely different use of CADI and CATI. With CADI, we want to copy a form and check what has gone wrong; with CATI, we want to avoid that anything goes wrong.

The CATI-machine applies dynamic error checking. As soon as a consistency error is encountered, an error message is displayed on the screen, together with a list of all questions involved. Via a menu, the user may select the question to jump to in order to correct the error.

The BLAISE System can also produce a CAPI-program, an interviewing program to be used on a hand-held computer. CAPI combines the advantages of face-to-face interviewing with those of CATI interviewing. The program in the hand-held computer is in control of the interview, i.e. it determines the routing and checks consistency of answers. Since checking takes place during the interview, errors can be corrected on the spot. After successful completion of the interview, a clean record is stored in the hand-held computer. Later on, data is transmitted to the office by a modem and telephone, or a diskette can be send to the office.

7. Conclusion

The advent of the microcomputer has a considerable impact on the work of national statistical offices. More and more the unexperienced subject matter statistician is making use of it, and for his work he needs software like the BLAISE System. The power of this system lies on the one hand in the consistency it enforces in the various steps of data collection and data processing. This makes the whole process more easy to manage and to control. On the other hand it also promotes standardization between different departments. Since all departments use the same software for carrying out their surveys, exchange of information between departments is more easy and less error prone.

In some sense the BLAISE System might be called an expert system. The BLAISE questionnaire description contains all information necessary for subsequent steps in data processing. Therefore it serves as a knowledge base used by the system to generate applications.

The BLAISE System has been tested since the middle of 1986, in its various stages of development. The BLAISE environment, the CADI-machine, and the SPSS setup facility have been used succesfully for a substantial number of surveys of the Netherlands Central Bureau of Statistics. The other options are just completed and serious testing is scheduled for this year.

The BLAISE System runs on microcomputers under MS-DOS, both stand-alone machines and local area networks (running under Novell). For more details, see Bethlehem et al. (1987[a]-1987[d]).

Literature

Van Bastelaer, A.M.L., F.A.M. Kerssemakers and D. Sikkel (1987a): A test of the Continuous Labour Force Survey with hand-held computers: Interviewer behaviour and data quality. In: CBS select 4, Automation in Survey Processing (Netherlands Central Bureau of Statistics, Voorburg).

Van Bastelaer, A.M.L., L.M.P.B. Hofman and J,K, Jonker (1987b): Computer Assisted Coding of Occupation. In: CBS select 4, Automation in Survey Processing (Netherlands Central Bureau of Statistics, Voorburg).

Bemelmans-Spork, E.J. and D. Sikkel (1985[a]): Observation of prices with hand-held computers, Statistical Journal of the United Nations Economic Commission for Europe, vol. 3, no.2.

Bemelmans-Spork, E.J. and D. Sikkel (1985[b]): Data Collection with hand-held computers, Proceedings of the 45th session, International Statistical Institute, Book III, topic 18.3.

Bethlehem, J.G. (1985): LINWEIGHT User Manual. CBS report, Netherlands Central Bureau of Statistics, Voorburg.

Bethlehem, J.G., D. Denteneer, A.J. Hundepool, and W.J. Keller (1987[a]): BLAISE 1.1 / A first aquaintance, CBS report, Netherlands Central Bureau of Statistics, Voorburg.

Bethlehem, J.G., D. Denteneer, A.J. Hundepool, and W.J. Keller (1987[b]): BLAISE 1.1 / Language reference manual, CBS report, Netherlands Central Bureau of Statistics, Voorburg.

Bethlehem, J.G., D. Denteneer, A.J. Hundepool, and W.J. Keller (1987[c]): BLAISE 1.1 / The data entry machine, CBS report, Netherlands Central Bureau of Statistics, Voorburg, to appear.

Bethlehem, J.G., D. Denteneer, A.J. Hundepool, W.J. Keller, and M.H. Schuerhoff (1987[d]): BLAISE 1.1 / System description. CBS report, Netherlands Central Bureau of Statistics, Voorburg.

Bethlehem, J.G. and W.J. Keller (1987): Linear Weighting of Sample Survey Data. Journal of Official Statistics 3, pp. 141-154.

Bethlehem, J.G. and D. Sikkel (1985): CORAN User Manual, CBS report, Netherlands Central Bureau of Statistics, Voorburg.

Keller, W.J., A. Verbeek and J.G. Bethlehem (1985): ANOTA: Analysis of Tables. CBS report, Netherlands Central Bureau of Statistics, Voorburg.

Nicholls II, W.L, and R.M. Groves, (1986[a]), The status of computer assisted telephone interviewing: Part I- Introduction and impact and cost and timeliness of survey data. Journal of Official Statistics 2, pp 93-115.

Nicholls II, W.L, and R.M. Groves, (1986[b]), The status of computer assisted telephone interviewing: Part II- Data quality issues. Journal of Official Statistics 2, pp 117-134.

Multiple Imputation for Data-Base Construction

D. B. Rubin, Cambridge, Mass.

ABSTRACT

Multiple imputation is a technique for handling missing values in shared data-bases. The technique replaces each missing value by a vector of possible values that reflect uncertainty about which value to impute. The resultant multiply-imputed data base consists of the original data set with missing values replaced by pointers to the rows of a supplemental matrix of multiple imputations, each row being the vector of multiple imputations for a missing value. This data base can be analyzed by standard complete-data analysis techniques to obtain valid inferences.

1. INTRODUCTION

Anyone with experience in the field of data-base construction knows that essentially every data set suffers from some missing values. Nevertheless, the data-base constructor often feels obligated to create a fixed-up data-base for distribution -- one without missing values. If missing values are left blank, complete-data statistics that would have been used in the absence of missing data can no longer be calculated, and data analysts can no longer use standard complete-data methods to draw inferences.

1.1 Imputation

It is not surprising, therefore, that a very common method of handling missing values is to fill them in -- impute them. That is, with imputation, each missing value is replaced by a real value. Many different procedures have been proposed for imputation, for instance, filling in the respondents' mean for that variable or a value predicted from the modeling of the missing variable given observed variables using respondents' data; as a specific example, when the missing value is personal income, a linear regression model predicting log(income) from demographic characteristics such as age, sex, education and occupation might be regarded as reasonable. A common method of imputation in the United States federal government statistical system is "hot-deck" imputation; see Madow et al. (1983) for relevant definitions and references.

Compstat 1988

1.2 Advantages of Single Imputation

In addition to the obvious advantage of allowing complete-data methods of analysis, imputation by the data collector (e.g., the Census Bureau) also has the important advantage of allowing the use of information available to the data collector but not available to an external data analyst such as a university social scientist analyzing a public-use file. This information may involve detailed knowledge of interviewing procedures and reasons for nonresponse that are too cumbersome to place on public-use files, or may be facts, such as street addresses of dwelling units, that cannot be placed on public-use files because of confidentiality constraints. This kind of information, even though inaccessible to the user of a public-use file, can often improve the imputed values.

A third advantage of imputation by the data-base constructor is that the missing data problem is handled once, rather than many times by the users. This implies consistency of the data-bases across users, and a consequent consistency of answers from identical analyses. Too often the apparently same analysis (e.g., least squares regression) when applied to the apparently same data base will result in different outcomes because of differences in the way users and programs handle missing data. This situtation leads to unnecessary confusion and wasted resources. Imputation by the data-base constructor leads to greater consistency and thereby to reduced costs of this type.

1.3 Disadvantages of Single Imputation

Just as there are obvious advantages to imputing one value for each missing value, there are obvious disadvantages of this procedure arising from the fact that the one imputed value cannot itself represent any uncertainty about which value to impute: if one value were really adequate, then that value was never missing. Hence, analyses that treat imputed values just like observed values generally systematically underestimate uncertainty, even assuming the precise reasons for nonresponse are known. Equally serious, single imputation cannot represent any additional uncertainty that arises when the reasons for nonresponse are not known.

1.4 Multiple Imputation to the Rescue

Multiple imputation, first proposed in Rubin (1977,1978), retains the three major advantages of single imputation and rectifies its major disadvantages. As its name suggests, multiple imputation replaces each missing value by a vector composed of $M \geq 2$ possible values. The M values are ordered in the sense that the first components of the vectors for the missing values are used to create one completed data set, the second components of the vectors are used to create the second completed data set and so on. Display 1 depicts a multiply-imputed data set.

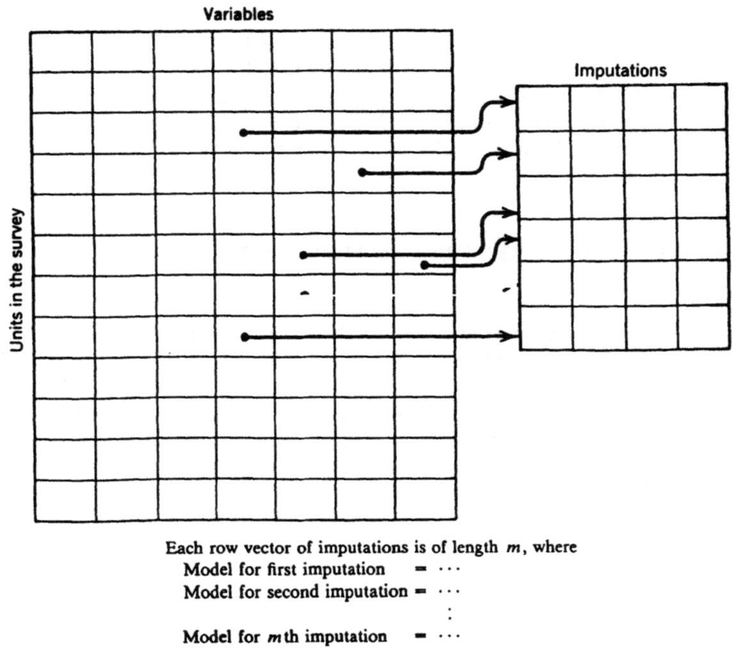

Display 1: Data set with m imputations for each missing **datum**. (From Rubin, 1987, p. 3).

The first major advantage of single imputation is retained with multiple imputation, since standard complete-data methods are used to analyze each completed data set. The second major advantage of imputation, that is, the ability to utilize data collectors' knowledge when handling the missing values, is not only retained but actually enhanced. In addition to allowing data collectors to use their knowledge to make point estimates for imputed values, multiple imputation allows data collectors to reflect their uncertainty as to which values to impute. This uncertainty is of two types: sampling variability assuming the reasons for nonresponse are known, and variability due to uncertainty about the reasons for nonresponse. Under each posited model for nonresponse, two or more imputations are created to reflect sampling variability under the model; imputations under more than one model for nonresponse reflect uncertainty about the reasons for nonresponse. The multiple imputations within one model are called repetitions and can be combined to form a valid inference under that model; the inferences under different models can be contrasted to reveal sensitivity of answers to posited reasons for nonresponse. The third advantage of single imputation -- consistency of answers across users -- is also retained, since the same set of multiple imputations are being passed on to all users, and all users applying one analysis method will obtain the same answer.

The unique advantage of multiple imputation is, as already suggested, that the multiple values can represent uncertainty due to both (a) sampling uncertainty -- by using repeated values within one model, and (b) reasons for nonresponse -- by using more than one model.

Rubin (1987) is a comprehensive treatment of multiple imputation. Other references on multiple imputation include Rubin (1977, 1978, 1979, 1980a, 1980b, 1980c, 1983a, 1983b, 1983c, 1986a, 1986b), Herzog and Rubin (1983), Li (1985), Schenker (1985), Glynn, Laird, and Rubin (1986) Rubin and Schenker (1986, 1987), Heitjan and Rubin (1986), Weld (1987), Raghunathan (1987), Little and Rubin (1987, Chapt. 11), and Treiman, Bielby and Cheng (1988).

2. GENERAL PROCEUDRES FOR CREATING A MULTIPLY-IMPUTED DATA SET

Multiple imputations ideally should be drawn according to the following general scheme. For each model being considered, the M imputations of the missing values, Y_{mis}, are M repetitions from the posterior predictive distribution of Y_{mis}, each repetition being an independent drawing of the parameters and missing values under appropriate Bayesian models for the data and the posited response mechanism. In practice, three important issues arise: explicit vs. implicit models, ignorable vs. nonignorable models, and proper imputation methods.

2.1 Explicit vs. Implicit Models

Explicit models are the ones usually used in mathematical statistics: normal linear regression, binomial, poisson, multinomial, etc. Implicit models are ones that can be thought of as underlying procedures used to "fix up" specific data structures in practice; often they have a "nonparametric", "locally linear", or "nearest neighbor" flavor to them. Although explicit models are the theoretical ideal precisely justifying multiple imputation technology, often implicit models can be used in place of explicit models. Both types of models are illustrated in Herzog and Rubin (1983), where repeated imputations are created using an explicit regression model and an implicit matching model, which is a modification of the Census Bureau's hot-deck. Many other examples of both kinds of models appear in Rubin (1987). Section 2.4, here, provides an example of an implicit model.

2.2 Ignorable vs. Nonignorable Models

The models underlying imputation methods, whether implicit or explicit, can be classified as assuming either ignorable reasons for missing data or nonignorable reasons. The term "ignorable" is coined in Rubin (1976) and is fully explicated in the context of multiple imputation in Rubin (1987). The basic idea is conveyed by a simple example in which X is observed for all units in the data base, and Y is missing for the nonrespondents but observed for the respondents. An ignorable model asserts that a nonrespondent is only randomly different from a respondent with the same value of X. A nonignorable model asserts that even though a respondent and nonrespondent appear identical with respect to X, their Y values systematically differ (e.g., the nonrespondent's Y is typically 20% larger than the respondent's Y with the same value of X). There is no <u>direct</u> evidence in the data to address the veracity of any such assumption, which is a good reason to consider several models and explore resultant sensitivity, whenever possible.

2.3 Proper Imputation Methods

Imputation procedures, whether based on explicit or implicit models, or ignorable or nonignorable models, that incorporate appropriate variability among the repetitions within a model are called *proper*, which is defined precisely in Rubin (1987). The essential reason for using proper imputation methods is that they properly reflect sampling variability when creating repeated imputations under a model, and as a result lead to valid inferences. For example, assume ignorable nonresponse so that respondents and nonrespondents with a common value of X have Y values only randomly different from each other. Even then, simply randomly drawing imputations for nonrespondents from matching respondents' Y values ignores some sampling variability. This variability arises from the fact that the sampled respondents' Y values at X randomly differ from the population of Y values at X. Properly reflecting this variability leads to repeated imputation inferences that are valid under the posited response mechanism. Ideal imputation methods (i.e., fully Bayesian ones) are automatically proper.

In the context of simple random samples and ignorable nonresponse, Rubin and Schenker (1986) study hot-deck imputation (i.e., simply randomly drawing imputed values from respondents), which is *not* proper, and a variety of proper imputation methods based on both explicit and implicit models, including a fully normal model, the Bayesian Bootstrap (Rubin, 1981), and an approximate Bayesian Bootstrap. The Approximate Bayesian Bootstrap (ABB) can be used to illustrate how an intuitive imputation method based on an implicit model, such as the simple random hot-deck, can be modified to be proper.

2.4 Example of a Proper Imputation Method - The ABB

Consider a collection of n units with the same value of X where there are n_R respondents and $n_{NR} = n - n_R$ nonrespondents. The ABB creates M ignorable repeated imputations as follows. For $\ell = 1, \ldots, M$, first create n possible values of Y by drawing n values at random with replacement from the n_R observed values of Y, and second, draw the n_{NR} missing values of Y at random with replacement from those n values. The drawing of n_{NR} missing values from a <u>possible</u> sample of n values rather than the <u>observed</u> sample of n_R values generates appropriate between imputation variability, at least assuming large simple random samples at X, as shown by Rubin and Schenker (1986). The ABB approximates the Bayesian Bootstrap by using a scaled multinomial distribution to approximate a Dirichlet distribution.

The ABB can be made nonignorable in many ways. For example, at the first step, instead of drawing n values of Y at random from the n_R observed values, independently draw n values of Y with probability proportional to Y^2 (or some other function of Y). This will skew the nonrespondents to have typically larger values of Y than respondents with the same values of X.

In practice, I believe that proper implicit models, both ignorable and nonignorable, will be the most useful from the perspective of the data-base constructor.

3. GENERAL PROCEDURES FOR ANALYZING A MULTIPLY IMPUTED DATA SET

The proper multiple imputations within each model are called repetitions and are combined (in ways to be described shortly) to create one inference under each model. The inferences across models are not combined but are contrasted to reveal sensitivity of inference to assumptions about the reasons for the missing data. The critical issue then is how to analyze the repetitions within one model to yield a valid inference under the posited reasons for missing data. The key idea is that M repetitions yield M completed data sets, each of which can be analyzed by standard complete-data methods just as if it were the real data set. The M complete-data analyses based on the M repeated imputations are then combined to create one repeated imputation inference.

3.1 The Repeated Imputation Inference For Point and Interval Estimation

Let $\hat{\theta}_\ell, U_\ell$, $\ell = 1, \ldots, M$ be M complete-data estimates and their associated variances for a parameter θ, calculated from the M data sets completed by repeated imputations under one model for nonresponse. For instance, for a regression analysis, $\theta = \beta$, $\hat{\theta}_\ell$ = the least squares estimate of β, and U_ℓ = (residual mean square) x $(X^T X)^{-1}$, in the standard notation. The final estimate of θ is

$$\bar{\theta} = \sum_{\ell=1}^{M} \hat{\theta}_\ell / M \quad .$$

The variability associated with this estimate has two components: the average within-imputation variance,

$$\bar{U} = \sum_{\ell=1}^{M} U_\ell / M \quad ,$$

and the between-imputation component,

$$B = \sum (\hat{\theta}_\ell - \bar{\theta}_M)^2 \,/\, (M - 1) \ ,$$

where with vector θ, $(\bullet)^2$ is replaced by $(\bullet)^T(\bullet)$. The total variability associated with $\bar{\theta}$ is then

$$T = \bar{U} + (1 + M^{-1})B \quad .$$

With scalar θ, the approximate reference distribution for interval estimates and significance tests is a t distribution:

$$(\theta - \bar{\theta})\, T^{-1/2} \sim t_v \quad ,$$

where the degrees of freedom,

$$v = (M - 1)\ \{1 + [(1 + M^{-1})B/\bar{U}]^{-1}\}^2 \quad ,$$

is based on a Satterthwaite approximation (Rubin and Schenker 1986, and Rubin 1987). The within to between ration $r=\bar{U}/B$ estimates the population quantity $(1 - \gamma)/\gamma$, where γ is the fraction of information about θ missing due to nonresponse. In the case of ignorable nonresponse with no covariates, γ equals the fraction of data values that are missing, but typically γ is less than this because of dependence between variables with the attendant ability to improve prediction of missing values from observed values.

3.2 Significance Levels for Multicomponent θ from $\{\hat{\theta}_\ell, \hat{U}_\ell; \ell=1,\ldots,M\}$

For θ with k components, significance levels for null values of θ can be obtained from M repeated complete-data estimates, θ_ℓ, and variance-covariance matrices, U_ℓ, using multivariate analogues of the previous expressions.

A simple procedure described in Li (1985) and Rubin (1987) that works well for M large relative to k is to let the p-value for the null value θ_o of θ be

$$\text{Prob } \{F_{k,v} > D\}$$

where $F_{k,v}$ is an F random variable, and

$$D = (\theta_o - \bar{\theta})T^{-1}\ (\theta_o - \bar{\theta})^T/k$$

with v defined by generalizing $r=B/\bar{U}$ to be the average diagonal element of $B\bar{U}^{-1}$, $r=\text{trace}(B\bar{U}^{-1})/k$.

A better procedure when M is modest, advocated in Rubin (1987), is to let the p-value be given by

$$\text{Prob}\{F_{k,v(k+1)/2} > \tilde{D}\}$$

where F and v are as previously defined, and

$$\tilde{D} = (\theta_o - \bar{\theta})\ \bar{U}^{-1}\ (\theta_o - \theta)/[(1+r)k] \quad .$$

This procedure is very accurate, except for large k when it tends to be a bit too conservative due to the approximate nature of the reference distribution.

More accurate procedures in this case are also somewhat more complicated, and are described by Raghunathan (1987).

3.3 Significance Levels from Repeated Significance Levels

With large data sets and large models, such as occur often with multiway contingency tables in social science research, a complete-data analysis may only produce a p-value or equivalently the χ^2 statistic on each completed data set:

$$d_\ell = (\theta_0 - \theta_\ell)\, U_\ell^{-1} (\theta_0 - \theta_\ell)^T \quad .$$

The problem of directly combining the $\{d_\ell, \ell=1,\ldots,M\}$ is tricky because each d_ℓ typically leads to a p-value that is too extreme (i.e., too significant). The representation that makes progress possible is to note that in large samples

$$\hat{D} \doteq \tilde{D}$$

where

$$\hat{D} = \frac{\left[\dfrac{\bar{d}}{k} - \dfrac{M-1}{M+1}\, r\right]}{(1+r)}$$

and $\bar{d}$ is the average d_ℓ. Replacing r in $\hat{D}$ with estimates obtained from the d_ℓ yields procedures, say D_*, that are acceptable in many cases (see Rubin, 1987; Raghunathan, 1987, and Weld, 1987). Truly satisfactory procedures for this case are the object of current research.

4. FREQUENCY EVALUATIONS

Although repeated imputation inferences are most directly motivated from the Bayesian perspective, they can be shown to possess good frequency properties. In fact, the definition of proper imputation methods means that in large samples infinite-M repeated imputation inferences will be valid. Since the finite-M adjustments are derived using approximations to Bayesian posterior distributions, however, some deficiencies can arise with finite M.

4.1 Relative Efficiency of Point Estimation

The large sample relative efficiency of the finite-M repeated imputation estimator using proper imputation methods relative to the infinite-M

estimator, in units of standard errors is

$$(1 + \gamma/M)^{-1/2} .$$

Even for relatively large γ, modest values of M result in estimates of $\bar{\theta}$ that are nearly fully efficient. For example, for γ=30% (usually a fairly extreme case), and M=3 (a reasonable choice for many applications), the relative efficiency of $\bar{\theta}$ is approximately 95%.

4.2 Confidence Coverage

In large samples, the confidence coverage of proper imputation methods using the t reference distribution can be tabulated as a function of M, γ and the nominal level, $1 - \alpha$. Extensive results are given in Rubin and Schenker (1986) and Rubin (1987). (With single imputation, the between component of variance is automatically set to zero, since it cannot be estimated, and the reference distribution is the normal, since v cannot be estimated without B.) Table 1 gives illustrative values for the improvements when going from single to multiple imputation with γ=30%. Three repeated imputations yield essentially valid confidence coverages, which is in striking contrast to using only one imputation. Even worse coverages for single imputation would have been obtained using best prediction methods, such as "fill in the mean".

Table 1

Actual large sample coverage percentages of confidence intervals based on single (M=1) vs. multiple (M=3) imputation with 30% missing information.

	Nominal Coverage		
M	90%	95%	99%
1	77	85	94
3	90	95	99

4.3 Significance Levels

Work on accurately obtaining significance levels is at a relatively early stage of development, but much effort has been expended and many tables are given in Li (1985), Rubin (1987), and Raghunathan (1987). Table 2 here indicates the tremendous improvements that can accrue when using multiple rather than single imputation. Notice that $\tilde{D}$ is essentially perfect for the values of γ, k and M presented; it is generally very good and clearly preferable to D. When the θ_ℓ and U_ℓ are not available, D_* can be used, and although not entirely satisfactory, it is still a dramatic improvement over single imputation, and would show an even more dramatic improvement if a "best prediction" method had been used for single imputation. Improvements over D_* are the topic of work in Raghunathan (1987) and current efforts.

Table 2

Actual large sample levels (in %) of tests based on single (M=1) vs. multiple (M=3) imputation with 30% missing information and a ten-component hypothesis being tested (i.e., k=10).

	Nominal Level		
	1%	5%	10%
$M = 1$	25	45	57
$M = 3$ using D	1	6	14
$M = 3$ using $\tilde{D}$	1	5	10
$M = 3$ using $\hat{D}_*$	4	10	17

5. CONCLUSION

In conclusion, multiple imputation is a very promising new tool for helping to handle missing values in shared data-bases. Although much work remains to be done before it will become a commonplace method, many interesting theoretical and practical results suggest effort expended in its development will be well rewarded by important contributions to applied work. Certainly, it gives data base constructors a new tool to address the difficult practical problem of how to handle missing data.

ACKNOWLEDGEMENTS

This work was partially supported by United States National Science Foundation Grants SES-83-11428 and DMS-85-04332. A similar but earlier summary is "Basic Ideas of Multiple Imputation for Nonresponse", Survey Methodology, 12, pp. 37-47.

REFERENCES

GLYNN, R., LAIRD, N., and RUBIN, D.B. (1986). Selection modelling versus mixture modelling with nonignorable nonresponse, *Drawing Inferences from Self-Selected Samples*, H. Wainer, ed., Springer-Verlag, pp. 119-146.

HEITJAN, D.F., and RUBIN, D.B. (1986). Inference for coarse data using multiple imputation. *Proceedings of the 18th Symposium on the Interface of Computer Science and Statistics*, pp. 138-143.

HERZOG, T.N., and RUBIN, D.B. (1983). Using multiple imputations to handle nonresponse in sample surveys. In *Incomplete Data in Sample Surveys, Volume 2: Theory and Bibliography*, New York: Academic Press, 209-245.

LI, K.H. (1985). *Hypothesis Testing in Multiple Imputation - with Emphasis on Mixed-up Frequencies in Contingency Tables*. Ph.D. Thesis, Department of Statistics, University of Chicago.

LITTLE, R.J.A. and RUBIN, D.B. (1987). *Statistical Analysis with Missing Data*, New York: Wiley.

MADOW, W.G. NISSELSON, H., and OLKIN, I. (1983). *Incomplete Data in Sample Survey, Volume 1, Report and Case Studies*, New York: Academic Press.

MADOW, W.G. and OLKIN, I. (1983). *Incomplete Data in Sample Surveys, Volume 3, Proceedings of the Symposium*, New York: Academic Press.

MADOW, W.G., OLKIN, I., and RUBIN, D.B. (1983). *Incomplete Data in Sample Surveys, Volume 2, Theory and Bibliographies*, New York: Academic Press.

RAGHUNATHAN, T.E. (1987). *Large Sample Significance Levels from Multiply-Imputed Data*. Ph.D. Thesis, Department of Statistics, Harvard University.

RUBIN, D.B. (1976). Inference and missing data. *Biometrika*, 63, 581-592.

RUBIN, D.B. (1977). The design of a general and flexible system for handling nonresponse in sample surveys. Unpublished paper prepared for the U.S. Social Security Administration.

RUBIN, D.B. (1978). Multiple imputations in sample surveys - a phenomenological Bayesian approach to nonresponse. *Proceedings of the Survey Research Methods Section of the American Statistical Association*, 20-34. Also in *Imputation and Editing of Faulty or Missing Survey Data*, U.S. Dept. of Commerce, 1-23.

RUBIN, D.B. (1979). Illustrating the use of multiple imputations to handle nonresponse in sample surveys. *Proceedings of the 1979 Meetings of the ISI-IASS, Manila*.

RUBIN, D.B. (1980a). *Handling Nonresponse in Sample Surveys by Multiple Imuputations*. U.S. Dept. of Commerce, Bureau of the Census Monograph.

RUBIN, D.B. (1980b). Illustrating the use of multiple imputations to handle nonresponse in sample surveys, *42nd Session of the International Statistical Institute*, 1979, Book 2, pp. 517-532.

RUBIN, D.B. (1980c). Introduction to papers on the CPS hot deck: evaluations and alternatives and related research, *Economic and Demographic Statistics*, U.S. Dept. of Health and Human Services, Social Security Administration, Office of Research and Statistics, pp. 67-70.

RUBIN, D.B. (1981). The Bayesian bootstrap. *Annals of Statistics*, 9, 130-134.

RUBIN, D.B. (1983a). Imputing income in the CPS: comments on 'Measures of Aggregate Labor Costs in the United States', In *The Measurement of Labor Cost*, Triplett, ed. University of Chicago Press, pp. 333-343.

RUBIN, D.B. (1983b). Conceptual issues in the presence of nonresponse, *Incomplete Data in Sample Surveys (Volume 2): Theory and Bibliography*, New York: Academic Press, pp. 123-142.

RUBIN, D.B. (1983c). Discussion of "Statistical Record Matching for Files" by M. Woodbury, *Incomplete Data in Sample Surveys (Volume 3): Proceedings of the Symposium*, New York: Academic Press, pp. 203-212.

RUBIN, D.B. (1986a). Basic Ideas of Multiple Imputation. *Survey Methodology*, 12, pp. 37-47.

RUBIN, D.B. (1986b). Statistical matching using file concatenation with adjusted weights and multiple imputations. *Journal of Business and Economic Statistics*, 87-94.

RUBIN, D.B. (1987). *Multiple Imputation for Nonresponse in Surveys*. New York: John Wiley & Sons.

RUBIN, D.B. and SCHENKER, N. (1986). Multiple imputation for interval estimation from simple random samples with ignorable nonresponse. *Journal of the American Statistical Association*, 81, 366-374.

RUBIN, D.B. and SCHENKER, N. (1987). Interval estimation from multiply-imputed data: a case study using agriculture industry codes, *Journal of Official Statistics*, 3, 375-387.

SCHENKER, N. (1985). *Multiple Imputation for Interval Estimation from Surveys with Ignorable Nonresponse*. Ph.D Thesis, Department of Statistics, University of Chicago.

TREIMAN, D.J., BIELBY, W.G., and CHENG, A.T. (1988). Evaluating a multiple imputation method for recalibration of 1970 U.S. Census detailed industry codes to the 1980 standard, *Sociological Methodology*.

WELD, L.H. (1987). *Significance Levels from Public Use Data With Multiply-Imputed Industry Codes*. Ph.D Thesis, Department of Statistics, Harvard University.

GRASP: AComplete Graphical Conceptual Language for Definition and Manipulation of Statistical Databases[1]

T. Catarci and G. Santucci, Roma

ABSTRACT

Recent research activities show a growing interest in statistical databases. Such interest is motivated by the increasing number of statistical applications and the inadequacies of traditional database languages and interfaces for statistical applications.
An outstanding open problem in statistical database development is how to express aggregate data starting from a description of elementary data. We propose a solution to this problem by means of a new language, called **GRASP** (**GRA**phical **S**tatistical **P**ackage). The language allows to query uniformly both elementary and aggregate data; and is, therefore, suitable for several different purposes: a) posing queries on elementary data; b) building aggregate data from elementary data; c) computing new aggregate data using statistical operators.

1. INTRODUCTION

An outstanding open problem in statistical database development is how to express aggregate data (i.e. statistical data) starting from a description of elementary data.
In this paper we deal with this problem, by proposing a new language, called **GRASP**: **GRA**phical **S**tatistical **P**ackage, which allows to build aggregate data starting from elementary data. This operation is executed by means of three different phases: 1) posing queries on elementary data, extracting the desired information for successive statistical applications (the result of each query is called class of objects); 2) defining the aggregation law by means of category attributes, that are properties of classes of objects, and by means of statistical classifications, that describe the relationship between a class of objects and the set of category attributes; 3) performing some computations on the partitions of the classes of objects, in order to obtain a class of data, that is a set of statistical data.
Moreover it is also possible to compute new statistical data from existing ones, by means of statistical operators.

[1] Work supported by Istituto Nazionale di Statistica (ISTAT)

GRASP makes use of two conceptual models, for elementary and statistical data respectively: elementary data are represented by the Entity-Relationship (E-R) model [Chen 76], augmented by the introduction of hierarchical structures [Bat 80]; statistical data are represented by the CSM model, described in [DiBatt 86], [DiBatt 87].

GRASP is part of a system to query uniformly both elementary and aggregate data, using diagrams as standard user interface. The simple graphic formalism used to represent conceptual schemes by means of diagrams allows the user to interact friendly with the system. The whole system is based on three basic ideas: 1) to represent the database by means of two conceptual models (E-R and MCS); 2) to allow an unambiguous definition of statistical data starting from elementary data; 3) to employ a fully graphical environment as user friendly interface with the system.

Such choices allow several benefits: a) the user has at his disposal, a simple yet powerful formalism to understand the data content; b) all different levels of aggregation of data, starting from elementary data, are defined precisely, and it is possible to check the completeness, the coherence and the minimality of the conceptual schema at different levels; c) the graphic formalism used to represent conceptual schemes, makes easier the execution of typical activities involved in query formulation, both at the level of the elementary data and of statistical data.

The paper is organized as follows. In section 2 we define the part of GRASP concerning elementary data; section 3 describes statistical classification; in section 4 the statistical operators are investigated.

2. QUERY LANGUAGE FOR ELEMENTARY DATA

The main idea of GRASP is to allow the user to define statistical data by means of queries on the conceptual schema of elementary data. In fact, in [DiBat 87] a class of objects is defined as a couple S=<E,Q>, where E is an entity in the E-R schema and Q is a generic query on it. We apply this definition, with the following extension: S = <E/R,Q>, where R is a relationship with cardinalities M:N. This is due to the information content of this kind of relationship that is equal to the one of an entity.

The language defined in this section is mainly a navigation language on E-R schemes provided with a set of operators giving it a property commonly called "more than completeness" (see [Atz 81]); the formal proof of relational completeness is in [Ange 88].

The main idea is to decompose a query into its elementary steps, expressed by a limited set of syntactic constructs. In the interaction with the user these constructs are expressed as selections of graphic symbols, made with a mouse; conditions on attributes are expressed textually.

2.1 Fundamental operators

The general query structure is based on the location of a central concept that can be seen as the entry point of one or more different subqueries; these subqueries express possible navigations from the central concept to all other concepts in the schema. The subqueries can be linked by the usual operators of union and intersect.

The instruction corresponding to the location of a central concept is:

SELE <ent rel name> WITH [<condition>]

where <condition> is:

{<attr>op<attr>/<costant> } / { DEL <attr> } and op $\in$ {=,<,>,<=,=>,<>,AND,OR,NOT,(,)}.

This is always the first instruction in a query. In addition to the usual conditions on attributes, the user may also delete some attributes in the query by means of the option DEL <attr>. This may be necessary, because all the attributes of the central entity are added to the final class of objects, differently from the other attributes met during the query.

At the end of the query, after the keyword ENDSELE, the user may store the query in a query library, in order to use it afterwards.

Once a concept has been selected, two different types of primitives are available in order to navigate in the schema: FIND and BRIDGE. It is possible to nest these two primitives as you like. The syntax of the primitives is the following:

1. FIND <ent rel name> [THROUGH <rel name>] [WITH <condition>] [ADD <ls attr>]

FIND is used to follow paths of concepts. The concept specified after the keyword is the final one in the path, while the starting entity (or relationship) is the last selected in the previous operation.

If the user wants to add an attribute to the final class of objects, he must explicity use the operation ADD <ls attr>.

2. BRIDGE <expression> BUILD WITH [<teta condition>] <expression>{SELE <ent rel name> [WITH <condition>] [ADD <ls attr>] [<nav>],ENDSELE [WITH <condition>]}

This operator compares concepts not directly connected, as the Θ-join in relational algebra [Codd 72]. In particular, by means of the clause denoted <teta condition> it is possible to perform directly, in addition to the condition on attributes, the cartesian product and the division as in relational algebra (for example it is possible to solve immediatly the query: "Find suppliers of all parts of a given product ").

At this point, we must specify something about hierarchical structures. If the previous constructs are applied on generalization hierarchies or on subsets, the following rules hold: if the query involves a lower entity, this entity inherits all links and attributes coming from the higher entity, in addition to its own links. On the contrary, if the entity involved is the higher entity, it doesn't inherit the properties of lower entities; however, a discriminant attribute is always defined among the attributes of the higher entity.

2.2 Attributes manipulation

During a query formulation, the user may express conditions on the attributes of selected entities. Moreover it is available a set of operators, as COUNT, SUM, AVG, ecc., in addition to the usual mathematical operators, that allow the user to build new attributes through the manipulation of existing ones, or to rename them. The construct for renaiming is:

<attr_name> REN OF <attr_name> OF <ent_rel_name> .

On the contrary, a new attribute may be defined by means of one of the previous navigation primitives and of one or more operators. The expression is:

<attr_name>NEW{FOR/STORED}<ent_rel_name>[<nav>,[<eta>,<nav>]]FORMULA{<operator >,(,<ent_attr_name>,) / <attr_name> [<op4><attr_name>] }

The clause FOR/STORED means that an attribute may be created temporary or permanently.

We have splitted the operators into 4 groups, according to their co-domain (we denote with D_i the attribute domains):

First group: D_1 -> D_1; (i.e. SUM/MINOR [(parameter)] /MAJOR [(parameter)])

Second group: D_1 -> Integer; (i.e. COUNT())

Third group : D_1 -> Real; (i.e. SD/AVG/VAR/SQM/%)

Fourth group : D_1 X D_2 X ... D_n -> Real; (i.e. +/-/*/:)

Now we give an example of attribute creation, referring to fig. 1.

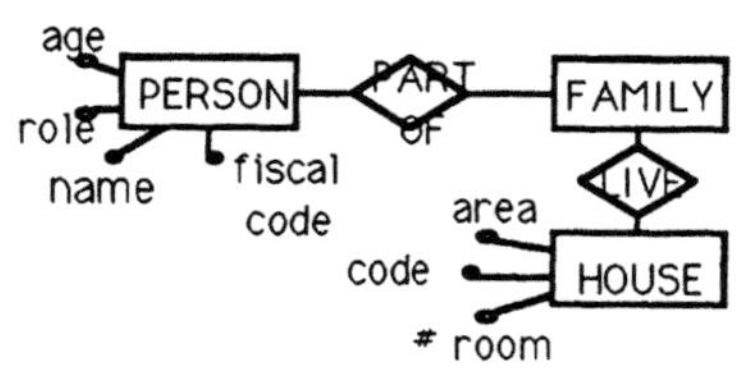

Fig. 1: query schema

The query: 'Find the families with at least two children' is expressed as:

SELE family WITH childrent_number (NEW FOR family
FIND person THROUGH part_of WITH role='C'
FORMULA COUNT (person)) >= 2
ENDSELE

3. DEFINITION OF STATISTICAL CLASSIFICATIONS

The result of the above phase of quering elementary data, is the specification of a set of classes of objects, each one with a label, automatically generated during the query formulation. Such a label, expressed in a language similar to natural language, represents the semantic content of the class of objects; it is therefore like an echo of the query that the user would express using a textual language and a check for the result of the query produced by the system. The corrispondence between the query and the label is based on set of correspondence rules.

With the language constructs shown in this section, the user may define, among all the ones proper of the class of objects, the category attributes involved in the statistical classification and their domains, building a partition on the instances of the class of objects. The domains of the category attributes may be the same proper of the corresponding attributes in the E-R schema.

The language syntax for the definition of statistical classifications is:

<statistical_classification>::=CSTAT <label_c> WITH <an_ad>1^n

<an_ad>::={<attribute_name> [DOMAIN MADE WITH <exp1>]/[DOMAIN <new_d_name>]/ DOMAIN AS <attr_name> OF {<ent_rel_name>/<label_c>}}1^n {GROUPING <an-ad> INTO <grouping_name>}

Several mechanism are available to define a domain, for example by means of a grouping on a domain existing in the E-R schema, with regular or enumerative intervals. This is shown by means of an example: we want to build a partition of the instances of the class PERSON (with attributes AGE, SEX, and STATE), based on the category attribute SEX and AGE, with the domain of AGE splitted into AGE CLASSES. The correspondent language construct is:

CSTAT person WITH sex, age DOMAIN MADE WITH (18, 30, 45, 60)

4. QUERY LANGUAGE FOR STATISTICAL DATA

Once classes of objects and statistical classifications have been produced, the user has at his disposal the conceptual model CSM and a fully graphical environment as a friendly interface with the system, in order to compute classes of data.

A class of data in CSM denotes a set of statistical data that we are interested to represent in the conceptual schema; it may be computed performing statistical computations on the subset of the partition of the class of objects. In our language it is defined an uniform mechanism to build a class of data starting from a partition of a class of objects or other classes of data and using different operators. The syntax is:

<cl_dat>::= <op> BUILD ON <label_c> USING <statistical_classification>/ <op> BUILD ON <ls_cl_dat> [WITH SUMMARIZE ON <ls_attr>]

The term "op" corresponds to many statistical operators: a) operators on class of objects, as COUNT, %, SUM, AVG, SUMM ON, and so on. This kind of operators has the additional possibility to perform the computation on a sample of data ; b) operators on a single class of data, as INDEX and CHANGE_RATE. c) poliadic operators (*,/,+,-, etc.) that allow to build a class of data starting from two or more classes of data; d) operators defined from the user, by means of the previous operators and the usual mathematical operations. Moreover, it is allowed to define also at statistical level groupings or aggregate of category attribute: in this case, if an attribute is involved in a classification, its grouping or aggregate attributes are involved as well.

Finally, in order to extend the concept of generalization at the classes of data, it is possible to define a generalization on statistical data with the following construct:

<gen>::=DATA ON <gent_name>GENERALIZATION OF {<cl_dat> / <gent_name>} $_1^n$

References

[Ange 88] M.Angelaccio - Completeness of Graphical Query Language - preliminary report, Consorzio Roma Ricerche, Roma, April 1988.

[Atz 81] P.Atzeni, P.P.Chen - Completeness of Query Languages for the Entity Relationship Model - in P.P. Chen (ed.) Proc. of the 2nd Intl. Conference on the Entity Relationship Approach to Information Modelling and Analysis, Washington D.C., North Holland, 1981.

[Bat 80] C.Batini, G.Santucci - Top-down design in the Entity-Relationship Model - International Conference on the E - R Approach to System's Analysis and Design, Los Angeles, 1980.

[Codd 72] E.F.Codd - Relational completeness of database sub-languages - pp.65-98 in Data Base Systems, R.Rustin (ed), Prentice Hall, Englewood Cliffs, New Jersey, 1972.

[DiBatt 86] G.Di Battista, G.Ferranti, C. Batini - A Methodology for Conceptual Design of Statistical Databases - Proc. of COMPSTAT Conference, Roma, 1986.

[DiBatt 87] G.Di Battista, C.Batini - Design of Statistical Databases: a Methodology for the Conceptual Step - Technical Report, Dipartimento di Informatica e Sistemistica, Universita' di Roma " La Sapienza", submitted for publication, 1987.

[DiLeva 87] A.Di Leva, P.Giolito - A Two Level data model for distributed statistical databases - pp. 19-54 in F.Ricci, L.Meoevoli (ed.) Proceedings of the Conference on Statistical Data Bases, Roma, 1987.

[Ozso 85] G.Ozsoyoglu, Z.Meral Ozsoyoglu - Statistical Database Query Language - in IEEE Transaction on Software Engineering, October 1985.

[Sho 85] A. Shoshani, H.K.T. Wong - Statistical and Scientific Database Issues - in IEEE Transaction on Software Engineering, October 1985.

[Wong 82] H.K.T.Wong, I.Kuo - GUIDE : Graphical User Interface for Database Exploration - pp. 22-31 in Proc. of the 8th VLDB Conference, Mexico City, September 1982.

Experimental Design

New Algorithmic and Software Tools for D-Optimal Design Computation in Nonlinear Regression

J. P. Vila, Jouy-en-Josas

Summary : The aim of this paper is to present

1. A new theoretical result for a problem of long standing (Atkinson et al. 1968, Box 1968) in the Theory of Optimal Design in Nonlinear Regression : the determination of optimality conditions for designs made of replications of the minimal D-optimal designs points (Vila 1986).

2. A new easy-to-use powerful software for optimal design computation in nonlinear regression, following the D-optimality criterion and other allied criteria.

Keywords : Nonlinear Regression; Optimal Designs; D-optimality; Constrained Optimization.

1 Introduction

The efficient fitting of a general regression model by least squares needs beforehand the determination of an optimal design for the model parameters estimation (Silvey 1980). The D-optimality criterion (Kiefer 1961) is most often used for this preliminary step. For a nonlinear regression model this criterion is based upon an implicit model linearization about a priori values for the unknown parameters. The designs can be considered either as probability measures on a compact set, or from a more practical point of view, as discrete designs of chosen size n (exact designs), in a given experimental region. In the first case Fedorov (1972) showed that the D-optimal design can be searched for in the set of all discrete measures supported on at most $p(p+1)/2$ distinct points, when the regression model has p parameters. This support is incidentally found with exactly p points. A similar replication phenomenon is frequently observed for exact designs (D_n-optimal designs) : the optimal design is found as made of "equitable" replications (r_i, $i = 1,\dots,p : \sum_{i=1}^{p} r_i = n$ and $| r_i - r_j | \leq 1 \ \forall i,j$) of p distinct points which constitute themselves the D_p-optimal design (the minimal D-optimal design). This phenomenon is non systematic (counter-example : incomplete polynomial models). Its prediction would allow reduction of the problem dimension by n/p, as well as numerical conditioning improvement.
Atkinson et al. gave a sufficient condition based on geometric considerations for the particular case $n = rp$. For exact designs obtained by rounding off weights of D-optimal measures of minimum support, Gaffke (1987) gave a sufficient condition of D_n-optimality. He proved an other condition to be necessary for D-optimality of a given exact design, in the special case in which the experimental region is an interval.
Constrained optimization theory helped us to determine a couple of very close sufficient condition and necessary condition, for the local optimality (at least) of designs made of general replications

Compstat 1988

of the p points of the minimal D-optimal design. These conditions are applicable whatever the size n of the searched design, the number ℓ of independent variables in the regression model, and for general types of constraints defining the experimental region. This "sufficient and quasi-necessary condition" can often be tested analytically (current simple models) and otherwise numerically in every case.
This result can easily be implemented in any software for D-optimal design computation, and used to spare computing time, sometimes drastically on micro-computers.

2 Properties of replicated minimal D-optimal designs

2.1 Notations

Let the regression model

$$y_i = f(x_i, \theta^*) + e_i \qquad i = 1, \ldots, n \tag{1}$$

$\theta^* \in \Theta$ a compact set of R^p : θ^* is the unknown p-vector of parameters,
$x \in E$ a compact set of R^ℓ : x is the ℓ-vector of independent variables.
x_{i_j} will be the jth component of the ith observation point x_i, $j = 1, \ldots, \ell$, $i = 1, \ldots, n$
$f : E \times \Theta \longrightarrow R$, the second-order-differentiable response fonction,
$\{e_i\}$: independent and identically distributed errors with $E(e_i) = 0$ and $Var(e_i) = \sigma^2$
$\xi_n = (x_1, \ldots, x_n)$, the n-experimental design
$X_n = [\{\partial f(x_i, \theta)/\partial\theta_j\}]$ $i = 1, \ldots, n$ $j = 1, \ldots, p$
$v_\alpha = (\partial f(x_\alpha, \theta)/\partial\theta_1 \ldots \partial f(x_\alpha, \theta)/\partial\theta_p)$, $\dot{v}_{\alpha_h} = \partial v_\alpha/\partial x_{\alpha_h}$, $\ddot{v}_{\alpha_{hk}} = \partial^2 v_\alpha/\partial x_{\alpha_h}\partial x_{\alpha_k}$
$\Delta_n = | X_n^T X_n |$, the D_n-optimality criterion, to be maximized with respect to ξ_n in a particular experimental region E_n, for a priori parameters values θ^0.

2.2 Constraints set and experimental region

The constraint functions defining the experimental region E_n, $\phi_k \leq 0$ or $\phi_k = 0$, $k = 1, \ldots, m$ play a crucial role in the optimality conditions to be proposed in the following. For brevity's sake in this paper, let us simply say that these constraint functions can be much more general than classic inequality or equality constraints which have to be satisfied by each of the n points of the design in turn.
We shall suppose that the gradients of the nonlinear active constraints at the optimal ξ_n, are linearly independent, which is the most usual *constraints "qualification condition"*.

2.3 Illustration

We shall just present here a couple of simple example and counter-example.

- Model 1 : Michaelis-Menten model
  $f(x, \theta) = \theta_1 x/(\theta_2 + x)$, $\theta_1 > 0, \theta_2 > 0$, $0 \leq x \leq x_{max}$
  For $n = p = 2$, analytical calculation gives $\Delta_2 = [x_1 x_2(x_2 - x_1)\theta_1/\{(\theta_2 + x_1)^2(\theta_2 + x_2)^2\}]^2$ and the minimal D-optimal design :
  ξ_2 : $x_1 = \theta_2 x_{max}/(2\theta_2 + x_{max})$, $x_2 = x_{max}$.
  For $n > 2$, optimal designs are always found as equitable replications of x_1 and x_2.

- Model 2 : Incomplete fourth degree polynomial
  $f(x, \theta) = \theta_1 + \theta_2 x + \theta_3 x^4$ $-1. \leq x \leq 1.$
  For this linear model, the optimal designs will be independent of the parameters values.
  - For $n = p = 3$: ξ_3 = (-1., 0., 1.) is the optimal design ($\Delta_3 = 4$).

- For $n = 3r$ with $r \in N^+$, numerical computations show that replications of ξ_3 are never optimal even locally (they are saddle points of the Δ_n criterion). The optimal designs are always supported on four points varying with n. For example for $n = 6$ the optimal design is :
  ξ_6 = (-1., -1., -0.3413, 0.3413, 1., 1.), with $\Delta_6 = 32.9509$.
  We observed equivalent phenomena with all incomplete polynomial models we tested.

2.4 A sufficient and a necessary conditions for the optimality of replications of the points of a minimal D-optimal design

Constrained optimization theory led us to develop these two conditions from the usual necessary condition for a local maximum and sufficient condition for a strong local maximum of a second order differentiable function (Gill et al. 1981).
Let $\xi_p = (x_1, \dots, x_p)$ be the D_p-optimal design of model (1) for the a priori parameters values θ^0, in the experimental region defined by the m constraints $\phi_k \leq 0 \quad k = 1, \dots, m$.
Let $\xi_n = (\underbrace{x_1, \dots, x_1}_{r_1}, \dots, \underbrace{x_p, \dots, x_p}_{r_p})$ with $\sum_{\alpha=1}^{p} r_\alpha = n$
Is ξ_n locally D_n-optimal?

2.4.1 A sufficient condition

We shall suppose the following set H of assumptions is satisfied by ξ_p :

a. ξ_p satisfies the usual sufficient condition for a strong local constrained maximum of $\log \Delta_p$.

b. There are $0 \leq t \leq m$ active constraints. These t constraints are supposed to be differentiable.

c. There are $0 \leq s \leq t$ active constraints associated with *strictly negative Lagrange multipliers.*

For $\alpha = 1, \dots, p$:

d. Among the s preceding constraints there are s_α constraints involving, among others, some of the ℓ components of the x_α point, $\quad 0 \leq s_\alpha \leq s$.

e. If $s_\alpha > 0$ let $\dot{\Phi}_\alpha$ be the $s_\alpha \times \ell$ - matrix whose lines are the partial derivatives of these s_α constraints with respect to the ℓ components of x_α.

f. Let $\rho_\alpha = rank(\dot{\Phi}_\alpha)$ if $s_\alpha > 0$, and $\rho_\alpha = 0$ if $s_\alpha = 0$.

g. If $\rho_\alpha < \ell$, let Q_α be a $\ell \times (\ell - \rho_\alpha)$ - matrix, whose columns form a basis for the set of vectors orthogonal to the rows of $\dot{\Phi}_\alpha$.

h. Let B_α be the $\ell \times \ell$ - symmetric matrix whose general term is :

$$\left\{(1/r_\alpha)\, v_\alpha \left(X_p^T X_p\right)^{-1} \ddot{v}_{\alpha_{jk}}^T + \dot{v}_{\alpha_j} \left[\sum_{h=1}^{p} (1/r_h) \left(X_p^T X_p\right)^{-1} v_h^T v_h \left(X_p^T X_p\right)^{-1}\right] \dot{v}_{\alpha_k}^T\right\}_{jk} \qquad 1 \leq j,k \leq \ell$$

Then :

ξ_n will satisfy the usual sufficient condition for a strong local constrained maximum of $\log \Delta_n$, *if and only if*:
for each $x_\alpha \in \xi_p$ such that $r_\alpha > 1$ and $\rho_\alpha < \ell$, the $(\ell - \rho_\alpha) \times (\ell - \rho_\alpha)$-symmetric-matrix $Q_\alpha^T B_\alpha Q_\alpha$ is *strictly negative definite.*

When $\ell = 1$ and $r_\alpha = r > 1\ \forall\alpha = 1,\ldots,p$ (the most frequent case), this condition reduces to

$$v_\alpha\left(X_p^TX_p\right)^{-1}\ddot{v}_\alpha^T + \dot{v}_\alpha\left(X_p^TX_p\right)^{-1}\dot{v}_\alpha^T < 0 \tag{2}$$

which has to be satisfied by each $x_\alpha \in \xi_p$ not involved in any of the s active constraints associated with strictly negative Lagrange multipliers.

Remarks :

1. The preceding condition applies to *all types* of replications $r_1,\ldots,r_p$ of the p points of ξ_p, and in particular to *"the equitable ones"* (such that $\mid r_i - r_j \mid \leq 1\ \forall i,j$), which are *the most interesting* for Δ_n maximization (Box 1968), when $E_n \equiv E^n$.
   The generality of this condition allows the test of the local optimality of a design in which some experimental failures destroy the initial equitability of replications.
2. Condition (2) is free from the value of the balanced replication factor r.
3. Lagrange multipliers estimates are obtained from the necessary condition of Kuhn and Tucker, which ξ_p must satisfy.
4. One of the most practical choice for Q_α is given by a variable reduction technique.

2.4.2 A necessary condition

Suppose now the design ξ_p satisfies the set H' of assumptions :

a'. ξ_p satisfies the usual necessary condition for a local constrained maximum of $\log\Delta_p$.

b. as previously

For $\alpha = 1,\ldots,p$:

d'. There are $0 \leq t_\alpha \leq t$ active constraints involving, among others, some of the ℓ components of the x_α point.

e'. If $t_\alpha > 0$ let $\ddot{\Phi}_\alpha$ be the $t_\alpha \times \ell$ - matrix whose lines are the partial derivatives of these t_α constraints with respect to the ℓ components of x_α.

f'. Let $\bar{\rho}_\alpha = rank(\ddot{\Phi}_\alpha)$ if $t_\alpha > 0$, and $\bar{\rho}_\alpha = 0$ if $t_\alpha = 0$.

g'. If $\bar{\rho}_\alpha < \ell$, let $\bar{Q}_\alpha$ be a $\ell \times (\ell - \bar{\rho}_\alpha)$ - matrix, whose columns form a basis for the set of vectors orthogonal to the rows of $\ddot{\Phi}_\alpha$.

Then :
ξ_n will satisfy the usual necessary condition for a constrained local maximum of $\log\Delta_n$, *if and only if* :
for each point $x_\alpha \in \xi_p$, such that $r_\alpha > 1$ and $\bar{\rho}_\alpha < \ell$, the $(\ell - \bar{\rho}_\alpha) \times (\ell - \bar{\rho}_\alpha)$-symmetric-matrix $\bar{Q}_\alpha^T B_\alpha \bar{Q}_\alpha$ is *negative semi-definite*.

When $\ell = 1$ and $r_\alpha = r > 1\ \forall\alpha = 1,\ldots,p$, this condition reduces to

$$v_\alpha\left(X_p^TX_p\right)^{-1}\ddot{v}_\alpha^T + \dot{v}_\alpha\left(X_p^TX_p\right)^{-1}\dot{v}_\alpha^T \leq 0 \tag{3}$$

which has to be satisfied by each $x_\alpha \in \xi_p$ not involved in any of the t active constraints.

Remark : It is not difficult to see that this necessary condition is simply a slight attenuation of the preceding sufficient condition. These two conditions are close enough for *condition 2.4.1 being necessary and sufficient for most practical purposes.*

2.5 Application examples

We shall limit to the balanced replications case ($r_\alpha = r \ \forall \alpha = 1, \ldots, p$).

- Model 1 : Michaelis-Menten model ($\ell = 1$)
  for $n = p = 2$ the minimal D-optimal design is $\xi_2 : x_1 = \theta_2 x_{max}/(2\theta_2 + x_{max})$, $x_2 = x_{max}$. Among the four simple bound-constraints defining the experimental region, only the fourth one is active : $\phi_4 = x_2 - x_{max} = 0$.
  Simple calculation gives the associated Lagrange multiplier, λ_4 :
  $\lambda_4 = -(\partial \log \Delta_2/\partial x_2) \mid_{\xi_2} = -6\theta_2/(x_{max}(\theta_2 + x_{max})) < 0 \qquad \forall \theta_2 > 0, \quad \forall x_{max} > 0.$
  This shows that condition (2) must be tested for x_1 only : after some algebraic calculation we get
  $B_1 = v_1(X_2^T X_2)^{-1}\ddot{v}_1^T + \dot{v}_1(X_2^T X_2)^{-1}\dot{v}_1^T = -7(2\theta_2 + x_{max})^4/(16\theta_2^2 x_{max}^2(\theta_2 + x_{max})^2) < 0$
  This ensures that for the Michaelis-Menten model any balanced replication ($r_1 = r_2 = r$) of the minimal D-optimal design is itself D_n-optimal among all possible designs of size $n = 2r$.

- Model 2 : Incomplete fourth degree polynomial ($\ell = 1$)
  Let us test condition (2) for the minimal optimal design $\xi_3 = (-1., 0., 1.)$. For the inner point $x_2 = 0.$, we get $B_2 = 0.5 > 0$. Even the necessary condition (3) is not satisfied. This result explains the non optimality of the $3r$-designs replicated from ξ_3. We observed the same violation of the necessary condition for all incomplete polynomial models we tested.

2.6 Extensions

To conclude this part, let us say shortly that the preceding conditions adapt to the case of variance error depending on the unknown parameters and the independent variables, and to more general regression model of the g.l.m. types.

3 Software Tools

The previous optimality conditions are to be implemented in PLAND, an already fully operative software for optimal design computation :

PLAND is an interactive software and menu-driven program well adapted to the demands of the professional statistician wishing to undertake a complete experimental design study, as well as to the needs of the more occasional user wanting to compute without too much prerequesite background one or a series of optimal designs with varying conditionings (a priori parameter values, fixed points,...).
Three optimality criteria are presently available :

1. *D-optimality* : for subsequent optimal parameters estimation by Ordinary Least Squares or by Modified Least Squares (heteroscedastic case with parameters-dependent variance error, Huet 1986).

2. D_s *– optimality* : for subsequent optimal estimation of a subset of the model parameters, by OLS or MLS.

3. D_f *– optimality* : for subsequent optimal estimation of one or several general functions of the model parameters.

The following problem characteristics are supported :

1. The model considered can be a multiresponse nonlinear regression model.

2. The model can include several independent variables. They can be free or constrained to lie between bounds defining the experimental region.
3. The optimal design can be constrained to include an a priori fixed sub-design.
4. In the case of a multiresponse model, the optimal designs for each of the response functions can be constrained to be identical to each other.

Algorithmic and structural characteristics :

1. Well efficient-proved numerical algorithms have been used, particularly for the computations of the information matrices determinants (through singular values decomposition of the model jacobian) and the optimisation of the criteria (quasi-Newton method).
2. The Fortran code is presently embedded in a well known statistical host system (Consistent System - Klensin, 1980), whose macro facility allows easy definition of the model function and variance error function, on the same basis as that of the related statistical package NL for general nonlinear regression problems (Huet et al. 1986).

4 References

Atkinson, A.C. and Hunter,W.G. (1968). The design of experiments for parameter estimation. *Technometrics, 10, 271-289.*

Box, M.J. (1968). The occurrence of replications in optimal designs of experiments to estimate parameters in nonlinear models. *J.R. Statist. Soc., B, 30, 290-302.*

Fedorov, V.V. (1972). *Theory of Optimal Experiments.* Academic Press, New York.

Gaffke, N. (1987). On D-optimality of exact linear regression designs with minimum support. *J. Statist. Plann. and Inference, 15, 189-204.*

Gill, P.E., Murray, W. and Wright M. (1981). *Practical Optimization.* Academic Press, New York.

Huet, S. (1986). Maximum Likelihood and Least Squares Estimators for a Non-Linear Model with Heterogeneous Variances. *Statistics, 17, 4, 517-526.*

Huet, S. and Messéan, A. (1986). A Statistical Package for General Nonlinear Regression Problems. *COMPSTAT 1986, Physica-Verlag, Heildelberg, 326-331.*

Kiefer, J. (1961). Optimum designs in regression problems, II. *Ann. Math. Statist. 32, 298-325.*

Klensin, J. (1980). The Consistent System. *The American Statistician, 34, 3, 169-176.*

Silvey, S.D. (1980). *Optimal Design. An introduction to the theory for parameter estimation.* London : Chapman and Hall.

Vila, J.P. (1986). Local Optimality of Replications from a Minimal D-optimal Design in Nonlinear Regression : A sufficient and quasi-necessary condition. *INRA. Dpt. de Biométrie, - R.T. n⁰ 86 - 13.*

Econometric Computing

Model-Building on Micro-Computers: Spreadsheets or Specific Software

J.-L. Brillet, Paris

In recent years, the growth of micro-computer power has led to the development of specific modelling packages, trying to combine the advantages of a dedicated software with the features of the micro-computer. But spreadsheets have also improved, both in speed and in versatility, and might now be considered as a modelling tool. Thus, one might question the necessity of buying and learning a specific software, especially users which are already familiar with spreadsheets, or will have to be anyway.

This paper will try to provide an answer, by testing the production of the same small economic model on four packages, two dedicated to modelling: AREMOS and SORITEC, two spreadsheet-oriented ones: Lotus 123 and JAVELIN.

1 The steps in the modelling process

Let us start with a description of the modelling process.

Data gathering and updating : data for our model can come from (1) an ASCII file, with the data as a sequence of numerical characters and labels,(2) a file with a standard format used for interfacing (PRN, DIF, WKS), or (3) be entered directly by hand (in general more than one mode will be used). Then the data will have to be updated, by adding series, adding values, or replacing erroneous ones.

Model specification and estimation : the equations for the model must now be defined. Some of them, describing behaviours, will generally include unknown parameters, which will have to be estimated using more or less sophisticated methods.

Compstat 1988

Model check and simulation : the completed model can now be solved. For this time-consuming and frequently performed step, much attention must be given to the efficiency of the method.

Presentation of the results : results will be included in tables and graphs (both on the monitor and on an external device, printer or plotter).

2 Comparison of the two processes

General options : the first big difference comes rather from historical reasons : for micro-computers, most spreadsheets will use menus and full-screen data editing, while modelling packages adapted from mainframes are generally command driven.

The menu technique might look more convenient. However, once the user is reasonably experienced, specifying a task as a line will be probably faster than a sequence of menu-driven answers. And having the commands in text form allows batch mode (delaying the execution of a set of commands), memorizing on an external editable (and executable) file the commands performed, exporting and importing formulas or whole models, wildcards (symbols representing any single character or sequence of characters) and naming lists of variables. Perhaps one should consider both modes, like AREMOS for some features.

Data management : Simulation packages can import data using ASCII files, or several specific formats, like Lotus 123 (WKS) spreadsheets. Then selected amount of this data is stored in specific banks, from which they will be retrieved through implicit calls. User or system-defined comments (date) will be associated with each series. Export follows the same lines as imports.

Spreadsheets also use standard formats. Once read in, the series are kept in a worksheet, coexisting with text and formulas. Thus full screen editing can be done simultaneously on several series.

Here JAVELIN shows originality, coming from its multi-mode presentation : one can move the line or histogram display of a series (through the up and down arrows) to change its values. But this proves more spectacular than useful.

On the whole, the time series philosophy used by the modelling packages gives them a significant advantage.

Model specification : the approach is quite similar: each package allows definition of variables as formulas, using mathematical and conditional operators. JAVELIN in particular looks here like a modelling package, using lagged operators and named variables.

But for Lotus 123 and most spreadsheets, the formula must be introduced first in one cell, then copied in a specific zone. This can be tedious, but allows versatility in defining variables: each cell in the line can have its individual formula (or contain just data).

However, the modelling packages havea specific advantage: the equations are considered as parts of a model, which will be treated by ordering algorithms to reduce the cost of the solving process, first by looking for recursive sub-models, then for the best reordering within each sub-block (AREMOS). The resulting code can be compiled, to avoid reinterpretation for several simulations of the same code.

Model estimation : Here the single variable (JAVELIN) or multi-variable (Lotus 123) least squares cannot compete with the numerous methods offered by the modelling packages, especially SORITEC. Moreover, spreadsheets require equations to be linear in the variables, introducing intermediary variables. In practice, estimation will have to be made outside the package, which leads to problems as formulas can not be imported.

Model simulation : here large differences will be found. First, it is advised (especially for big models) to check first for consistency the set of equations, data, and coefficients. Each equation is computed separately, using as explanatory variables the observed values. Results are compared to estimated values (or to actual ones for identities). For modelling packages, this is a simulation option, but for spreadsheets copying the formulas in a different part of the sheet (123) or edited the formulas (JAVELIN) will be necessary.

As to normal simulation, worksheets face two connected problems : solving a non-recursive system ; taking into account the time dimension. Obviously, recursivity is the issue : if solution for one period can be reached in one pass, the whole set can be also. But this is not the general case.

The aspects of the problem : evaluating and improving the accuracy of the solution ; minimizing the time it takes ; treating the cases of non-convergence, are dealed with by the simulation package, which succeed more or less. For worksheets, the user must provide its own solutions, by organizing the sheet to order computations ; then building macro-instructions allowing the repetition of passes, and testing for convergence.

Of course, are can iterate by hand, and control convergence visually. But asks for constant attention from the user, and thus can become tedious for large models.

The time dimension can be introduced in worksheets. The problem remains of solving the model one period at a time: for dynamic models (formulas include previous values of variables), it is in practice more efficient to solve the model one period at a time than to include the time dimension in each iteration. One can feel that global solving, by multiplying the sources of variation (dynamic and simultaneous) will increase the time taken in reaching convergence in the last periods. Reducing the computing process to a single column of formulas proves surprisingly easier for 123, which allows versatility for the nature of cells in a given column, than for JAVELIN.

And options like exogeneizing a set of variables, introducing add-factors, or changing the process from static to dynamic, will never be available.

Presentation of results : the quality of results presentation, both in graphs and tables, is essential to both types of package. It is treated in a rather similar way. The main potential differences are in the mode of access to the computed values. The sheet format (which allows labels and lines) can be used for table presentation, either directly (if the model is small enough) or using a dedicated sheet with formulas depending on the solution variables. For graphs, the only difference is in the selection of data. The advantage of AREMOS and JAVELIN is only due to the better handling of this function.

Intermediate solutions : of course, one can also consider using both kinds of software in the same modelling process, using standard interfaces. Let us consider the possible solutions.

Clearly, no spreadsheet is adequate for estimation. This in practice forbids also simulation, as entering formulas by hand will lose the essential (as every experienced modeller knows) garantee of simulating the actual equations one has estimated. This limits spreadsheets to the data management and presentation phases. Even then, there is no obvious advantage : if some modelling software offer inadequate graphic features, they should rather be completed by a specialized graphic package.

Conclusion : the conclusion of this study is not surprising : a specialized packaged is much better than a spreadsheet at producing and managing a model. The use of a spreadsheet can be considered only if several of these cases are met :

- the user has already bought one, and has too tight a budget to consider buying an additional package.
- the model will be used for a short time, and the user is not prepared to invest intellectually in learning about a new package.
- the model is very simple to solve: fully recursive, or non-dynamic.
- the user is an expert at using the spreadsheet, and will build macro-instructions reproducing most of the features of the modelling packages. Solving iteratively the model, until a condition is met on the difference between two consecutive values, can very well be written as a 123 macro. This macro can be also built in the same working team or institution, or obtained from some expert. However, some of the specific simulation features, like automatic reordering of equations, information on convergence problems or using sophisticated algorithms for model solution, will never be available. Thus the macros built using the specialized packages will be much more efficient and complete.

It is obvious, however, that most of the present modelling packages have not evolved enough from mainframe philosophy, and will look a little outdated, especially to newcomers used to the micro-computer environment. They would gain much from reproducing some of the features of the most sophisticated spreadsheets, like simultaneous access to different presentations and the user-friendliness of menus and full screen editing.

Three Examples of Computer - Intensive Statistical Inference

B. Efron, Stanford

Abstract

We discuss three recent data analyses which illustrate making statistical inferences (finding significance levels, confidence intervals, and standard errors) with the critical assistance of a computer. The first example concerns a permutation test for a linear model situation with several covariates. We provide a computer-based compromise between complete randomization and optimum design, partially answering the question "how much randomization is enough?"

A problem in particle physics provides the second example. We use bootstrap both to find a good estimator for an interesting decay probability, and then to obtain a believable confidence interval.

The third problem involves a long-running cancer trial in which the z-value in favor of the more rigorous treatment wandered extensively during the course of the experiment. A dubious theory, which suggests that the wandering is just due to random noise, is rendered more believable by a bootstrap analysis. All three examples illustrate the tendancy for computer-based inference to raise new points in statistical theory.

Compstat 1988

MOUSE DATA: ORDINARY PERMUTATION TEST

[1] Two groups of mice, "A"=Treatment (7 mice)
"B"=Control (9 mice)

[2] for each mouse measured survival time in days after surgery:

group A:	94	197	16	38	99	141	23			Mean: 86.9
group B:	52	104	146	10	50	31	40	27	46	Mean: 56.2
										Diff: 30.6

[3] 1000 random divisions of the 16 numbers into groups of 7 and 9 gave 1000 corresponding values Diff*=mean A* -mean B*. [In other words we permute labels "A" and "B".]

[4] of these 126 exceeded Diff=30.6, for an attained sig level of
asl = .126

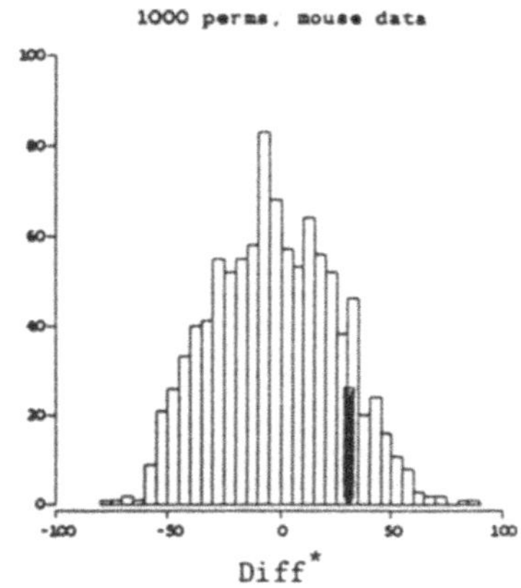

The 14 Scoliosis Patients

PATIENT	y	T	age	ht	wt	sex	health	constant
			X					
[1,]	14.84	1	14.33	175.25	54.35	1	2	1
[2,]	15.65	1	14.67	169.45	53.67	2	0	1
[3,]	19.47	1	16.58	179.35	59.25	1	1	1
[4,]	11.59	-1	13.75	169.85	50.95	1	2	1
[5,]	12.25	-1	13.58	154.75	34.28	1	2	1
[6,]	18.55	1	14.17	173.05	46.45	1	1	1
[7,]	20.33	1	17.17	177.35	54.25	1	2	1
[8,]	12.88	-1	14.42	173.65	59.57	1	0	1
[9,]	10.54	-1	12.58	165.05	39.27	2	0	1
[10,]	17.57	-1	15.67	192.25	64.25	2	1	1
[11,]	16.03	-1	20.17	183.75	62.75	2	2	1
[12,]	18.75	-1	19.33	187.95	56.51	2	1	1
[13,]	9.19	1	13.75	169.05	42.78	1	2	1
[14,]	17.37	1	14.75	177.15	54.35	2	2	1

DATA MATRIX

Usual Linear Model

$$\underset{14\times1}{y} = \underset{14\times1}{T}\beta + \underset{14\times6}{X}\ \underset{6\times1}{\alpha} + \underset{14\times1}{\varepsilon}$$

where $\underset{\sim}{\varepsilon} \sim (0, \sigma^2 I)$.

Usual ANOVA test for $H_0{:}\beta=0$

Rejects H_0 for large values of

$$S = \overset{\perp}{T}{}'\overset{\perp}{y}/(|\overset{\perp}{T}||\overset{\perp}{y}|)$$

where $\overset{\perp}{T}$ and $\overset{\perp}{y}$ are projections orthogonal to $\mathcal{L}(X)$ [equivalent to t-test for $\beta = 0$].

Data was actually generated from

$$y = \text{age} + .667\cdot T + \varepsilon$$

where ε =

-0.16 0.31 2.22 -1.49 -0.66 3.71 2.49 -0.87 -1.37 2.57
-3.47 0.09 -5.23 1.95

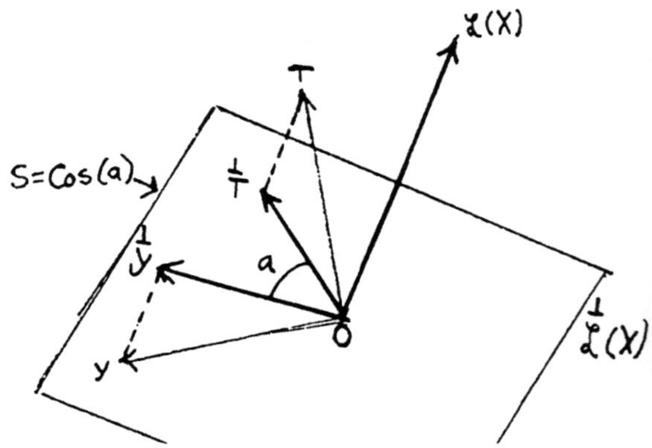

- Gave S = 0.557 Reject H_0?
- Usual t-test gave
  asl = $P\{t_7 > 1.77\}$ = .060

- Compare S with values of S^* obtained by permuting T to T^* (i.e., permuting -1's and 1's).
- Choose 400 T^* vectors randomly from $\binom{14}{7} = 3432$ possibilities.
- Would like $\overset{\perp}{T}{}^*$ to be uniformly distributed in $\overset{\perp}{\mathcal{L}}(X)$.
- For $v_1, v_2, \ldots, v_8$ an orthonormal basis for $\overset{\perp}{\mathcal{L}}(X)$, looked at projections of $\overset{\perp}{T}{}^*$ vectors along each v_j ; counted # projections in deciles of "perfectly uniform" distribution.

decile	v_1	v_2	v_3	v_4 ↓	v_5	v_6	v_7	v_8
1	33	38	43	31	42	51	45	43
2	37	39	51	66	40	36	33	39
3	31	45	38	57	40	40	43	46
4	38	49	37	29	46	46	36	43
5	39	45	34	12	40	25	46	45
6	45	36	39	11	49	33	36	30
7	47	27	45	28	43	39	45	42
8	42	38	44	67	34	43	37	36
9	34	40	36	67	31	46	34	38
10	54	43	33	32	35	41	45	38
				↑				

counts for 400 Tvecs actually used

v_1	v_2	v_3	v_4 ↓	v_5	v_6	v_7	v_8
47	40	32	44	41	34	49	33
45	36	39	61	40	49	35	32
43	52	45	57	52	44	37	35
27	36	49	31	41	38	32	44
37	37	37	15	34	38	41	48
44	31	40	11	36	43	32	53
34	42	35	31	38	50	45	32
37	45	51	49	35	35	48	49
44	43	34	56	43	38	38	32
42	38	38	45	40	31	43	42
			↑				

- Projection along v_4 very non-uniform, so chose another 400 T^* vectors.
- Not much better, but these are the ones I decided to use.

Each T^* gives

$$S^* = \overset{\perp}{T}{}^{*\prime}\overset{\perp}{y}/(|\overset{\perp}{T}{}^*|\,|\overset{\perp}{y}|)$$

(with y fixed as shown). Of the 400 S^* values, 25.5 exceeded S = 0.557, giving

$$\text{asl} = \frac{25.5}{400} = .064 .$$

400 permutation values

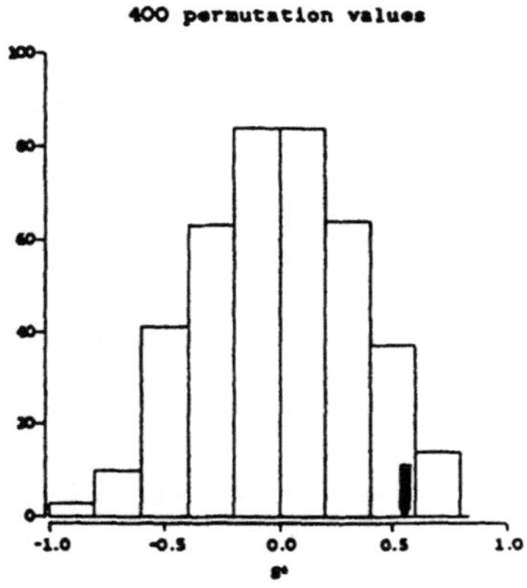

"top 40" perm values

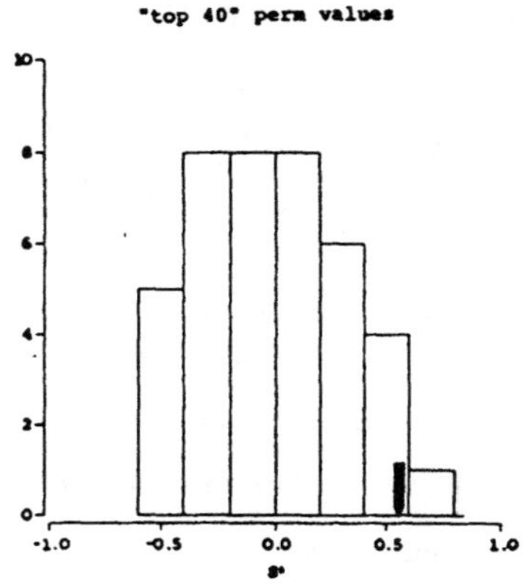

- Ideal T vector would have $\tau(T) = |\overset{\perp}{T}|^2 = 14$ (that is, $T \perp \mathcal{L}(X)$).
- $\mathrm{Var}\{\hat{\beta}|T\} = \sigma^2/\tau(T)$ in usual model.
- If T chosen randomly from 400 , mean $\{\tau(T)\} = 8.43$.
- I chose T randomly from "top 40", i.e. those 40 T vectors having greatest $\tau(T)$ values.
  mean $\{\tau(T)|\text{Top } 40\} = 12.53$
- Actually got T_{372} , with $\tau_{372} = 12.28$.

- 1.5 of the 40 S^* values for "top 40" reference set exceeded $S = 0.557$,

  asl = .038

permutation asl, all 400:	.064	[=25.5/400]
permutation asl, top 40:	.038	[=1.5/40 NOTE: Binomial SD for asl is .04!]
anova asl:	.060	[=Prob{$t_7 > 1.77$}]

Maybe top 40 T^* vectors all point in the same direction! No, their direction counts are reasonably uniform in $\mathcal{L}^{\perp}(X)$, except near v_4.

We actually got lucky: $\overset{\perp}{y}$ is not very close to v_4. Here are the cosines of angle $(\overset{\perp}{y}, v_j)$:

1	2	3	4	5	6	7	8
.31	.19	.01	-.02	.30	.26	-.56	.63

	[,1]	[,2]	[,3]	↓ [,4]	[,5]	[,6]	[,7]	[,8]
[1,]	5	4	4	0	8	2	2	3
[2,]	3	4	4	13	4	8	3	3
[3,]	3	4	3	12	3	4	6	4
[4,]	1	3	5	0	3	3	4	2
[5,]	3	3	4	0	4	4	6	7
[6,]	6	5	2	0	3	3	2	7
[7,]	5	3	6	1	2	6	1	5
[8,]	2	3	3	6	5	4	6	4
[9,]	5	5	4	8	4	5	7	0
[10,]	7	6	5	0 ↑	4	1	3	5

The Tau Data

All τ events

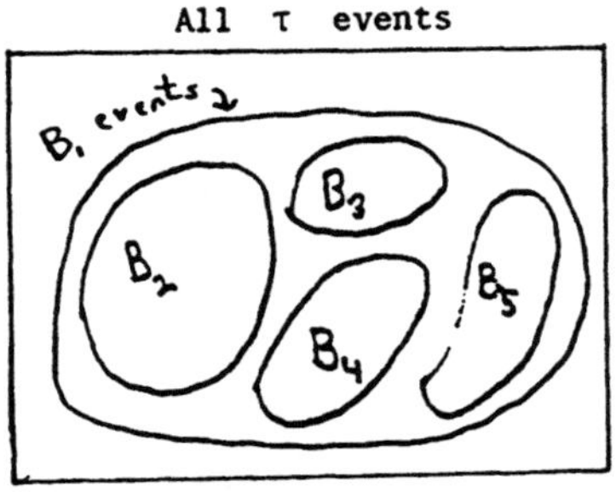

- Occurrence rates of five different tau decay events estimated,

  B1, B2, B3, B4, B5.

  (also the "estimated" SD for each).

- Should have

  $D = B1 - (B2+B3+B4+B5) = 0$.

- In fact $\hat{D} = 18.25$ (or 16.90).

- Wanted: a central 98% confidence interval for D.

- Normal Theory: $D \in [15.41, 21.09]$.

B1 ESTIMATES (13 LABS)

	[, 1]	[, 2]	[, 3]	[, 4]	[, 5]	[, 6]	[, 7]	[, 8]	[, 9]	[,10]
EST	84	86.0	85.2	85.2	85.1	87.8	84.7	86.7	86.9	86.1
SD	2	2.2	1.7	2.9	3.1	4.1	1.9	0.7	0.4	1.0

	[,11]	[,12]	[,13]
EST	87.9	87.2	84.7
SD	1.3	0.9	1.0

B2 ESTIMATES (6 LABS)

	[,1]	[,2]	[,3]	[,4]	[,5]	[,6]
EST	24	20.5	22.1	22.3	22.3	22.6
SD	9	4.1	2.5	1.5	2.1	1.5

B3 ESTIMATES (7 LABS)

	[,1]	[,2]	[,3]	[,4]	[,5]	[,6]	[,7]
EST	9.0	8.0	11.7	9.9	11.8	10.7	10.0
SD	3.8	3.5	1.8	2.1	1.3	0.9	1.8

B4 ESTIMATES (14 LABS)

	[, 1]	[, 2]	[, 3]	[, 4]	[, 5]	[, 6]	[, 7]	[, 8]	[, 9]	[,10]
EST	18.9	22.4	16.0	18.2	19	17.6	18.3	20.4	13.0	18.2
SD	3.0	5.5	1.3	3.1	9	1.3	3.1	3.3	3.5	0.9

	[,11]	[,12]	[,13]	[,14]
EST	17.4	17.0	18.4	19.1
SD	0.9	1.1	1.6	1.4

B5 ESTIMATES (20 LABS)

	[, 1]	[, 2]	[, 3]	[, 4]	[, 5]	[, 6]	[, 7]	[, 8]	[, 9]	[,10]
EST	18.3	17.5	22	22.4	15	21	22	18.2	35	17.8
SD	3.0	4.0	10	5.5	3	6	7	3.1	14	2.7

	[,11]	[,12]	[,13]	[,14]	[,15]	[,16]	[,17]	[,18]	[,19]	[,20]
EST	17.1	17.6	12.9	19.4	18.0	17.7	17.4	18.8	17.7	18.3
SD	1.3	3.3	1.8	2.3	1.2	0.9	1.0	1.1	1.4	1.2

POINT ESTIMATES OF B1,B2,B3,B4,B5 AND D=B1-(B2+B3+B4+B5)

	D	B1	B2	B3	B4	B5
weighted mean:	18.25	86.72	22.31	10.78	17.68	17.70
25%trimmed mean:	16.90	85.88	22.32	10.22	18.26	18.18

- Tried different trimmed means: 0, .1, .2, .25, .3, .5.
- For each trim, evaluated bootstrap SD estimate.
- ".3" gives lowest SD estimate for $\hat{D}$, but ".25" easier to explain.
- <u>Chose ".25" for remainder of Analysis.</u>

matrix of bootstrap sds

TRIM	$\hat{D}$	$\hat{B1}$	$\hat{B2}$	$\hat{B3}$	$\hat{B4}$	$\hat{B5}$	
0	1.39	0.35	0.43	0.46	0.55	0.94	
.1	1.18	0.41	0.41	0.54	0.54	0.69	
.2	1.04	0.46	0.40	0.60	0.37	0.54	
.25	1.04	0.43	0.39	0.64	0.35	0.54	
.3	1.01	0.48	0.38	0.61	0.38	0.49	•
.5	1.11	0.66	0.37	0.73	0.40	0.36	
theo	1.22	.27 small	.86 big	.60	.41	.38 small	•

- Bootstrap SD estimates based on 200 bootstrap replications each for $\hat{B1}, \hat{B2}, \ldots, \hat{B5}$.
- Then $\widehat{SD}(\hat{D}) = [\widehat{SD}(\hat{B1})^2 + \ldots + \widehat{SD}(\hat{B5})^2]^{\frac{1}{2}}$.

observed data

$$F_1 \xrightarrow{\text{iid}} (\hat{B1}_1, \hat{B1}_2, \ldots, \hat{B1}_{13}) \rightarrow \hat{B1} \quad \text{(25\% trimmed mean)}$$

$$\hat{F}_1 \xrightarrow{\text{iid}} (\hat{B1}^*_1, \hat{B1}^*_2, \ldots, \hat{B1}^*_{13}) \rightarrow \hat{B1}^* \quad \text{(25\% trimmed mean)}$$

bootstrap data

Repeat 200 times

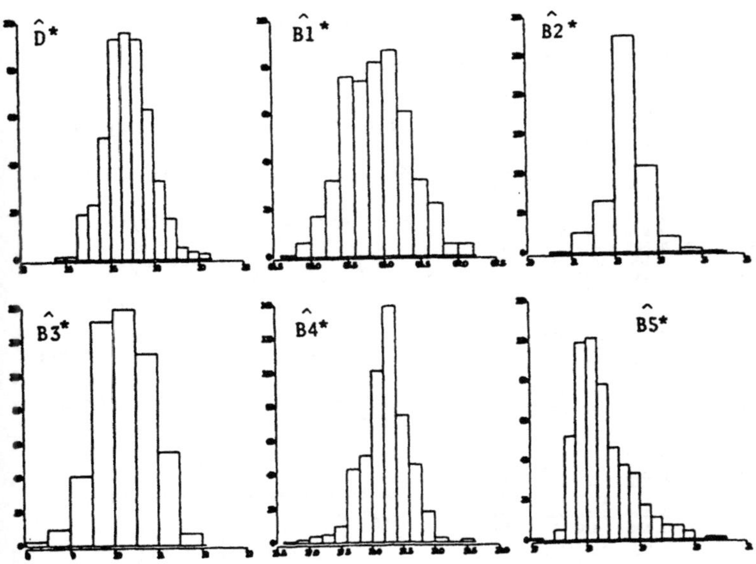

- Here are histograms of $\hat{B1}^*, \hat{B2}^*, \ldots, \hat{B5}^*$, and also $\hat{D}^* = \hat{B1}^* - (\hat{B2}^* + \ldots + \hat{B5}^*)$ (500 bootstraps for each).
- Percentiles of $\hat{D}^*$ were 14.20 (.01) and 19.34 (.99)
- Approximate Confidence Intervals for D:

	.01	.99
BC	14.29	19.53
BC_a	14.25	19.49
Boot-T	14.73	18.99
Boot-T (smooth)	14.22	19.20
Normal Theory	15.41	21.09

<u>Reference</u>:

Efron, B. and Tibshirani, R. (1986). Bootstrap methods for standard errors, confidence intervals, and other measures of statistical accuracy. <u>Statistical Science</u> 1, 54-77.

Bootstrap T

- Let $T = \frac{\hat{D}-D}{\hat{\sigma}}$ ← jackknife est. of $SD(\hat{D})$.
- Use bootstrap to estimate $T^{(.01)}$, $T^{(.99)}$.
- 2000 bootstraps of $T^* = \frac{\hat{D}^*-\hat{D}}{\hat{\sigma}^*}$ gave estimated percentiles

  $\hat{T}^{(.01)} = -2.63 \qquad \hat{T}^{(.99)} = 2.73$

  $[\hat{D}-\hat{\sigma}\cdot 2.73, \hat{D}+\hat{\sigma}\cdot 2.63] = [14.73, 18.99]$. ← "Boot-T"

- Smoothed Bootstrap: Draw from $\tilde{F}_1 = \hat{F}_1 \oplus N(0,\sigma_1^2)$, etc. Gave

  $\hat{T}^{(.01)} = -2.97 \qquad \hat{T}^{(.99)} = 3.31$, so

  $[\hat{D}-\hat{\sigma}\cdot 3.31, \hat{D}+\hat{\sigma}\cdot 2.97] = [14.22, 19.20]$. ← "Boot-T smooth"

- Then "T interval" is $[\hat{D}-\hat{\sigma}T^{(.99)}, \hat{D}-\hat{\sigma}T^{(.01)}]$

Boot-T for taudata

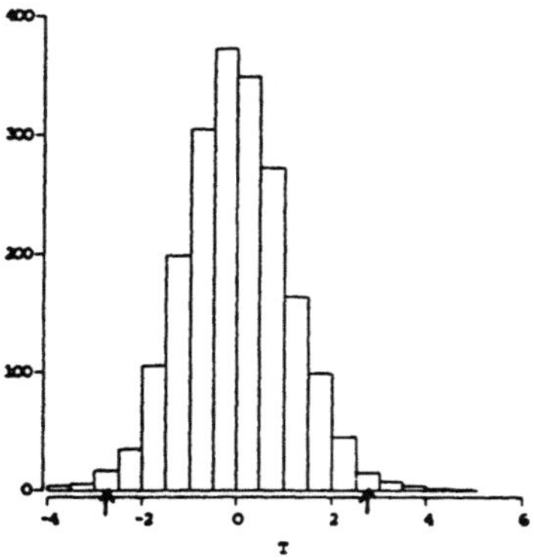

Cancer Treatment Data

- Randomized clinical trial for head and neck cancer.
- Data as of June 30, 1985.
- 51 patients in "A", radiation.
- 46 patients in "B", rad. + chemo.
- y = time to relapse (days).
- $d = \begin{cases} 1 & \text{if relapse observed} \\ 0 & \text{censored} \end{cases}$
- km = Kaplan-Meier survival curve.
- Logrank (Mantel-Haenszel) test for equality of survival was

  $z = 2.29$

  for an attained significance level of

  $1-\Phi(2.29) = .011.$

NCOG7h61 SEPARATED AND ORDERED

	group A					group B				
	y	d	km	ent	ind	y	d	km	ent	ind
1	7	1	0.980	1131.188	24	37	1	0.978	160.190	54
2	34	1	0.961	1795.380	44	84	1	0.956	968.566	73
3	42	1	0.941	2096.754	48	92	1	0.933	1745.942	88
4	63	1	0.922	861.252	16	94	1	0.911	979.004	72
5	64	1	0.902	1210.064	27	110	1	0.889	1588.752	85
6	74	0	0.902	1117.750	25	112	1	0.867	605.316	61
7	83	1	0.882	1529.876	38	119	1	0.844	2316.252	96
8	84	1	0.862	849.814	15	127	1	0.822	831.814	71
9	91	1	0.842	472.564	6	130	1	0.800	439.126	58
10	108	1	0.822	972.566	12	133	1	0.778	1684.066	87
11	112	1	0.802	1818.818	43	140	1	0.756	720.068	66
12	129	1	0.782	1812.818	42	146	1	0.733	1333.816	78
13	133	1	0.762	1683.066	39	155	1	0.711	1264.940	77
14	133	1	0.742	880.252	17	159	1	0.689	1034.442	76
15	139	1	0.722	1431.568	34	169	0	0.689	472.564	60
16	140	1	0.702	690.630	10	173	1	0.666	1370.692	79
17	140	1	0.681	968.566	20	179	1	0.643	291.942	52
18	146	1	0.661	714.068	11	194	1	0.620	810.376	70
19	149	1	0.641	1390.692	35	195	1	0.597	2314.252	95
20	154	1	0.621	2068.316	45	209	1	0.574	1991.440	91
21	157	1	0.601	671.630	9	249	1	0.551	1567.314	84
22	160	1	0.581	349.818	2	281	1	0.528	618.754	63
23	160	1	0.561	797.376	13	319	1	0.505	720.068	65
24	165	1	0.541	1719.504	41	339	1	0.482	1677.066	86
25	173	1	0.521	2106.192	46	432	1	0.459	1006.442	74
26	176	1	0.501	1694.066	40	469	1	0.436	1381.692	81
27	185	0	0.501	1392.692	32	519	1	0.413	2105.192	92
28	218	1	0.480	892.690	18	528	0	0.413	438.126	57
29	225	1	0.459	2158.630	50	547	0	0.413	2153.630	94
30	241	1	0.438	528.440	7	613	0	0.413	2106.192	93
31	248	1	0.418	171.190	1	633	1	0.386	746.500	67
32	273	1	0.397	802.376	14	725	1	0.358	1857.688	89
33	277	1	0.376	321.380	4	759	0	0.358	1968.002	90
34	279	0	0.376	2380.128	51	817	1	0.328	410.688	59
35	297	1	0.354	1409.130	33	1092	0	0.328	1526.876	83
36	319	0	0.354	202.628	3	1245	0	0.328	1357.254	80
37	405	1	0.330	887.690	19	1331	0	0.328	1398.692	82
38	417	1	0.307	2201.500	49	1557	1	0.287	605.316	62
39	420	1	0.283	1517.438	37	1642	0	0.287	1084.318	75
40	440	1	0.259	349.818	5	1771	0	0.287	818.376	68
41	523	0	0.259	2153.630	47	1776	1	0.230	407.688	53
42	523	1	0.233	1041.880	21	1897	0	0.230	774.938	69
43	583	1	0.208	1480.000	36	2023	0	0.230	621.754	64
44	594	1	0.182	546.440	8	2146	0	0.230	338.818	55
45	1101	1	0.156	1037.880	23	2297	0	0.0	431.126	56
46	1116	0	0.156	1259.940	28					
47	1146	1	0.125	1320.816	30					
48	1226	0	0.125	1040.880	22					
49	1349	0	0.125	1293.378	31					
50	1412	0	0.125	1314.816	29					
51	1417	1	0.0	1172.626	26					

y=days observed
d=failure or not
km=Kaplan-Meier
ent=entry date (from 1/1/78)
ind=index number in original data

• Kaplan-Meier estimated survival curves for the two treatments →

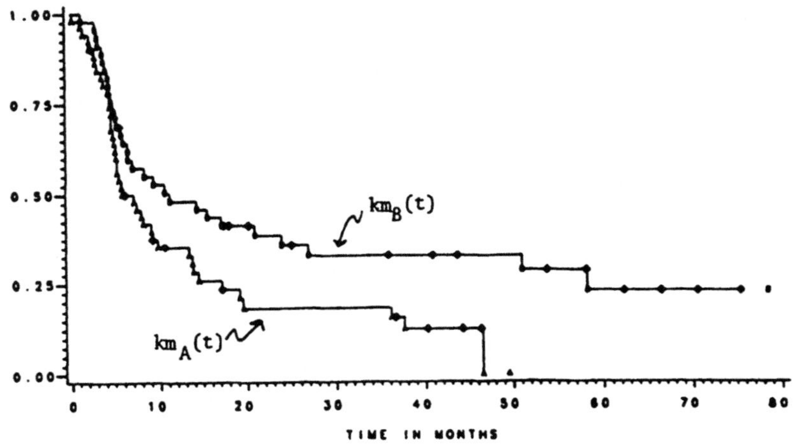

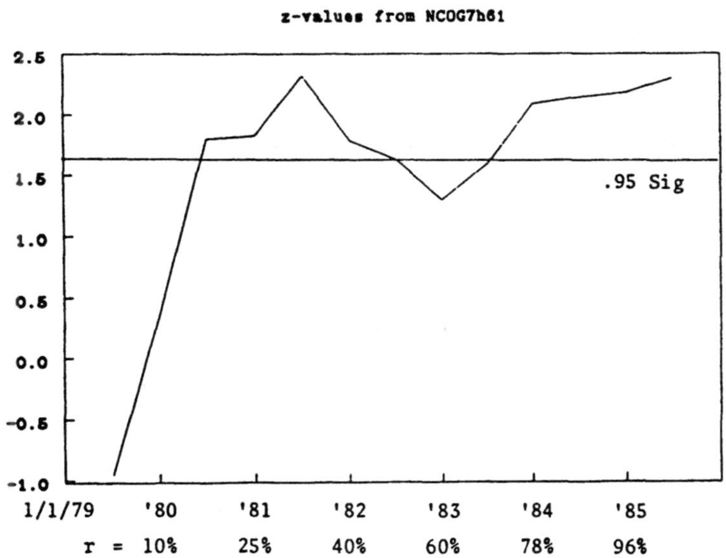

• z-values from log-rank test, at various calendar times.

• z = 2.31 on 6/30/81. Experiment nearly halted.

• "r" = proportion of total experience (Σy for all patients at that date compared to Σy on June 30, 1985).

• Question: Was treatment B relatively more effective early in experiment?

Some dubious theory: Let "z_r" be the z-value when proportion r of the total data is available (so z_1 = final z-value). Then

(a) $E\{z_r\}/E\{z_1\} \doteq \sqrt{r}$ (b) $z_r \mathrel{\dot\sim} N(\sqrt{r}\, E\{z_1\},1)$ (c) $z_r | z_1 \mathrel{\dot\sim} N(\sqrt{r}\, E\{z_1\},1-r)$

(d) z_{r_1} and z_{r_2} are approximately bivariate normal, with corr $= \sqrt{\frac{r_1}{r_2}}$.

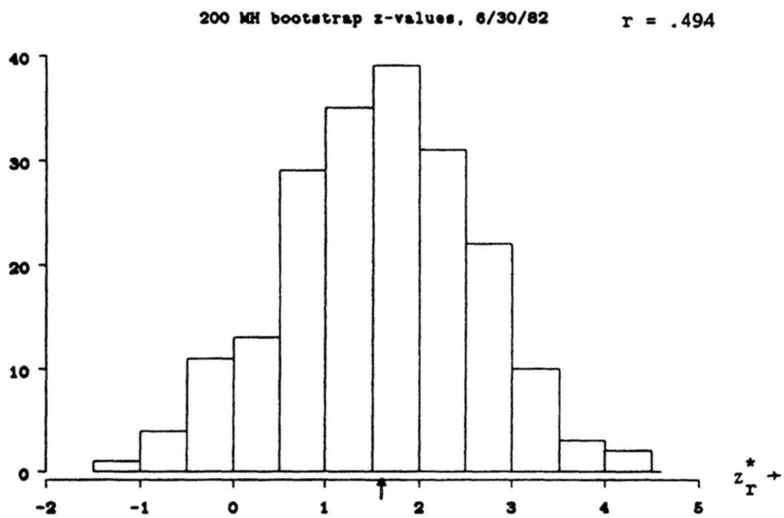

Bootstrap investigation of z-value on June 30, 1982: $r = .494$.

- Consider as fixed the 72 entry dates (38 for A, 34 for B) occurring before June 30, '82.
- For entry date e_i compute c_i = # of days from e_i until June 30, '82.
- Let $Y_1^*, Y_2^*, \ldots, Y_{38}^*$ be i.i.d. draws from km_A, the final Kaplan-Meier curve (as of June 30, '85).
- For each Y_i^*, let $y_i^* = \min(Y_i^*, c_i)$ and $d_i^* = 1$ or 0 as y_i^* equals Y_i^* or c_i.
- Likewise draw $Y_1^*, Y_2^*, \ldots, Y_{34}^*$ from km_B, the final Kaplan-Meier curve (as of June 30, '85).
- Then compute z_r^*, the log-rank z-value for the bootstrap data.
- 200 z_r^*'s $\mathrel{\dot\sim} N(1.56, 1.04^2)$ $\qquad$ • $\sqrt{r}\, E\{z_1\} = 1.61$.
- Compare with $N(1.61,1)$ from (b)!

- Did corresponding z_r^*'s for $r = .246$ (Jan. 1, '81).
- $\operatorname{cor}(z_{r_1}, z_{r_2}) = .727$.
- Compare with (d), $\operatorname{corr} = \sqrt{\frac{r_1}{r_2}} = .706$.

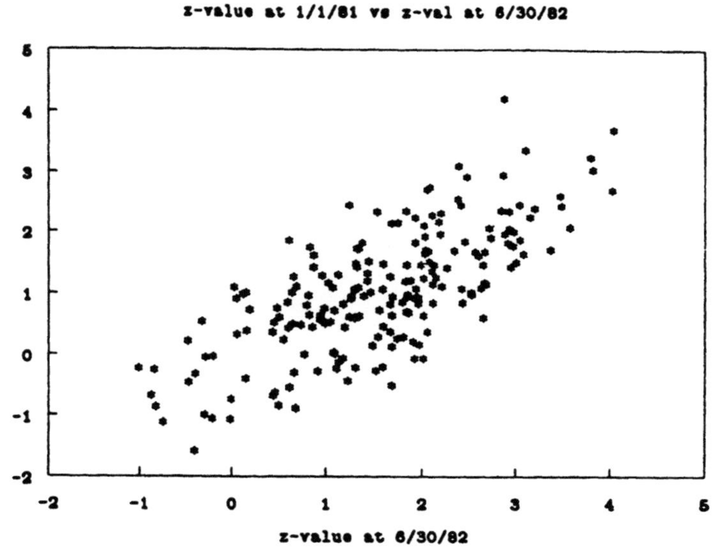

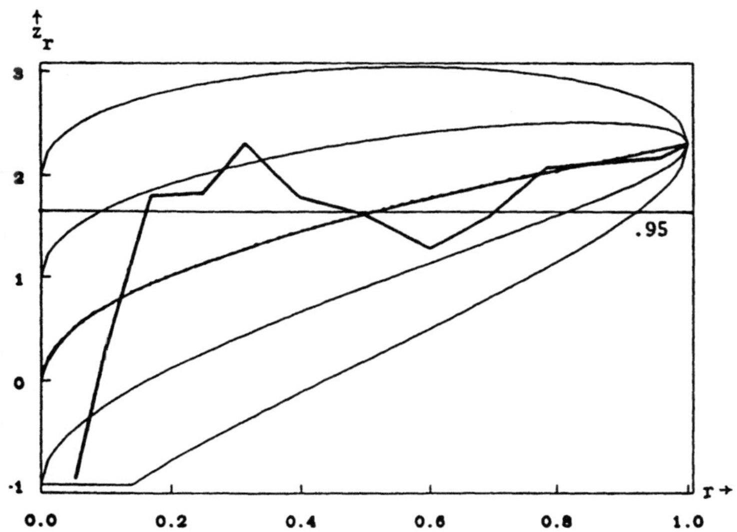

- Jagged line: z_r vs. r.
- Smooth curves:

  $\sqrt{r}\, z_1 + c\sqrt{1-r}$ c=.2,-1,0,1,2.
- Middle curve is $E\{z_r | z_1\}$.
- Others show 1 and 2 conditional standard deviation excursions.
- Nothing unusual happened! Maximum excursion is less than 1.3 conditional standard deviations.

New Computer Procedures for Generating Optimal Mixture Designs on Finite Design Spaces

H. A. Yonchev, Sofia

1. Introduction

The investigations for developing industrial products with better properties are often based on experiments with different kinds of mixtures. In a q-component mixture the sum of the proportions of the components is unity. If the proportion of the i-th component is denoted by x_i then

$$\sum_{i=1}^{q} X_i = 1, \text{ and} \tag{1}$$

$$0 \leq X_i \leq 1, \ i = 1,q \tag{2}$$

The experimental region formed by the constraints (1) and (2) is (q-1)-dimensional regular simplex. In many cases by economic or technological considerations additional constraints A_i and B_i are imposed on the mixture components:

$$0 \leq A_i \leq X_i \leq B_i \leq 1, \quad i = 1,q \tag{3}$$

They form a subarea of the (q-1)-dimensional simplex, which is in the general case an irregular hyperpolyhedron. The proportions X_i are called "related" or "mixture" variables. The temperature, the pressure or the reaction time of the technological process of manufacturing the product are called "process" variables. They are not related and in coded form are varied in the interval [-1,1]:

$$-1 \leq X_i \leq 1, \ i = q+1,q+2,\dots,m \tag{4}$$

The number of process variables is $r = m-q$, where by m is denoted the total number of variables.

The success of the experimental investigation using statistical methods depends on the information obtained by the experimental design. Specific problems arise at desigining optimal experimen-

Compstat 1988

tal plans including both mixture and process variables due to the necessity of working with mixture variables in a simplex or its subarea while the process variables are varied in multidimensional cube in Euclidean factor space. In this paper a procedure for generating D-optimal experimental designs including both kinds of variables is presented. Its application is considered especially for generating designs in constrained subarea of the simplex defined by (3) for the mixture variables and by (4) for the process variables.

The concept of D-optimality is based on the efficiency of the design to estimate the parameters of an assumed model. Assume that the model is expressed in matrix notation in the following usual form:

$$Y = F\beta + \varepsilon, \tag{5}$$

where Y is N×1 vector of observations; F is N×k augmented design matrix with ith row f(X) depending on the model; X is m×1 vector of predictor variables presenting the ith design point; β is the k×1 vector of coefficients to be extimated; ε is N×1 vector of uncorrelated random variables identically distributed with mean zero and variance σ_ε^2 assumed to be equal to 1. The vector of the least squares estimates b of the parameters β is given by

$$b = (F^TF)^{-1}F^TY \tag{6}$$

with covariance matrix $(F^TF)^{-1}\sigma_\varepsilon^2$. The variance of the predicted response $\hat{Y}(x)$ at point x is

$$d(Y(x)) = f^T(x)(F^TF)^{-1}f(x)\sigma_\varepsilon^2 \tag{7}$$

A design with matrix F_N^* is said to be an exact N-point D-optimal one if it maximizes the determinant of the information matrix $M = F_N^TF_N$ over all possible N-point designs in the factor space χ

$$|F^{*T}F^*| \rightarrow \max_{\chi}|F^TF| \tag{8}$$

2. Designs in mixture and process variables in the presence of the constraints (3)

The inclusion of process variables in the constrained mixture problem is considered first by Cornell (1971). The proposed designs are constructed in the space obtained as a combination of spherical subspace of the simplex in the mixture related variables and the rectangular space of r-process variables.

A similar approach is used by Li (1971) who transforms the region of interest into a cuboidal in the space of mixture variables. That space is then combined with the cuboidal space of the process variables and rotatable designs are developed in the combined space. Very near to the approach of Li is that of Hare (1979), who also suggests working in mixture variables in a cuboidal region of interest which is the largest inscribed cuboidal in the simplex or in the constrained mixture space. Then the cuboidal in mixture variables is linked to a cuboidal factor space for the process variables. The experimental design is developed in the combined space of the mixture related variables and the process variables.

The methods described above have the following main shortcomings:

1. The methods of inscribing ellipsoidal or cuboidal regions in the mixture space cause a significant loss of information in that the full region of interest is not explored.
2. The best criteria for constructing designs, such as D-optimality, G-optimality etc. are not used.
3. Different kinds of transformations of the mixture variables are used making the use of the designs inconvenient for the experimenters.

To overcome these shortcomings in the paper is proposed the following three-stage procedure for developing designs of high efficiency for constrained mixture variables and process variables:

1. A set S_1 of points consisting of the vertices and the centroids of the various faces of the hyperpolyhedron defined by (3) is generated. In that case the entire region of interest in the simplex is explored.
2. A set S_2 of points consisting of the points of support of the continuous D-optimal design in r-process variables is formed. The set S_2 can also be generated to consist of 2^r or 3^r points (levels (-1,1 or -1,0,1) in the case of including first or second order terms in process variables in the model (5).

3. Combining the coordinates of the points of S_1 with the coordinates of the points of S_2 a set S of L points called "candidate points" is constructed.

The goal is to find N points in S that maximize the determinant of $M = F_N^T F_N$ or come close to its maximum. The obtained N-point designs are exact D-optimal designs found on the discrete set of points S constituting the finite design space χ.

The main problem arising at generating exact designs using the given above three-stage procedure is to develope a highly effective algorithm for searching the N design points on the set of candidate points S. The main difficulties arise especially at solving problems of high dimension.

In the next section a new simple algorithm with improved efficiency which can be used for constructing exact D-optimal designs in mixture and process variables in finite factor space is suggested. It is called FSDOP.

3. The algorithm FSDOP

The general idea of the algorithm FSDOP is shown in the following sequence of steps:

Step 1. Using the three-stage procedure described above a finite set of L "candidate points" S is generated and stored.

Step 2. The points of S are distributed randomly into R groups consisting of p points, where $k/2 < p < 3k$.

Step 3. An initial random N-point design $\xi_o(N)$ with nonsingular matrix $M(\xi_o(N))$ is generated. The value of $\det M(\xi_o(N))$ is computed and saved.

Step 4. The first group is taken from S and added to the initial design forming a "candidate list" of H = p+N points. Then the augmented matrix F_H of the "candidate list" is formed and $M_H^{-1} = (F_H^T F_H)^{-1}$ and det M_H are calculated and saved.

Step 5. Using formula (7) at each point X_i, $i = 1,H$ of the "candidate list" the values of the variance of prediction $d_H(X_i)$ are calculated.

Step 6. The point X_{min} with a minimal value of d_H is found and deleted from the "candidate list". The values of M^{-1}, det M and $d(X_i)$ are updated for H-1 points using the formulae

$$M_{H-1}^{-1} = M_H^{-1} + (M_H^{-1}X_{min})(M_H^{-1}X_{min})^T/(1-X_{min}^T M_H^{-1}X_{min}) \quad (9)$$

$$d_{H-1}(x_i) = d_H(x_i) + (X_i^T M_H^{-1}X_{min})^2/(1+X_{min}^T M_H^{-1}X_{min}) \quad (10)$$

$$\det M_{H-1} = \det M_H(1-X_{min}^T M_H^{-1}X_{min}) \quad (11)$$

Step 7. The procedure is repeated from step 5 on the "candidate list" with H-1 points, H-2 points and so on until N points are left. The last N points are considered as a new N-point design $\xi_1(N)$ and one "short excursion" on H is counted out. The design $\xi_1(N)$ is taken as a new initial design if the following condition is satisfied

$$\det M(\xi_1(N)) > \det M(\xi_o(N)) \quad (12)$$

Otherwise $\xi_o(N)$ is used and the short excursion is considered unsuccessful.

Step 8. The next short excursion begins with constructing a new "candidate list" taking the second group with P points from S and adding it to the new initial design. It is obvious that R "short excursions" are possible until the set S is exhausted. At the Rth "short excursion" a design $\xi_R(N)$ with $\det M(\xi_R(N))$ is obtained.

The sequence of R "short excursions" is denoted as an "excursion" over the whole set of candidate points S. It is considered successful if

$$\det M(\xi_R(N)) > \det M(\xi_o(N)) \quad (13)$$

Step 9. The sequence of "excursions" terminates whenever a given number of unsuccessful excursions are performed. In some variants of FSDOP only the first excursion is performed over the whole Set S. For examples of high dimension ($m>7$) are used L/t candidate points chosen from S to be with maximal values of d(x) and calculated with the best F_N obtained in the first excursion. The value of t is taken in the limits t = 2÷10.

4. Examples

To compare the performance of FSDOP with that of another algorithms some examples are given below. A FORTRAN-77 program for VAX11/750 computer was used to conduct the examples.

Johnson and Nachtsheim (1983) compared different algorithms using the following mixture model in three variables $Y = b_1x_1 + b_2x_2+b_3x_3+b_{12}x_1x_2+b_{13}x_1x_3+b_{23}x_2x_3+b_{-1}x_1^{-1}+b_{-2}x_2^{-1}+b_{-3}x_3^{-1}$. An exact

D-optimal design with N=12 points was searched in a subarea of the simplex defined by the constraints $0{,}05<x_1<0{,}95$, $0{,}05<x_2<0{,}95$, $0{,}05<x_3<0{,}95$.

Using FSDOP for searching on a grid constructed with a step 0,025 in the given subarea a value of $\mathrm{Det}F_N^TF_N=0{,}42E-7$ was obtained for CPU=20 s. The same value was found by the best known algorithms of Fedorov (1972) Mod. Fedorov, k-exchange and Mitchell (1974) for 165 s, 91 s, 35 s and 94 s correspondingly.

Li (1971) presented an example of D-optimal design constructed for experiments with three mixture variables x_1, x_2 and x_3 and three process variables x_4, x_5 and x_6 to evaluate a second order polynomial model. The variables are varied in the constraints: $0<x_1<0{,}6$, $0<x_2<0{,}6$, $0<x_3<0{,}8$, $-1<x_4<1$, $-1<x_5<1$, $-1<x_6<1$ and a design with N=29 is searched. Transforming the region of interest into a cuboidal in the space of mixture variables Li obtained a design with $\det(F_N^TF_N)=0{,}93E-26$. For the same example using FSDOP was generated D-optimal design with $\det(F_N^TF_N)=0{,}162E-22$. The big difference in the results obtained is due to the significant loss of information in that the full region of interest was not explored by the algorithm of Li.

Better results were obtained also using FSDOP for the example of Hare (1979) generating 27-point D-optimal design for second order model in three mixture variables x_1, x_2 and x_3 and one process variable x_4 varied in the constraints: $0{,}43\le x_1\le 0{,}645$, $0{,}35<x_2<0{,}55$, $0{,}005<x_3<0{,}02$ and $-1<x_4<1$. The value of $\mathrm{Det}(F_N^TF_N)$ corresponding to the design of Hare is 0,186E-34 while FSDOP obtained 0,5E-34.

References

Cornell, J.A. (1971), J. Amer. Statist. Assoc., 66, 42-48.

Fedorov, V.V. (1972), Theory of optimal experiments, translated and edited by W.J. Studden and E.M. Klimko, New York, Academic Press.

Hare, L.B. (1979), Technometrics, 21, No 2, 159-173.

Johnson, M.E., and Nachtsheim, C.J. Technometrics, 25, No 3, 271-277.

Li, J. (1971), Designs of experiments with mixtures and independent variable factors, unpublished Ph.D. dissertation, Rutgers/ The State University, New Brunswick, NJ 08903.

Mitchell, T.J. (1974), Technometrics, 16, No 2, 203-210.

Screening Based Exclusively on Experts Opinions

Y. Ćwik and E. Pleszczyńska, Warsaw

Key words: latent trait, conditional independence, performance measure, positive dependence, statistical consulting system.

1. Introduction.

Let U be a performance variable which cannot be observed neither for the classified item nor for items from a learning sample. Let Z be a screening variable positively dependent on U. Besides, a learning sample is available in which the value of U for each item is evaluated by $(Z, \tilde{Z})$ where Z and $\tilde{Z}$ are the opinions of two experts. We assume that Z and $\tilde{Z}$ are conditionally independent given U and that for any u the conditional distributions of $(Z|U=u)$ and of $(\tilde{Z}|U=u)$ are identical. Models of this type are thoroughly discussed by Holland and Rosenbaum (1986).

The distribution of $(Z, \tilde{Z})$ is estimable from a learning sample, and we propose to use it for constructing a suitable screening rule $\delta(z)$. This will be exemplified by our approach to Owen's screening problem. Suppose that an item is considered <u>acceptable</u> if and only if the value of U is greater than the quantile of U of a <u>chosen</u> order π. Owen's problem consists in minimizing the expected proportion γ of items to be rejected under condition that the expected proportion τ of acceptable items among the selected ones is not less than a prescribed value τ_{min} (Birnbaum & Chapman (1950), Owen et al.(1975), Kowalczyk (1987)). Under some kind of positive dependence between Z and U, τ is a strictly increasing function of γ. Thus, Owen's problem is solved by the rule $\delta(z)$ which selects the item if and only if z is greater than the quantile of Z of order γ corresponding to τ_{min}.

But $\tau(\gamma)$ depends on the distribution of (U,Z) which is neither known nor estimable. To remove this difficulty, we construct an approximation of $\tau(\gamma)$

Compstat 1988

basing on the estimable distribution of $(Z,\tilde{Z})$. This is done as follows. For any (U,Z) and π, we use a decreasing convex function from (0,1) to (0,1), called screening curve (SC), which characterizes the discrepancy between the distributions of $(Z|U\le u_\pi)$ and $(Z|U>u_\pi)$ (Gafrikova & Kowalczyk (1988)). Then we show that $\tau(\gamma)$ may be determined from SC. Next, we consider the screening curve for $(\tilde{Z},Z)$ and π. Finally, we construct a mapping of SC for $((\tilde{Z},Z),\pi)$ onto a curve which is a good approximation of SC for $((U,Z),\pi)$. The mapping is <u>distribution-free</u> in a model consisting of $(U,Z,\tilde{Z})$ such that U, Z and $\tilde{Z}$ are continuous and suitably positively dependent.

This paper is devoted to a short presentation of the main ideas. In Sec.2 we consider a two-valued performance variable, say I, and show that in this case the mapping of the estimable SC into SC for (I,Z) (which is <u>not</u> directly estimable), is given analytically. In Sec.3 we present some Monte Carlo results, according to which the case of a two-valued latent trait provides a good approximation to the case of a continuous latent trait. The scheme of a related statistical consulting system is mentioned in Sec.4.

2. Screening concerning two-valued latent traits.

Consider (I,Z) consisting of I with values 1 and 2 and of a continuous r.v. Z such that $F_1 \ge F_2$, where $F_i(z)=P(Z<z|I=i)$, $i=1,2$. Moreover, let $P(I=2|Z=z)$ be an increasing function of z and let the supports of (Z|I=1) and (Z|I=2) be identical.

An item will be considered <u>acceptable</u> if I=2. Let δ_γ be a screening rule which rejects an item if $Z<z_\gamma$, where z_γ denotes the quantile of Z of order γ. We recall that (under the above assumptions concerning (I,Z)) the set $\{\delta_\gamma, \gamma\in[0,1]\}$ consists of rules admissible with respect to the ordering such that $\delta<\delta'$ if each conditional error rate for δ is not smaller than the respective one for δ'.

Let $\pi=P(I=1)$. The conditional error rates for δ_γ are $x_{\gamma,\pi}=F_2(z_\gamma)$ and $y_{\gamma,\pi}=1-F_1(z_\gamma)$ where $\pi(1-y_{\gamma,\pi})+(1-\pi)x_{\gamma,\pi}=\gamma$. Given (I,Z), the performance of the set of screening rules $\{\delta_\gamma, \gamma\in[0,1]\}$ is characterized by a convex curve

$$C_{I,Z} = \{(x_{\gamma,\pi},y_{\gamma,\pi}),\ \gamma\in[0,1]\},$$

which will be called <u>screening curve</u>. Obviously, $C_{I,Z}$ does not depend on π since it can be rewritten as $\{(F_2(z),1-F_1(z)),\ z\in R\}$. But the performance of δ_γ changes with π since $(x_{\gamma,\pi},y_{\gamma,\pi})$ is the point at which $C_{I,Z}$ intersects the line $\pi(1-y)+(1-\pi)x=\gamma$ (cf. Fig.1).

Let $\tau(\gamma) \underset{df}{=} P(I=2|Z>z_\gamma)$ and let τ_{min} be the smallest tolerable level of $\tau(\gamma)$. A screening problem expressed by Owen is to find $\gamma_{min}=\inf(\gamma|\tau(\gamma)\geq\tau_{min})$. Since $\tau(\gamma)$ can be expressed as $(1-x_{\gamma,\pi})(1-\pi)/(1-\gamma)$ then γ_{min} is easily derived from $C_{I,Z}$ (cf. Fig.2).

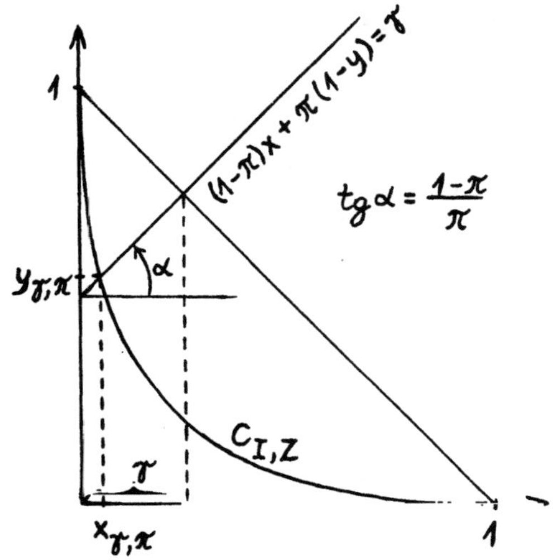

Fig.1.
$(x_{\gamma,\pi}, y_{\gamma,\pi})$ for $\gamma=0.25$, $\pi=0.5$.

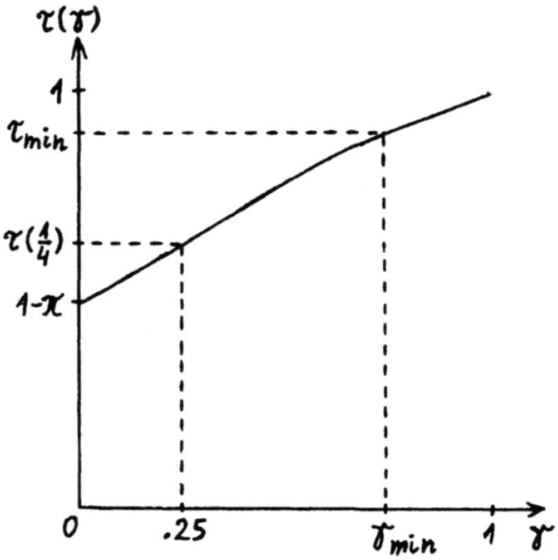

Fig.2.
Graph of $\tau(\cdot)$ for $C_{I,Z}$ from Fig.1.

Now suppose that I is a latent trait so that F_1 and F_2 are not known and learning samples for I=1 and I=2 are not available. Thus, γ_{min} is not directly estimable, and some additional information is needed. Such information will be exemplified in this paper by parallel experts' opinions.

Let Z express the degree of expert's belief that the classified item is acceptable. Suppose that a learning sample is available with <u>two</u> experts involved so that Z and $\tilde{Z}$ are respectively observed for each item from this sample. Thus, the distribution of $(Z,\tilde{Z})$ is directly estimable. We assume that Z and $\tilde{Z}$ are conditionally independent under I and that $(Z|I=i)\sim(\tilde{Z}|I=i)$ for i=1,2. Let

$$\tilde{Z}^{(\pi)} = \begin{cases} 1 & \text{if } \tilde{Z}<z_\pi, \\ 2 & \text{otherwise,} \end{cases}$$

and let $x^*_{\gamma,\pi}$, $y^*_{\gamma,\pi}$ be the conditional error rates of δ_γ for the pair $(\tilde{Z}^{(\pi)},Z)$. Then the screening curve for $(\tilde{Z}^{(\pi)},Z)$, namely

$$C^* = \{ (x^*_{\gamma,\pi}, y^*_{\gamma,\pi}),\ \gamma\in[0,1] \},$$

lies "above" $C_{I,Z}$ and is related to it by the formulas

$$x^{*}_{\gamma,\pi} = x_{\gamma,\pi}(1-\frac{x_{\pi,\pi}}{\pi}) + \frac{\gamma}{\pi}\, x_{\pi,\pi}\,, \qquad \gamma\in[0,1] \tag{1}$$

(cf. Fig.3). In particular, for $\gamma=\pi$ $\quad x^{*}_{\pi,\pi} = x_{\pi,\pi}(2-\frac{x_{\pi,\pi}}{\pi})$. Consequently,

$$x_{\gamma,\pi} = f_{\gamma,\pi}(x^{*}_{\gamma,\pi}, x_{\pi,\pi}), \qquad x_{\pi,\pi} = f_{\pi}(x^{*}_{\pi,\pi}),$$

where $f_{\gamma,\pi}$ and f_{π} are obtained from (1). Thus $C_{I,Z}$ is easily reconstructed from C^{*}. It follows that the solution of the Owen's problem is indirectly estimable from parallel experts' opinions.

It can be argued that the knowledge of π is not directly available in case of a latent trait. However, π can be deduced from the distribution of $(\tilde{Z},Z)$ in a way, rather complex, which will be not discussed here.

In the next section we exploit the fact that the transformation of C^{*} onto $C_{I,Z}$ depends only on π (and not on F_1 and F_2).

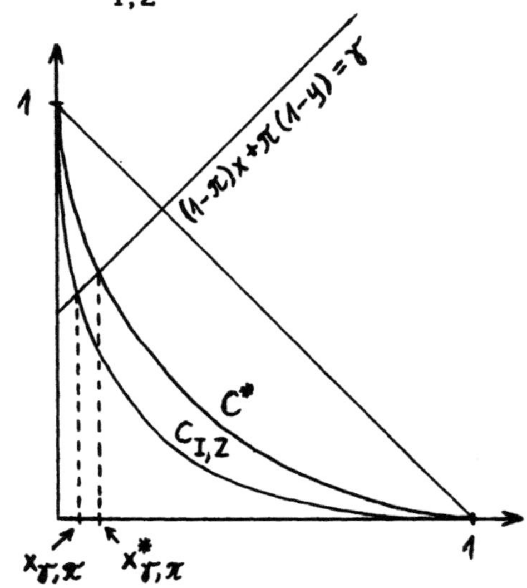

Fig.3. $x^{*}_{\gamma,\pi}$ and $x_{\gamma,\pi}$ for $\gamma=0.25$, $\pi=0.5$.

3. Screening concerning continuous latent traits.

Consider (U,Z) consisting of a continuous latent trait U and a scalar continuous screening variable Z such that for any z $P(Z<z|U=u)$ is a decreasing function of u, and for any u $P(U<u|Z=z)$ is a decreasing function of z. We select a number π from (0,1) and treat as acceptable items with the value of U exceeding u_{π} where u_{π} is the π-th quantile of U. Let

$$U^{(\pi)} = \begin{cases} 1 & \text{if } U<u_{\pi}, \\ 2 & \text{otherwise,} \end{cases}$$

and let Owen's screening problem be formulated for $(U^{(\pi)},Z)$. Clearly, its solution is derivable from the screening curve $C=\{(x_{\gamma,\pi},y_{\gamma,\pi}),\ \gamma\in[0,1]\}$ which corresponds to $(U^{(\pi)},Z)$. But, as previously stated in Sec.2, C is not directly estimable because U is latent. So we try to make use of parallel expert's opinions, available from a learning sample. Consequently, we deal with $(U,Z,\tilde{Z})$.

Assume that Z and $\tilde{Z}$ are conditionally independent under U and that $(Z|U=u)$ and $(\tilde{Z}|U=u)$ are identically distributed. The assumptions made on $(U,Z,\tilde{Z})$ imply that the distribution of $(\tilde{Z},Z)$ is symmetric and that for any z $P(Z<z|\tilde{Z}=\tilde{z})$ is a decreasing function of $\tilde{z}$ (Gafrikova & Kowalczyk (1988)). As previously in Sec.2, we take into account $(Z^{(\pi)},Z)$ and its screening curve $C^*=\{(x^*_{\gamma,\pi},y^*_{\gamma,\pi}),\ \gamma\in[0,1]\}$. Our aim is now to map C^* onto C. This mapping should be distribution-free in the considered model of $(U,Z,\tilde{Z})$, i.e. no more information about the distribution of (U,Z) must be used. But, obviously, Z and $\tilde{Z}$ are not conditionally independent under $U^{(\pi)}$, and formulas (1) from Sec.2 do not hold. It could be only hoped that Z and $\tilde{Z}$ would be "almost" conditionally independent under U, so that (1) holds approximately.

The question is now: how serious is the discrepancy and what mapping of C^* on C should be used? A partial but encouraging answer results from the Monte Carlo experiments performed by us on two families of distributions of (U,Z):

$\{N(0,0,1,1,\rho),\ \rho=0.20(0.05)0.90\}$ and $\{M(1,1,\rho),\ \rho=0.20(0.05)0.90\}$, $M(1,1,\rho)$ being the distribution of $(0.5(\xi_1^2+\eta_1^2,\xi_2^2+\eta_2^2))$ where (ξ_1,ξ_2) and (η_1,η_2) are independent and equally distributed according to $N_2(0,0,1,1,\sqrt{\rho})$. The pairs $(x^*_{\gamma,\pi},x_{\gamma,\pi})$ were obtained by simulation for various combinations of (γ,π,ρ) for both families. Then, the values of $x_{\gamma,\pi}$ were compared with $f_{\gamma,\pi}(x^*_{\gamma,\pi},x_{\pi,\pi})$ obtained according to (1). To illustrate, let $\gamma=\pi=0.5$. Then for a two-valued latent trait $x_{0.5,0.5}=f_{0.5}(x^*_{0.5,0.5})$. The graph of $f_{0.5}$ is presented in Fig.4, together with the set of crosses and circles corresponding to pairs $(x^*_{0.5,0.5},x_{0.5,0.5})$ for (U,Z) from the first and second family, respectively. It is evident that, notwithstanding the diversity of respective distributions of $(U^{(0.5)},Z)$, crosses and circles follow approximately the graph of a smooth increasing function, which is near but above the graph of $f_{0.5}$.

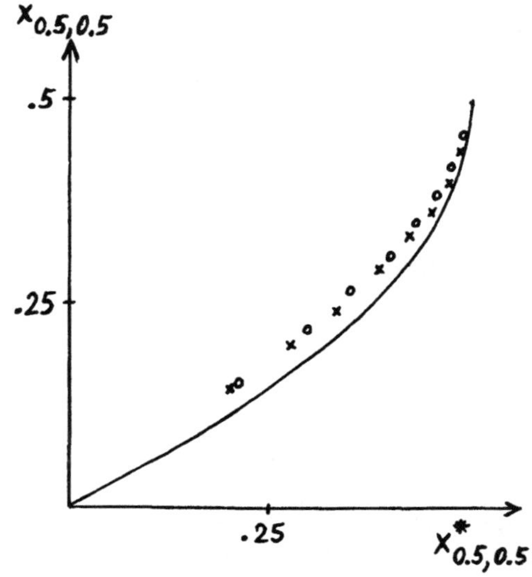

Fig.4. Graph of $f_{0.5}$ and simulated points $(x^*_{0.5,0.5},x_{0.5,0.5})$.

This graph, together with its analogues obtained by simulation for othes values of γ and π, seems to be a proper tool to transform C^* on C for any $(U,Z,\tilde{Z})$ satisfying the assumptions listed at the beginning of this section.

4. Final remarks.

A statistical consulting system, concerning screening and discrimination based on parallel experts' opinions, is now being prepared in our institute. The system consists of three modules: basic, simulatory and educational. The latter summarizes the theoretical bases of statistical inference on a latent performance variable. The contents of the educational module justifies the choice of rules proposed to users in the basic module.

The educational module exploits an ample set of examples, ranging from simple to rather complex. Each example concerns some family of distributions of (U,Z) and is used to illustrate successive stages of reasoning. The user can choose any example at any stage. He is also able to investigate his own examples in the simulatory module.

According to Sec. 2 and 3, solutions of screening problems can be derived from the screening curve corresponding to (U,Z) and π. Therefore, our main task is to estimate this curve. In Sec. 2 and 3 we aimed at a form of the screening curve which could be directly estimated from data consisting of parallel experts' opinions. We are now investigating properties of the respective substitute rules combining theory and simulation, and taking into account suitable bootstrap or cross-validation approach.

Acknowledgement. Thanks are due to Teresa Kowalczyk for helpful comments.

References:

Birnbaum Z. W., Chapman D. G. (1950), On optimum selections from multinormal populations, Ann. Math. Statist. 21, 443-447.

Gafrikova W., Kowalczyk T. (1988), Discriminant analysis based on expert opinions, Submitted for publication.

Holland P. W., Rosenbaum P. R. (1986), Conditional association and unidimensionality in monotone latent variable models, Annals of Statistics 14, 1523-1543.

Kowalczyk T. (1987), On screening problems linked with discriminant analysis, Proceedings DIANA II Conference on Discriminant Analysis, Cluster Analysis and other Methods of Data Classification, Liblice 1986, 175-180. Mathematical Institute of the Czechoslovak Academy of Sciences. Praha.

Owen D. B., McIntire D. M., Seymour E. (1975), Tables using one or two screening variables to increase acceptable product under one-sided specificatons, J. Quality Technology 7, 127-138.

Address list of Authors

Adorf, H.-M.
European Southern Observatory
Space Telescope
Karl-Schwarzschild-Straße 2
D-8046 Garching bei München
West Germany

Alvoni, Elisabetta
Università di Bologna
Facoltà di Economia e
Commercio.
Istituto di Matematica
generale e Finanzia
Piazza Scaravilli, 2
40126 Bologna
Italy

Baldessari, Bruno
Università 'La Sapienza'
Dipartimento di Statistica
Piazzale Aldo Moro
I- 00185 Roma
Italy

Bartlett, Geoff
The Univ. of Western Australia
Dept. of Management
University of Western Australia
Nedlands 6009
Australia

Bell, Edwina
Thames Polytechnic
Wellington St.
London, SE18 6PF
England

Belsley, David A.
Massachusetts Inst. of Techn.
Center for Computational
Research in Economics and
Management Science
1 Amherst Street- E40-111
Cambridge, MA 02139
USA

van den Berg, G.M.
University of Leiden
Department of Human and
Computer Studies
P.O.Box 9605
2300 RC Leiden
The Netherlands

Berzuini, Carlo
Università di Pavia
Dipartimento di Informatica e
Sistematica.
Via Abbiategrasso 209
27100 Pavia
Italy

Bethlehem, Jelke G.
Central Bureau of Statistics
P.O. Box 959
2270 A2 Voorburg
The Netherlands

Boomsma, Anne
Rijksuniversiteit Groningen
Vakgroep Stat. & Meettheorie
Oude Boteringestraat 23
9712 GC Groningen
The Netherlands

Brillet, Jean-Louis
National Institute
Statistics and Economics Studies
18 blvd Adolphe PINARD
75675 PARIS Cedex 14
France

Brown, Philip
Department of Statistics and
Computational Mathematics
University of Liverpool
U.K.

le Calvé, G.
Université de Rennes 2
6 Avenue Gaston Berger
35043 Rennes Cedex
France

Catarci, Tiziana
Università di Roma
'La Sapienza'
Dip. Informatica e Sistemistica
Via Buonarroti 12
00185 Roma
Italy

Ciampi, Antonio
Montreal Children's Hospital
Research Institute
4060 St.Catherine St.West 2nd
Montreal, Quebec
Canada H3Z 2Z3

Cleave, Nancy
Department of Statistics and
Computational Mathematics
University of Liverpool
U.K.

Cox, Michael A. A.,
Univ. of Newcastle upon Tyne
Dept. of Engin. Mathematics
Newcastle upon Tyne NE1 7RU
U.K.

Cox, Trevor
Univ. of Newcastle upon Tyne
Dept. of Statistics
Newcastle upon Tyne NE1 7RU
U.K.

Cwik, Jan
Inst. of Computer Science
Polish Academy of Sciences
PKIN
00-901 Warsaw
Poland

Deville, J.C.
Division Méthodes de Sondages
75675 Paris
Cedex 14
France

Dobrić, V.
University of Zagreb
University Computing Centre
Engelsdva B.B.
41000 Zagreb
Yugoslavia

Duijkers, Thomas J.
University of Leiden
Institute for Social Medicine
P.O.Box 9605
2300 RC Leiden
The Netherlands

Duijsens, I.J.
University of Leiden
Department of Human and
Computer Studies
P.O.Box 9605
2300 RC Leiden
The Netherlands

Dutter, Rudolf
Technische Universität Wien
Institute for Statistics
Wiedner Haupstraße 8-10
1040 Wien
Austria

Efron, Bradley
Stanford University
Department of Statistics
Sequoia Hall
Stanford, California 94305
U.S.A.

Eilers , Paul H.C.
Rijnmond Central
Central Environmental Agency
's-Gravelandseweg 565
3119 XT Schiedam
The Netherlands

Elguero, Eric
Unité de Biométrie
I.N.R.A.-U.S.T.L.-E.N.S.A.
9, Place Pierre Viala
34060 Montpellier
France

de Falguerolles, A.
Université Paul Sabatier
Lab.de Statistique e Probabil.
118, Route de Narbonne
31062 Toulouse Cedex
France

Felsenstein, Klaus
Technische Universität Wien
Institut für Statistik und
Wahrscheinlichkeitstheorie
Wiedner Haupstraße 8-10
1040 Wien
Austria

Gallo, Francesca
Università 'La Sapienza'
Dipartimento di Statistica
Piazzale Aldo Moro
I- 00185 Roma
Italy

Gilchrist, Robert
Polytechnic of North London
Holloway Road
London N7 8DB
England

Gordesch, Johannes
Free University Berlin
Babelsberger Str. 14-16
D-1000 Berlin 31
West Germany

Green, Mick
Lancatser University
Centre for Applied Statistics
Lancaster LA1 4YF
England

Grosbas, J.M.
Division Méthodes de Sondages
75675 Paris
Cedex 14
France

Hastie, Trevor
AT&T Bell Laboratories
600 Mountain Ave.
Murray Hill, New Jersey 07974
U.S.A.

Hietala, Paula
University of Tampere
Dept. of Mathematical Sciences
and Statistics
P.O.Box 607
SF-33101 Tampere
Finland

Holmes-Junca, Susan
Unité de Biométrie
I.N.R.A.-U.S.T.L.-E.N.S.A.
9, Place Pierre Viala
34060 Montpellier
France

Hougaard, Philip
NOVO Research Institute
Novo Alle
2880 Bagsværd
Denmark

Härdle, Wolfgang
Institut für Gesellschafts -
und Wirtschaftswissenschaften
Rheinische Friedrich-Wilhelms
Universität
D-5300 Bonn 1
West Germany

Hájek, Petr
Mathematical Institute
CSAV
Zitná 25
115 67 Prague
Czechoslovakia

Isogai, Takafumi
Osaka University
Dept. of Statistics
Toyonaka
Osaka 560
Japan

Ivánek , Jiří
Prague School of Economics
nám, A.Zápotockého 4
130 67 Prague
Czechoslovakia

Jirků, Petr
Center of Biomathematics
Czech. Academy of Sciences
Vídeňská 1083
CS - 142 20 Praha 4
Czechoslovakia

Jiroušek, Radim
Czechoslovak Academy
of Sciences
Institute of Information
Theory & Automation
Pod Vodárenskou Vézí 4
182 08 Prague 8
Czechoslovakia

Keller, Wouter J.
Central Bureau of Statistics
P.O. Box 959
2270 A2 Voorburg
The Netherlands

Kříž, O.
Czechoslovak Academy
of Sciences
Institute of information
Theory & Automation
Pod Vodárenskou Vézí 4
182 08 Prague 8
Czechoslovakia

Liski, Erkki P.
University of Tampere
P.O.Box 607
33101 Tampere
Finland

Luijben, Thom C.
Rijksuniversiteit Groningen
Vakgroep Stat. & Meettheorie
Oude Boteringestraat 23
9712 GC Groningen
The Netherlands

Marti-Recober, M.
Facultat d'informàtica
OPC
Pau Gargallo,5
08028 Barcelona
Spain

Mathieu, J.R.
Université Paul Sabatier
Lab.de Statistique e Probabil.
118, Route de Narbonne
31062 Toulouse Cedex
France

Mehta, Cyrus R.
Harvard University of
Public Health
Dana-Farber Cancer Institute
44 Binney Street
Boston, MA 02115
U.S.A.

Messéan, A.
INRA
CRJ-INRA
Laboratorie de Biométrie
Jouy en Josas 78350
France

Molenaar, Ivo W.
University of Groningen
Vakgroep S&M FPPSW
Oude Boteringestraat 23
9712 GC Groningen
The Netherlands

Momirović, Konstantin
University of Zagreb
University Computing Centre
Engelsdva B.B.
41000 Zagreb
Yugoslavia

Morin, Anne
Université de Rennes 1
IRISA
Rennes Cedex 35042
France

Muñoz-Gracia, Pilar M.
Facultat d'informàtica
OPC
Pau Gargallo,5
08028 Barcelona
Spain

Murphy, Brian P.
The Univ. of Western Australia
Dept. of Management
University of Western Australia
Nedlands 6009
Australia

Murtagh, F.
European Southern Observatory
Space Telescope
Karl-Schwarzschild-Straße 2
D-8046 Garching bei München
West Germany

Naeve, Peter
Universität Bielefeld
Fakultet für
Wirtschaftswissenschaften
Universitätsstraße, Postf. 8640
D- 4800 Bielefeld 1
West Germany

Nelder, John
Imperial College of
Science & Technology
Dept. of Mathematics
Huxley Bldg. 180 Queens Gate
London SW7 2 BZ
England

Nordbotten, Svein
University of Bergen
Dept. of Information Science
N-5000 Bergen
Norway

Nummi, Tapio
University of Tampere
Dept. of Mathematical Sciences
P.O. Box 607
SF- 33101 Tampere
Finland

Nurhonen, Markku
University of Tampere
Dept. of Mathematical Sciences
P.O. Box 607
SF- 33101 Tampere
Finland

Okur, Mehmet C.
Ege University
Faculty of Engineering
Bilgisayar Merkezi
35100 Bornova-Izmir
Turkey

Pages-Fita, J.
Facultat d'informàtica
OPC
Pau Gargallo,5
08028 Barcelona
Spain

Patel, Nitin R.
Harvard University of
Public Health
Dana-Farber Cancer Institute
44 Binney Street
Boston, MA 02115
U.S.A.

Payne, Clive
University of Oxford
Social Studies Faculty Centre
George Street
Oxford 2RL
U.K.

Payne, Roger W.
Rothamsted Experimental
Station,
Statistics Department
Harpenden
Herts AL5 2JQ
U.K.

Pleszczyńska, E.
Inst. of Computer Science
Polish Academy of Sciences
PKIN
00-901 Warsaw
Poland

Pötzelberger, K.
Technische Universität Wien
Institut für Statistik und
Wahrscheinlichkeitstheorie
Wiedner Hauptstraße 8-10
1040 Wien
Austria

Pregibon, Daryl
AT&T Bell Laboratories
600 Mountain Ave.
Murray Hill, New Jersey 07974
U.S.A.

Radaković, J.
University of Zagreb
University Computing Centre
Engelsdva B.B.
41000 Zagreb
Yugoslavia

Ross, G.J.S.
Rothamsted Exper. Station
Statistics Department
Harpenden
Herts AL5 2JQ
England

Roth, N.
Division Méthodes de Sondages
75675 Paris
Cedex 14
France

Rubin, Donald B.
Harvard University
Science Center
One Oxford Center
Cambridge, Mass. 02138
U.S.A.

Santucci, Giuseppe
Università di Roma
'La Sapienza'
Dip. Informatica e Sistemistica
Via Buonarroti 12
00185 Roma
Italy

Scallan, Antony
Polytechnic of North London
Holloway Road
London N7 8DB
England

Schimek, M.G.
University of Graz
Universitäts-Kinderklinik
Auenbruggerplatz
A-8036 Graz
Austria

Senchaudhuri, Pralay
Harvard University of
Public Health
Dana-Farber Cancer Institute
44 Binney Street
Boston, MA 02115
U.S.A.

Smith, Peter W.F.
University of Lancaster
Dept. of Mathematics
University of Lancaster
LA1 4YW
U.K.

Stejskal, Břetislav
Prague School of Economics
nám, A.Zápotockého 4
130 67 Prague
Czechoslovakia

Stewart, G.W.
The University of Maryland
Dept. of Computer Sciences
College Park
MD 20742
U.S.A.

Thiffault, Johanne
Montreal Children's Hospital
Research Institute
4060 St.Catherine St.West 2nd
Montreal, Quebec
Canada H3Z 2Z3

Venetoulias, Achilles
Massachusetts Inst. of Techn.
Center for Computational
Research in economics
and Management Science
1 Amherst Street- E40-111
Cambridge, MA 02139
USA

Vila, Jean-Pierre
Laboratoire de Biométrie
INRA
CRJJ
78350 Jouy-en-Josas
France

Watts, Peter
Thames Polytechnic
Wellington St.
London, SE18 6PF
England

Welsch, Roy E.
Massachusetts Inst. of Techn.
Center for Computational
Research in Economics and
Management Science
1 Amherst Street- E40-111
Cambridge, MA 02139
USA

Westerhoff, Thomas
Nixdorf Computer AG
Pontanusstraße 55
D- 4790 Paderborn
West Germany

Whittaker, Joe
University of Lancaster
Dept. of Mathematics
University of Lancaster
LA1 4YW
U.K.

Wittkowski, Knut M.
Eberhard-Karls-University
Dept. of Medical Biometry
Westbahnhofstr. 55
D-7400 Tübingen
West Germany

Yenyukov, I.
CEMI of the USSR Academy
of Science
Krasikova 32
117418 Moscow
USSR

Yonchev, Hristo A.
Higher Institute of
Chemical Technology
Dept. of Automation
Bul. Kliment Ohridski No 8
Sofia 1156
Bulgaria

Young, Alastair
University of Cambridge
Statistical Laboratory
16 Mill Lane
Cambridge CB2 1SB
England

GPSR Compliance

The European Union's (EU) General Product Safety Regulation (GPSR) is a set of rules that requires consumer products to be safe and our obligations to ensure this.

If you have any concerns about our products, you can contact us on ProductSafety@springernature.com

In case Publisher is established outside the EU, the EU authorized representative is:

Springer Nature Customer Service Center GmbH
Europaplatz 3
69115 Heidelberg, Germany

Batch number: 10584556

Printed by Printforce, the Netherlands